CONTEMPORARY MATHEMATICS

Titles in this Series

Volume

1. **Markov random fields and their applications,** Ross Kindermann and J. Laurie Snell
2. **Proceedings of the conference on integration, topology, and geometry in linear spaces,** William H. Graves, Editor
3. **The closed graph and P-closed graph properties in general topology,** T. R. Hamlett and L. L. Herrington
4. **Problems of elastic stability and vibrations,** Vadim Komkov, Editor
5. **Rational constructions of modules for simple Lie algebras,** George B. Seligman
6. **Umbral calculus and Hopf algebras,** Robert Morris, Editor
7. **Complex contour integral representation of cardinal spline functions,** Walter Schempp
8. **Ordered fields and real algebraic geometry,** D. W. Dubois and T. Recio, Editors
9. **Papers in algebra, analysis and statistics,** R. Lidl, Editor
10. **Operator algebras and K-theory,** Ronald G. Douglas and Claude Schochet, Editors
11. **Plane ellipticity and related problems,** Robert P. Gilbert, Editor
12. **Symposium on algebraic topology in honor of José Adem,** Samuel Gitler, Editor
13. **Algebraists' homage: Papers in ring theory and related topics,** S. A. Amitsur, D. J. Saltman and G. B. Seligman, Editors
14. **Lectures on Nielsen fixed point theory,** Boju Jiang
15. **Advanced analytic number theory. Part I: Ramification theoretic methods,** Carlos J. Moreno
16. **Complex representations of GL(2, K) for finite fields K,** Ilya Piatetski-Shapiro
17. **Nonlinear partial differential equations,** Joel A. Smoller, Editor
18. **Fixed points and nonexpansive mappings,** Robert C. Sine, Editor
19. **Proceedings of the Northwestern homotopy theory conference,** Haynes R. Miller and Stewart B. Priddy, Editors
20. **Low dimensional topology,** Samuel J. Lomonaco, Jr., Editor

Titles in this Series

Volume

21 **Topological methods in nonlinear functional analysis,** S. P. Singh, S. Thomeier, and B. Watson, Editors

22 **Factorizations of $b^n \pm 1$, $b = 2, 3, 5, 6, 7, 10, 11, 12$ up to high powers,** John Brillhart, D. H. Lehmer, J. L. Selfridge, Bryant Tuckerman, and S. S. Wagstaff, Jr.

23 **Chapter 9 of Ramanujan's second notebook—Infinite series identities, transformations, and evaluations,** Bruce C. Berndt and Padmini T. Joshi

24 **Central extensions, Galois groups, and ideal class groups of number fields,** A. Fröhlich

25 **Value distribution theory and its applications,** Chung-Chun Yang, Editor

26 **Conference in modern analysis and probability,** Richard Beals, Anatole Beck, Alexandra Bellow and Arshag Hajian, Editors

27 **Microlocal analysis,** M. Salah Baouendi, Richard Beals and Linda Preiss Rothschild, Editors

28 **Fluids and plasmas: geometry and dynamics,** Jerrold E. Marsden, Editor

29 **Automated theorem proving,** W. W. Bledsoe and Donald Loveland, Editors

30 **Mathematical applications of category theory,** J. W. Gray, Editor

CONTEMPORARY MATHEMATICS
Volume 29

Automated Theorem Proving:
After 25 Years

W. W. Bledsoe and
D. W. Loveland, Editors

AMERICAN MATHEMATICAL SOCIETY
Providence · Rhode Island

EDITORIAL BOARD

R. O. Wells, Jr.,
managing editor
Jeff Cheeger
Adriano M. Garsia

Kenneth Kunen
James I. Lepowsky
Johannes C. C. Nitsche
Irving Reiner

PROCEEDINGS OF THE SPECIAL SESSION ON AUTOMATIC THEOREM PROVING
89TH ANNUAL MEETING OF THE AMERICAN MATHEMATICAL SOCIETY

HELD IN DENVER, COLORADO

JANUARY 5–9, 1983

1980 *Mathematics Subject Classification.* Primary 68G15; Secondary 03B35.

Library of Congress Cataloging in Publication Data

Special Session on Automatic Theorem Proving (1983: Denver, Colo.)
    Automated theorem proving.
    (Contemporary mathematics, 0271-4132; v. 29)
    "Proceedings of the Special Session on Automatic Theorem Proving, 89th Annual Meeting of the American Mathematical Society, held in Denver, Colorado, January 5–9, 1983"–T. p. verso.
    Includes bibliographies.
    1. Automatic theorem proving—Congresses. I. Bledsoe, W.W. II. Loveland, Donald W. III. American Mathematical Society. Meeting (89th: 1983: Denver, Colo.) IV. Title. V. Series: Contemporary mathematics (American Mathematical Society; v. 29)
QA76.9.A96S64   1983   511.3   84-9226
ISBN 0-8218-5027-X

    Copying and reprinting. Individual readers of this publication, and nonprofit libraries acting for them, are permitted to make fair use of the material, such as to copy an article for use in teaching or research. Permission is granted to quote brief passages from this publication in reviews provided the customary acknowledgment of the source is given.
    Republication, systematic copying, or multiple reproduction of any material in this publication (including abstracts) is permitted only under license from the American Mathematical Society. Requests for such permission should be addressed to the Executive Director, American Mathematical Society, P. O. Box 6248, Providence, Rhode Island 02940.
    The appearance of the code on the first page of an article in this volume indicates the copyright owner's consent for copying beyond that permitted by Sections 107 or 108 of the U. S. Copyright Law, provided that the fee of $1.00 plus $.25 per page for each copy be paid directly to Copyright Clearance Center, Inc. 21 Congress Street, Salem, Massachusetts 01970. This consent does not extend to other kinds of copying, such as copying for general distribution, for advertising or promotion purposes, for creating new collective works or for resale.

Copyright © 1984 by the American Mathematical Society
Printed in the United States of America
All rights reserved except those granted to the United States Government
This volume was printed directly from author prepared copy.

The paper used in this book is acid-free and falls within the guidelines
established to ensure permanence and durability.

Table of Contents

Preface vii

Acknowledgments ix

D. W. Loveland. Automated Theorem Proving:
a Quarter Century Review. 1

Citation to Hao Wang. 47

Hao Wang. Computer Theorem Proving and
Artificial Intelligence. 49

Citation to Lawrence Wos and Steven Winker. 71

L. Wos and S. Winker. Open Questions Solved
with the Assistance of AURA. 73

W. W. Bledsoe. Some Automatic Proofs in Analysis. 89

R. S. Boyer and J. S. Moore. Proof-Checking,
Theorem-Proving, and Program Verification. 119

R. S. Boyer and J. S. Moore. A Mechanical Proof
of the Turing Completeness of Pure LISP. 133

P. B. Andrews, D. A. Miller, E. L. Cohen and F. Pfenning.
Automating Higher-order Logic. 169

D. Lankford, G. Butler and B. Brady. Abelian Group
Unification Algorithms for Elementary Terms. 193

Table of Contents

G. Nelson. Combining Satisfiability Procedures
by Equality Sharing. 201

Wu Wen-Tsun. On the Decision Problem and
the Mechanization of Theorem-Proving
in Elementary Geometry. 213

Wu Wen-Tsun. Some Recent Advances in
Mechanical Theorem-Proving of Geometries 235

Shang-Ching Chou. Proving Elementary
Geometry Theorems Using Wu's Algorithm. 243

D. B. Lenat. Automated Theory Formation
in Mathematics. 287

J. McDonald and P. Suppes. Student Use of
an Interactive Theorem Prover. 315

Preface

The genesis of the collection of papers presented in this volume is a special session on automated theorem proving (ATP), held in conjunction with the annual meeting of the American Mathematical Society in Denver, January, 1983, to honor several research leaders in the field of ATP. Hao Wang received the first Milestone Prize for his pioneering work in the early years of research in ATP and Lawrence Wos and Steven Winker received the Current Research Prize for recent and ongoing work. The citations appear in this volume with papers by the recipients. It is anticipated that both awards will be given periodically at a yet undetermined interval.

A secondary purpose for the special session was to bring to the attention of the mathematics research community the status of work in ATP. It is clear from the work of several research groups that the state has been reached where serious thought can be given by mathematicians to the possibilities of computer-aided proof discovery. (Fully automated proof, and theorem, discovery remains a goal but current success, in some ways quite impressive, is still very limited.) It is our hope that some active mathematicians will choose to accept the challenge this field extends and work, probably with an existing research group in ATP, to extend our capabilities towards frontier mathematics.

The papers in this volume derive primarily from the presentations by invited speakers at the special sessions, although invitations were extended to Wu Wen-Tsun and Patrick Suppes to submit papers for the volume. It was our feeling that the presentation in the special session collectively portrayed the state-of-the-art, with a sense of progress and high activity that exists in the field today. We also feel that the papers presented in this volume are even more successful at presenting the position and the momentum of the field of ATP in 1983.

The first paper in the volume serves as the introduction and overview to this volume as well as to the field of ATP. The papers of Bledsoe, Boyer and Moore, and Wos and Winker have a tutorial aspect and might be read next by those with little prior acquaintance with the field. A non-technical personal perspective of the endeavor to automate theorem proving is offered by Hao Wang. The papers by Lenat and Suppes outline substantial projects of considerable interest to our community but whose published reports have generally appeared in other settings. The remaining papers are more technical but report substantial contributions to the field.

We wish to thank the American Mathematics Society for their interest in this developing field of Automated Theorem Proving, both for the staging of the special session at the AMS Annual Meeting in Denver, January 1983, and for their interest in publishing this volume within the *Contemporary Mathematics* series. We thank Tony Palermino for competent stewardship at AMS in bringing this volume to press. Also, we thank David Mutchler at Duke University for his time and expertise in guiding several of these papers through a balking text-processor and output device, we thank Barbara Smith at the University of Texas, Austin, for typing several papers under time pressure, and indirectly we thank those who typed and supervised camera-ready copy preparation for individual papers.

W.W. Bledsoe
D.W. Loveland

February 15, 1984

Acknowledgments

Wu Wen-Tsun. On the decision problem and the mechanization of theorem proving in elementary geometry. Reprinted from *Scientia Sinica* 21 (2), 1978, with permission of the author.

Douglas Lenat. Automated theory formation in mathematics. Adopted from the article of the same title in the *Proceedings of the International Joint Conference on Artificial Intelligence*, August, 1977, with the permission of the International Joint Conferences on Artificial Intelligence, Inc. Copies of the proceedings are available from William Kaufmann, Inc., 95 First Street, Los Altos, CA 94022, USA.

AUTOMATED THEOREM-PROVING: A QUARTER-CENTURY REVIEW

Donald W. Loveland[1]

This opportunity to review the events and concepts that shape the present field of automated theorem proving (ATP) is particularly attractive because of the timing; twenty-five years ago (1957), the first publication appeared that reported results of a computer program that proved theorems. The paper, "The Logic Theory Machine" by Newell, Shaw and Simon [NS1], presented a program that could prove some theorems in the sentential calculus of *Principia Mathematica*. (A program to prove theorems in additive arithmetic was written by Martin Davis in 1954 but led to no publication.) The paper by Newell, Shaw and Simon, the stimulus of a 1956 Dartmouth conference on "artificial intelligence", and the increasing availability to researchers of then high-powered computers (i.e., substantial memory and speed) led to a spurt of activity that initiated the field. This activity has continued, slower than some anticipated yet faster than others expected, with an occasional shift in approach. It is this activity that we summarize.

A primary purpose of this paper is to provide an overview for the mathematics community of the history of ATP so as to better understand the contributions to this field by the first winners of

---
[1] Computer Science Dept., Duke University, Durham, NC 27706.

© 1984 American Mathematical Society
0271-4132/84 $1.00 + $.25 per page

the ATP "Milestone" prize (Hao Wang) and the "Current Research" prize (Lawrence Wos and Steven Winker). We do not dwell upon their contributions, for they have review papers in this volume, but seek to place their contributions, and those of others with work presented here, in overall perspective. This is a history summary, not a tutorial; the technical aspect is not developed. Unfortunately, space limitations prevent an encompassing document of history, so we apologize in advance for all the significant developments and (many) important contributors not mentioned in this review.

By automated theorem proving we mean the use of a computer to prove non-numerical results, i.e. determine their truth (validity). Often (but not always, e.g. decision procedures) we may also require human readable proofs rather than a simple statement: "proved". Two modes of operation are used: fully automatic proof searches and man-machine interaction ("interactive") proof searches. The label ATP will cover both.

The first question of interest to mathematicians is: how good are provers at this time? New mathematical results have been discovered by automated provers. In 1967 SAM's lemma was derived automatically during an interactive session with a mathematician concerned with modular lattices (see [GO1]). The mathematician recognized the lemma as the key to an open question. Other open questions have been answered within finitely axiomatizable theories. For example, in about 1980 Wos *et al*. answered positively the following open question, brought to their attention by Kaplansky (see [WW2]). Does there exist a finite semigroup which simultaneously admits of a nontrivial antiautomorphism without admitting a nontrivial involution? In set theory and analysis progress is slower (not surprisingly) but definitely nontrivial theorems such as Cantor's theorem (Andrews, this volume) and the Intermediate Value (IV) theorem (Bledsoe [Bl2], also this volume) have been proven by automated provers. (The IV theorem had the least upper bound principle as given.)

One of the most interesting ways to view the progression of this field is by identifying the two major approaches, or philosophic viewpoints adopted, and reviewing contributions in this light. We call the two approaches the *logic* approach and the *human-oriented* or *human simulation* approach. The *logic* approach is characterized by the presence of a dominant logical system that is carefully delineated and essentially static over the development stage of the theorem proving system. Also, search control is clearly separable from the logic and can be regarded as sitting "over" the logic. Control "heuristics" exist but are syntax-directed primarily. We shall include modular decision procedures augmenting systems as quite within the spirit of the logic approach. The *human simulation* approach generally is the antithesis of the above. Specific traits will be exhibited as we progress; the thrust is obviously simulation of human problem solving techniques.

The logic and human simulation approaches are admittedly not clearly delineated nor will ATP systems generally fit neatly in one extreme or the other. We make no further apology for this. The distinct philosophies represented in the two approaches have been very important historically and thus it is instructive to consider the influence of each approach even on systems not strictly in one camp.

A reader wishing a fuller review of the early history of ATP than space permits us may consult Davis' paper "The prehistory and early history of automated deduction" (see Davis [Da2]). The Davis paper covers to the mid-1960's. The late 1960's saw an intense investigation of the resolution principle of J.A. Robinson;

we give a very limited treatment of that energetic period (1965-1970) and refer the interested reader to Chang and Lee [CL1] or Loveland [Lo5] for (detailed) information on resolution theorem proving. We do not mean to imply that resolution is unimportant (quite the contrary, as we hope to convey) but allows us to concentrate more on the period 1970-1982. For those interested in the original papers of this field, through the early 1970's, see the Siekmann and Wrightson anthology [SW1].

Throughout this review we give a year associated with many of the contributions for relative perspective. As an exact year is sometimes hard to determine for an ongoing project or because of publication delay after the result has been announced informally, dates must be regarded as approximate.

Figure 1 lists a number of major contributions, listed according to their bias towards the logic or human simulation approach. The Presburger prover programmed by Martin Davis (1954), already mentioned, was a straight implementation of the classical Presburger decision procedure for additive number theory; roughly, first order sentences about linear equalities and inequalities over the domain of non-negative integers. A decision procedure for a theory is a precise recipe (procedure, algorithm) that halts for any well-formed sentence in the domain and outputs "true" (or "valid") if that indeed is the case. The complexity of the decision procedure is high (as is now better understood) so the program proved only very simple facts.

The Logic Theorist (LT) previously mentioned is the acknowledged first published exploration in automated theorem proving. Its goal was to mechanically simulate the deduction processes of humans as they prove propositional logic theorems. The methods used were, first, substitution into established formulae to directly obtain a desired result, or, failing that, to find a subproblem whose proof represents progress towards proving the goal problem. The iteration of the above led to a "chaining" of subproblems. LT was able to prove 38 theorems in the *Principia Mathematica*. More importantly, it introduced several basic ATP concepts (e.g. chaining), artificial intelligence (AI) techniques (list processing) and initiated the field of ATP.

The Geometry Theorem-proving Machine (GTM) (see Gelernter [Ge1], Gelernter *et al.* [GH1]) (1959) was a landmark project in its time in the AI field. At a 1956 conference Minsky made the observation that the diagram that traditionally accompanies plane geometry problems is a simple model for the theorem that could greatly prune the proof search. Inspired by this idea, Herbert Gelernter and Nathaniel Rochester at IBM organized a project to build the GTM. The diagram proved to be a powerful heuristic. The technique for its use is to work backwards ("backward chaining") from the conclusion (goal) towards the premises creating new subgoals whose truth imply (by the chain just created) the final goal. When working backwards from goal G one locates in memory the inference rules that conclude G and takes the hypotheses of each inference rule as a new subgoal. Of course, every hypothesis of some single inference rule must be established in order to establish G. (Note that this can be diagrammed by arcs emanating from a node marked G; this begins a proof search *tree*. This tree is an AND/OR tree in that from some nodes emanate arcs all of which must be pursued successfully and from other nodes emanate arcs only one of which must be pursued successfully.) Not all of the subgoals need be true. The geometry model was used to say which were true; the others were dropped as subgoals. This should be contrasted with forward chaining (reasoning forward from premises via *modus ponens*) where all conclusions are true. The GTM was capable of proving most high school exam problems within its domain (rectilinear diagrams). Incidentally, running time was often comparable to high school student time, e.g., about 20 minutes for a harder problem.

Simultaneously with the GTM effort the first efforts in the logic framework were occurring. Paul Gilmore and Hao Wang independently sought to mechanically prove theorems in the predicate calculus using not the heuristic methods of Newell and Simon but methods derived from classical logic proof procedures. Both were aware that known complete proof procedures need not be as clumsy as implied by Newell and Simon. Gilmore was motivated by Beth's semantic tableau technique. The procedure chosen was crude by any standard (and acknowledged so by Gilmore) but it was probably the first working mechanized proof procedure for the predicate calculus and did prove some theorems of modest difficulty (see Gilmore [Gi1,Gi2]). Besides the novel undertaking itself Gilmore was motivated by a desire to learn what computer programming was like. He satisfied all his motivations and effectively retired from the field (except for an enlightening analysis of the GTM in [Gi3]).

Hao Wang explicitly sought to counter the impression Newell and Simon imparted about the efficiency of proof procedures. Wang initiated the program at IBM in the summer 1958 and continued his work at Bell Labs in 1959-1960 (see Wang [Wa1], [Wa2], [Wa3]). Three programs were developed, one for propositional calculus, one for a decidable part of the predicate calculus, and in 1959-1960 a third for all of predicate calculus. The first program proved 220 theorems of *Principia Mathematica* in 37 minutes (3 minutes actual computing time); an effective display that using human-oriented techniques was not the only game in town. The program was organized around Gertzen-Herbrand methods rather than the semantic tableau approach of Gilmore. Of considerable interest was Wang's later attention to pattern analysis to overcome the explosion of formulae that resulted from enumerating all possible substitutions of terms for quantified variables. This anticipated somewhat the matching technique of Prawitz (see below) but also suggested heuristic exploration with shades of AI techniques.

The straightforward approach taken by Gilmore made his theorem prover a target for improvement. One of the most successful was the procedure introduced by Martin Davis and Hilary Putnam (see Davis, Putnam [DP1]). The D-P procedure (1960) greatly reduced the propositional processing. The question of validity of a predicate calculus formula can be reduced to a series of validity questions about ever-expanding propositional formulas. Herbrand's Theorem states that to each valid first-order formula C there is a propositional formula $C_1 \vee C_2 \vee C_3 \vee \cdots \vee C_n$ (truth-functionally) valid if C is valid. The C_i are essentially substitution instances over an expanded term alphabet of C with quantifiers removed. Thus one could test $C_1 \vee C_2 \vee C_3 \vee \cdots \vee C_i$, for ever-increasing i, by truth tables if one wished, and could proclaim the original a theorem if a tautology was shown to exist. Truth tables happen to be quite inefficient. The D-P procedure actually sought unsatisfiable formulas (the dual of the validity problem) and worked with conjunctive normal form (cnf form). Cnf form is a conjunction of *clauses*, each clause a disjunction of *literals*, atomic formulas (atoms) or their negations. The D-P procedure made optimal use of simplification by cancellation due to one-literal clauses or because some literals might not have their complement (the same atomic formula, but only one with a negation sign) in the formula. A simplified formula was *split* into two formulas so further simplification could recur anew.

The three years around 1960 was a significant time for the logic approach to ATP. One more notable event occurred at this time. In 1960 Dag Prawitz published a paper attacking the all-too-apparent major problem with the D-P procedure, the enumeration of substitutions (see Prawitz [Pr1]). Prior to this point substitutions were determined by some enumeration scheme that covered every possibility. Prawitz realized that the only important substitutions create complementary literals. Prawitz found substitutions by deriving a set of identity conditions (equations) that will lead to contradictory propositional formula if the conditions are met. The insight was profound but the implementation awkward. Davis later combined the idea with the D-P procedure to realize the advantages of both (see Davis [Da1]).

The idea of matching found its cleanest formulation in the resolution principle put forth by J.A. Robinson (1965) (see Robinson [Ro1]). The logic formulated by Robinson had no logical axioms and one inference rule that employed a substitution device called *unification*. The inference rule is as follows: if $A \vee l_1$ and $B \vee \sim l_2$ are variable-disjoint clauses (A and B are disjunctions of literals; l_1 and l_2 are literals) then from $A \vee l_1$ and $B \vee \sim l_2$ (the *parent* clauses) deduce clause $A\vartheta \vee B\vartheta$ (the *resolvent*) where ϑ is a substitution such that $l_1\vartheta = l_2\vartheta$. The propositional part of this inference rule is essentially the cut rule of Gentzen [Ge2], a generalization of *modus ponens*. The substitution ϑ is special in that it is the most general substitution that allows the equality to hold. Thus any substitution that would identify these two literals is an instance of this substitution. This is a powerful tool because it eliminates branching of search due to different possible substitutions that equate these two atoms but lead to different clauses. One uses the resolution inference rule to generate clauses (with some pruning possible) until for some literal l, literals l and $\sim l$

are generated. Since their conjunction is contradictory, the original formula is unsatisfiable.

The logic is one of extreme simplicity and, with unification, power. However, clauses accumulated at a rapid rate. The elegant simplicity of resolution led many researchers to invest time in trying to find restrictions and strategies of resolution to control the explosive clause growth and to provide resolution deduction formats amenable to heuristics for trimming clause growth. To even name all the refinements of resolution is beyond reason here (the appendix of Loveland [Lo5] summarizes twenty-five variations; more exist). We settle for naming some of the variations that have held continued interest beyond the 1965-1970 period when most variations were proposed. Figure 2 names the variations we summarize here. Of course, any short list must do injustice to useful variants omitted; for this we can only apologize.

The variants we consider here are complete proof procedures unless otherwise stated; that is, every unsatisfiable formula has a refutation within the proper format. By a *restriction of resolution* we mean a variant for which some clauses generated by the basic resolution procedure are not generated; a *strategy of resolution* only rearranges the order of generation to get likely useful clauses earlier. A *refinement of resolution* is either a restriction or a strategy. Recall that formulas are tested for unsatisfiability by resolution, that conjunctive normal form is used and that all quantifiers are removed by a process that involves the use of "Skolem functions". (See Chang and Lee [CL1] or Loveland [Lo5].)

Several refinements of resolutions are important because they are used within the successful theorem provers of Wos *et al.* to be mentioned later (and considered in the Wos and Winker paper in this volume). *Unit preference*, one of the earliest refinements, is a strategy favoring one-literal clauses. This guarantees that the resolvent is shorter than the longer parent clause, clearly beneficial when one seeks two one-literal clauses of complementary literals to conclude the proof. This strategy was introduced in Wos, Robinson and Carson [WC1].

A resolution restriction also introduced in Wos, Robinson and Carson [WC1] is the *set-of-support* restriction. A support set T, a subset of the set S of all the given clauses, is chosen and then two clauses from S-T are never resolved together. This means that every resolvent has in its deduction history some clause of T. T is usually chosen to be (a subset of) the clauses special to the problem at hand.

Hyperresolution (see Robinson [Ro2]), also used by Wos *et al.*, in its original form restricts resolutions to where one parent clause contains only positive literals (i.e. atomic formulas) and any resolvent containing a negative literal is immediately used in all permitted resolutions and then discarded. There now exist a number of important generalizations and variants of this restriction. The three refinements discussed all appeared within the first year of the introduction of resolution.

A proof procedure called *model elimination* (see Loveland [Lo1], [Lo2]) appeared about the same time. It is not a resolution procedure but is close to a very restricted form of *linear resolution*, which was not defined until several years later (and introduced simultaneously in Loveland [Lo3] and Luckham [Lu1], back-to-back at a 1968 conference). Linear resolution has each resolution of a deduction use the preceding clause of the deduction as one parent clause. One views the deduction as a series of modifications to the first clause, since (usually) only one literal is dropped and perhaps some added when passing from one clause to the next. (This ignores any further instantiation effect of a substitution.) Kowalski and Kuehner [KK1] provided the translation of model elimination into a very restricted linear resolution format (also done independently in Loveland [Lo4]). Their procedure, called *SL-resolution*, was used by several people in theorem proving projects, and by Colmerauer and Roussel in an early version of Prolog, a programming language we discuss later. Perhaps the primary importance of the model elimination procedure itself is in the MESON format, where it provides a natural completion to the problem reduction format (see Loveland and Stickel [LS2], Loveland [Lo5]). The problem reduction format is as old as the field itself, and probably is in its purest form in the Geometry Theorem-proving Machine discussed earlier. The problem reduction format is associated with "natural inference systems", considered later. (For a totally different completion of a problem reduction format see Plaisted [Pl1].)

A number of resolution refinements (including variants of model elimination) allow literals in clauses to be ordered to reduce multiple derivations of the same clause. The strongest ordering restrictions were based on the method called *locking* (Boyer [Bo1]) which allows a unique integer assignment to each literal occurrence of each input clause. Resolution need be done using only the lowest indexed integer of each clause. This does not allow complete freedom of order assignment for each resolvent (without endangering completeness) but is very close to that. This powerful ordering property still has its adherents (e.g., see Sandford [Sa1]).

Two important incomplete forms of resolution restrictions are *unit clause* resolution and *input clause* resolution. The former permits resolution only when one parent is a unit clause and the latter is a restricted form of linear resolution where one parent is always an input (given) clause. Chang [Ch1] shows these procedures to be of equal power; indeed, they are complete over the class of Horn formulas, a class receiving increased attention lately as we consider later. A Horn formula in conjunctive normal form has at most one positive literal per clause, but as many negative literals as desired.

Associated with resolution procedures is a means for handling equality called *paramodulation* by Wos and G. Robinson (see [WR2]). In essence this is the equality replacement rule with unification; it replaces all equality axioms except certain reflexivity axioms for functions. The term *demodulation* is used when paramodulation is restricted to replacement of the (usually) shorter term by the longer term with no instantiation allowed in the formula incorporating the replacement (see [WR1]). Paramodulation and demodulation continue to play a very important role in AURA, a replacement resolution theorem prover now used by Wos *et al.* at Argonne National Labs. We discuss AURA later.

The above refinements are discussed in detail in Chang and Lee [CL1] or Loveland [Lo5].

It is interesting to note that a proof procedure very close to resolution in spirit (given as a test for validity rather than unsatisfiability) was invented by Maslov in the USSR almost simultaneously with Robinson's invention of resolution. The original work (Maslov [Ma1]), called the *inverse* method, is reviewed together with resolution by Maslov [Ma2]; also see Kuehner [Ku1] for a direct comparison of resolution and the inverse method of Maslov.

One of the earliest and certainly the most successful of investigators to implement a resolution system is Larry Wos, one of the recipients of the first ATP prizes. We shall comment later on his long term development of resolution provers but the reader is advised to also see the paper in this volume.

While the logic approach received much of the attention in the early sixties and certainly most of the attention in the late sixties, there were other approaches to ATP, definitely more human-oriented. One already referenced was the Semi Automated Mathematics (SAM) project that spanned 1963 to 1967. Although much of this work did not get recorded in the scientific literature (however, see [GO1]) there was developed a succession of systems designed to interact with the mathematician. The systems had good interfacing to the human user, with good expressive power in its input/output language. The notation adopted was a many sorted ω-order logic with equality. λ-notation was employed. The SAM I system was primarily a proof checker, but the theorem proving power continued to increase through SAM V, which had substantial automatic capability. SAM's lemma mentioned earlier was proved on SAM V. Also, a matching procedure for ω-order logic was studied by Gould in a 1966 Ph.D. thesis. Interaction was aided by a natural deduction format that used subordinate proofs in the sense of Fitch 1952 [Fi1].

One bastion of dedication to the human oriented approach to theorem proving was (is) MIT. One project out of the MIT environment in the mid-sixties was ADEPT, the Ph.D. thesis of L.M. Norton [No1]. ADEPT was a heuristic theorem prover for group theory. Many heuristics were used, including a simple model to prune subset relations; the heuristics were deemed powerful enough that no backtracking was permitted. The limit of its capability was $x^2 = e \Longrightarrow$ *group is abelian*.

The fascination with resolution as the basis of theorem proving systems was overwhelming in the latter part of the 1960's. Not only did the theorem proving community adopt the principle but artificial intelligence researchers looked upon resolution as the key to their needs regarding reasoning systems. Question-answering (see Green and Raphael [GR1]) and robotics (see Fikes and Nilsson [FN1]) were two areas where resolution theorem proving formed the core of the reasoning component. The key to using resolution for problem solving (as compared to theorem proving) is answer extraction; such a procedure was given by Green (see [Gr1]).

The almost monolithic dedication to resolution during this period brought sharp criticism from some AI researchers, notably Marvin Minsky of MIT. By 1970, a Ph.D. student at MIT, Carl Hewitt, had formulated a specific rebuttal to resolution, and to the logic approach in general. He proposed PLANNER [He1], a human-oriented system not even a theorem prover *per se*, but a language in which a "user" was to write his own theorem prover, specifically tailored to the problem domain at hand. Besides this global commitment to flexibility, the language itself was structured for flexibility. For example, the inference rules could retain advice on when the rule was to be activated, thus incorporating search control at the heart of the inference mechanism. Because each step could be a procedure (a dynamic process) this approach has been called a *procedural* approach to theorem proving. Although the language was never fully implemented, a PLANNER subset was realized (microPLANNER). We comment on use of procedures again later.

At the time that the opponents of resolution were making themselves heard, a member of the resolution establishment, Woody Bledsoe, was questioning the capability of uniform logic systems to handle the mathematics of set theory and analysis in particular. A proof of a simple problem involving the intersection of families of sets proved awkward using resolution but fairly simple using a backward chaining methodology with some quite basic set theoretic inference rules. This observation began what is definitely the most fruitful and sustained effort in human oriented ATP to date. We return later to consider this approach to ATP.

An independent but parallel effort in human oriented theorem proving was undertaken about the same time by Arthur Nevins, then at George Washington University [Ne1,Ne2,Ne3]. Nevins built at least two provers; he received considerable attention for the ability of his provers to prove theorems which most resolution provers could not touch and was at the then outer limit of the prover of Wos *et al*. For example, Nevins could prove, fully automatically, $x^3 = e \implies f(f(a,b),b) = e$ where $f(x,y) = xyx^{-1}y^{-1}$, a result considerably harder than the "$x^2 = e \implies$ *the group is abelian*" proved (with guidance) by Norton with ADEPT. (A considerably more efficient automatic proof of this theorem was accomplished using the method of rewrite rules (see below) using an improvement due to Lankford. See [Bl1].)

Not all the activity and new ideas of 1970 were on the human simulation side of theorem proving. In 1970 Knuth and Bendix published a paper (see [KB1]) that refocussed the attention of many theoreticians in ATP on the potential of rewrite rules, a device familiar to logicians. (Effectively, the use of rewrite rules in ATP had been underway for several years by Wos via demodulation.) A rewrite rule is a replacement rule, usually written so that the left hand side is replaceable by the right hand side at any occurrence of the left hand side, after variable substitution. For example, $x + (-x) \implies 0$ is a rewrite rule of arithmetic; such rewrite rules that simplify an expression are called *simplification rules*. In equational theories, one converts equations to rewrite rules (primarily by choosing a direction for rewriting). Then one can repeatedly rewrite (reduce) terms via rewrite rules and equate terms that reduce to the same term (their "normal form") via rewrite rules. Knuth and Bendix gave an algorithm that for a class of equational theories permit computation of a *complete* set of rewrite rules; that is, a set of rewrite rules sufficient to check the truth of every equation of the theory by demanding that equal terms reduce to the same normal form. The theory of groups is one such theory while the theory of abelian groups is not. However, the problems of handling commutativity (and similar problems with associativity within some equational theories) have been surmounted in limited cases by researchers such as Lankford and Ballantyne, Huet and Peterson, Slagle, and Stickel. (For example, see [Sl2], [LB1], [Hu2], [PS1].) A key idea here has been the use of auxiliary equivalence classes; for example, one equates terms that reduce to terms within the same equivalence class. An unanswered question was (and is): what equational theories have complete sets of rewrite rules? One faces both the problems of finite termination of a sequence of terms under rewrite transformations and, if termination occurs, whether a normal form is reached whenever it should be. This problem has been investigated by a number of people (see Huet and Oppen [HO1] for a technical overview). This approach has been integrated with resolution (see Slagle [Sl1],[Sl2]).

A research area very close to rewrite (or reduction) systems is that of special unification algorithms. As previously mentioned, unification is the name J.A. Robinson gave to the process for finding a most general substitution for variables to make atomic formulas identical. Special unification also seeks to "unify" atomic formulas, but within special theories, usually equational theories. The usual unification process is augmented by equations or rewrite rules in such a manner as to make unnecessary the explicit listing of the equations as axioms. This idea was first formalized in Plotkin [Pl2] to deal with general equational theories. Equational theories with commutative and/or associative functions were pursued by others such as Stickel [St1] and Livesey and Siekmann [LS1]. The paper by Dallas Lankford in this volume presents a unification algorithm for elementary Abelian groups. A short survey of the research area (citing over ninety papers) appears in [RS1]. Even before Plotkins' 1972 paper, Nevins had incorporated some of these ideas in his theorem prover (see [Ne1]) to handle some associative and commutative functions. It should be noted that special unification does not always yield a single most general unifying substitution; indeed, there can be infinitely many unifiers, and the unification algorithm may not even be decidable.

The problem of unification of expressions has also been studied for higher order logic, some notable papers in this area being Gould [Go3], Huet [Hu1], and Jensen and Pietrzykowski [JP1]. Again, there may be one, finitely many or infinitely many unifiers. The unification algorithm is undecidable, even for the subcase of second-order logic (see Goldfarb [Go1]).

The advocates of the resolution approach have by no means been quiescent during the 1970's. About 1972, the theorem prover of Wos, Robinson and Carson was replaced by one developed by Ross Overbeek. The system has continued to develop with contributions from S. Winker, E. Lusk, B. Smith and L. Wos. The system has been named AURA, for *AU*tomated *R*easoning *A*ssistant. To the mechanisms introduced in the 1960's (unit preference, set-of-support for resolution; paramodulation and demodulation for handling equalities), they have added hyperresolution, more flexibility in demodulation, and preprocessors for preparation of input from a variety of formats. They have done an extensive amount of experimentation with the system, solving problems in circuit design and program verification as well as formal logic and (finite) mathematics. The result is a somewhat more human-oriented system but by our understanding of the term we leave the system firmly entrenched in the logic camp. This is not to say that domain-dependent information cannot influence search control. There is a weighting system that assigns values to terms, following user-proscribed rules, which is apparently very successful. However, the rules are purely syntax directed, although not limited to weights on nonlogical symbols. AURA now is viewed by its originators as a useful research tool for solving open problems subject to precise axiomatic formulations. (See Wos *et al.* [WW3].)

We should mention at this point an effort to make a flexible theorem prover available "to the people". More correctly, the purpose is to provide a structured set of tools, called LMA (for Logic Machine Architecture), that does include a default prover, named ITP, for those wishing to experiment with their own variation of theorem prover, not necessarily resolution based. ITP has the full capability of AURA. Lusk, McCune and Overbeek are designing the package in PASCAL for portability (see Lusk *et al.* [LM1]).

A very interesting effort that differs considerably from the resolution based prover of Wos *et al.* and the strong human-oriented prover of Bledsoe is the Computational Logic Theorem Prover (see [BM1], [BM2], [BM3]). This effort by R. Boyer and J. Moore, begun about 1972, focuses on inductive domains. The prover uses the language of quantifier-free first order logic with equality and features a general induction principle among the inference rules with the other inference rules, primarily rewrite rules. The latter are constantly undergoing change; for example, a proven equality can be converted to a rewrite rule for later use at the user's option. Besides working within traditional mathematics, primarily number theory using recursive arithmetic, they have proved properties of programs and algorithms, so-called "proofs of correctness", of particular interest within computer science. This is possible due to the inductive nature of many data structures, such as lists, used by programs. The program is capable of determining the variable of induction and also the induction predicate in many simple cases. A fairly sophisticated set of heuristics is used to guide the above tasks, particularly for the selection of an induction predicate. Besides its automatic mode, the prover can be used as a sophisticated proof checker (within its stylized domain) by supplying it with conjectures during a proof session. A conjecture is tackled and, if proved, added to the prover's library of proven results. Any result in the library may be used, generally as a rewrite rule. Through these conjectures and user-supplied recursively-defined functions which can help the prover find the right induction setting (with all such supplied functions checked by the prover for existence and uniqueness), quite deep theorems can be formally verified. Although Boyer and Moore and some others tend to think of their system as human-oriented, because of the heuristics used and the strong human interaction possible, we place the prover between the two approaches because the system also has a very dominant, carefully designed, logic with mostly syntactic heuristic rules, created directly to aid performance, not because humans use the principle.

In 1975-76 a flurry of interest occurred in proof procedures in the resolution flavor that utilized a graph representation, with literals or (conjunctive normal form) clauses as vertices and possible complementary literals (under substitution) connected by edges. (See Kowalski [Ko2], Sickel [Si1], Shostak [Sh1] and Andrews [An1].) Actually, the first explicit graph proof procedure in the resolution flavor was the "resolution graph" in 1970 (see [YR1]) and related ideas appeared earlier in Davis [Da1], Loveland [Lo1] and Prawitz [Pr2]. But the connection graph concept (to adopt Kowalski's term for the class of related procedures) prospered particularly well because a tight control on the introduction of new clauses was better appreciated after several years of attempting to constrain growth via resolution refinements, and because the associated deletion rules for links and clauses seemed possibly more effective than previous ideas for restraining redundant computation. A theorem prover being built by Jörg Siekmann and colleagues at Karlsruhe [BE1] (see later remarks) uses the Kowalski connection graph procedure as the basis of its syntactic processor and Wolfgang Bibel has used an alternative form of these ideas and those of Prawitz [Pr2] in constructing his "systematic proof procedure". (See [Bi1] and [Bi3]; see [Bi4] for comprehensive treatment of this procedure.) The ideas of Bibel have found independent and somewhat different realization in the work of Peter Andrews. The theorem prover of Andrews differs in an important way from all others we discuss in this paper (except the SAM project) in that it works within higher-order logic. We consider this theorem prover next.

In the late seventies, the development of TPS, a theorem prover for type theory, was undertaken by Peter Andrews and colleagues [An3], [MC1]. This prover, though firmly in the logic camps as regards approaches to theorem proving, also shows traits of the human simulation approach, but to date these traits are weak. One trait is much attention to a good proof development presentation to the user. The language of the prover is a type theory embodying λ-abstractions. As is essential to the provers of the logic school, it has a precise formal system -- the language just mentioned and a formal proof system. However, the proof system is complex, involving a natural deduction style proof format, a mating procedure used within the search process, and the ability to move from the mating format to the more decipherable natural deduction format. The mating procedure, like a connection graph procedure, links potentially complementary literals but here one uses a matrix format. A significant improvement is the freedom from conjunctive normal form. It is not necessary to use the distributive law to obtain clauses; rather, one can directly use expressions such as $A \wedge (B \vee (C \wedge D))$, a freedom that can significantly reduce a type of redundancy within the input expression (see [An2]). Arcs pass between potentially complementary literals; these are all identified before "unifying" substitutions are sought for certain adjacent nodes (the atomic formulas therein, actually). The order of finding complementary pairs (via unifiers) and simplifying greatly influence computation time and is subject to heuristic guidance. Andrews employs syntactic heuristics such as working with the most constrained path to minimize instantiation commitment. A key point of the mating method is the highly controlled rate of expansion of copies of quantified subformulas. The existing matrix is thoroughly explored for possible contradictions before an essential expansion is made in the number of clauses to be explored. Andrews uses the higher order unification process of Huet [Hu1], a nondeterministic algorithm often nonterminating, which also invites, or forces, heuristics to be incorporated in the processing, as Andrews has done.

The mode of operation of TPS is either automatic or interactive, so either the user or the mating process can guide the development of the natural deduction proof. In automatic mode TPS has proven Cantor's theorem as well as numerous first-order theorems.

The difficulty of finding the right substitution instances in higher order unification, where entire formulas can replace a variable, almost surely means that heuristics based on human "insight" (or use of the interactive mode) will play an increasingly important role in provers such as TPS.

An ambitious theorem prover now being developed at Karlsruhe, West Germany, the Markgraf Karl Refutation Procedure [BE1], promises to be a very significant prover if it approaches the capabilities planned for it by its designers. Early reports of problems run suggest that it compares favorably with other existing provers. It appears to split our classification, with a heuristic Supervisor (an overdirector) coupled to a logic engine built around a connection graph prover. Domain specific knowledge is available via a data bank; in the first version the chosen domain is automata theory.

The last decade also has seen the invention or amplification of new techniques for theorem proving. Included here is the connection graph method as an offshoot of resolution methods (this is mentioned in the preceding paragraphs). Two techniques not linked directly to resolution are rewrite rule systems, already discussed, and decision procedures. We comment on the latter briefly now. The rewrite rule systems and decision procedures have been featured in applications of theorem provers in non-mathematical roles, as discussed at the end of the paper.

By a decision procedure we mean a procedure that determines validity (or satisfiability) of a formula in a theory and halts on any formula in the theory. The Presburger decision procedure has already been mentioned. The decision procedures of recent note have tended to be tests for satisfiability and apply to quantifier-free first-order theories. Examples are Presburger arithmetic with uninterpreted function symbols (Shostak [Sh2]), augmented by array *store* and *select* (Suzuki and Jefferson [SJ1]), a set theory using $\cup, \cap, -, =, \in$ (Ferro, Omodeo, Schwartz [FO1]) and the theory of equality with uninterpreted function symbols (Ackerman [Ac1], Nelson and Oppen [NO1], Downey, Sethi and Tarjan [DS1], and others). An important paper in this regard is that of Nelson and

Oppen [NO2] where it is shown how to combine decision procedures for satisfiability applied to quantifier-free theories. Shostak has recently presented an interesting alternate method for combining decision procedures [Sh3]. Just recently, a very interesting decision procedure over a subset of elementary (plane) geometry has become known to the ATP community, although discovered by Wen-Tsün Wu in 1978, in China (see [Wu1]). The cleverness and novelty of the procedure within a field so well studied for so long illustrates how much we have yet to learn about decision procedures and the analytical approach to theorem proving in general.

We have deferred technical discussions of the human simulation approach represented by Bledsoe, Nevins and others. We have done so because the presentation is relatively lengthy and would disrupt our historical overview. However, because the human simulation approach is structurally complex even at the top (overview) level, and also a less familiar architecture to mathematicians than those systems following the logic approach, we do summarize some of the components and devices used in human-oriented theorem provers. To do so we adapt an overview list from Bledsoe [Bl1], which we recommend for a more in-depth survey of human-oriented systems (actually non-resolution systems in general). Some of the components are used in provers of the logic approach as well, as will be noted by the reader.

Concepts important to human-oriented theorem proving systems:

Knowledge base -- a data base (library) of facts, stored by context to the degree possible. This often means storing by object named rather than property named.

Rewrite rules -- often obtained from an equation by choosing a direction and then regarding the statement as a replacement rule, where variables can be instantiated to fit the application. We have discussed rewrite rule systems already in the context of equational theories. Rewrite systems are widely used, both as the major inference device and as a component system within larger systems. See Boyer-Moore [BM2] or Winker and Wos [WW1] (also Wos, this volume) for example systems using rewrite rules.

Algebraic simplification -- usually a special package which reduces algebraic expressions to a normal form to facilitate further processing. Such systems are usually implemented using rewrite rules. Algebraic simplifiers are only an example of simplification packages, but are the most important example and one having reached a high degree of sophistication (e.g., see [MF1]).

Natural inference systems -- predominantly backward chaining inference systems employing many inference rules. Such systems differ from many traditional logic systems (in particular, resolution) by not trying to economize on inference rules (the economy being useful when one wishes to prove theorems *about* logics) and by avoiding symmetric normal forms. Asymmetry is prized in human simulation systems: $A \to B$ and $\sim A \vee B$ are logically equivalent but the former often reveals a use preference over $\sim B \to \sim A$. In particular, conjunctive (disjunctive) normal form is avoided. The Bledsoe ATP prover [BT1] is an excellent example of a natural inference system. The problem reduction format, mentioned in our discussion of resolution refinements, is a basic format for structuring proof exploration for natural inference systems.

Control of search direction -- the ability to chain forward or backward as is warranted by the local situation. Although backward chaining from the goal towards the premises generally is the preferable choice of search direction, it is often wise to generate the consequences of certain statements using forward chaining. This is usually triggered directly by the statement itself. For example, if a premise employs symbols not in a common "vocabulary", it is often wise to seek consequences that employ the more common symbols. This increases the chance of a match with a subgoal to close off a branch of the search tree. In general, the combinatorics of node explosion with depth of search suggests bi-directional search. If twenty steps separate goal from premises, it is more efficient search to expand ten steps from each than twenty steps from the goal in all directions. Of course, one usually knows less about which premise to expand than which goal, so one biases the search in favor of the goal. However, Nevins in [Ne2] makes a case for a strong forward chaining bias.

Types -- to utilize different domains of objects over which variables range. If one distinguishes variables that range over integers from variables that range over reals (or variables over individuals from variables over sets) one realizes considerable efficiency in processing because restriction to the chosen domain is built-in rather than treated axiomatically, which requires deduction. By having the type built-in, the pattern matching (e.g., unification) on terms is considerably more powerful.

Procedures -- a mechanism of integrating search control with inference. Decision procedures are one extreme where a sentence is fully evaluated to true or false under control of the algorithm being used. At the other extreme are procedural heuristics that alter search flow in a minor way. As a simple example of procedural heuristics, statements that are frequently established in one step might be expanded first and established immediately if successful, so as to quickly reduce the number of unproven goals. This was done in the geometry theorem prover [GH1]. Hewitt's theorem proving language PLANNER [He1] allows for procedural information to be retained with each inference rule, information such as when to give preference to the inference rule, as previously mentioned.

Overdirector or executive program -- a portion of the system which is the top-level control. In the more sophisticated provers the overdirector is more than a statistics keeper and link between routines. An agenda is often maintained, which may be anything from a scoring mechanism to resolve conflicting alternatives to an active plan that develops the search along well prescribed lines. There can be overdirectors at several levels in a hierarchy. A sample task of an overdirector is to terminate a branch of search when the search effort is deemed "excessive" and backtrack to pursue a more promising path. The Bledsoe ATP prover [BT1] provides an excellent example of a significant overdirector.

Man-machine interaction -- the close involvement of the user in the proof search process. Although certain classes of problems can be proven fully automatically quite quickly, the user can be quite effective in cutting off "clearly" fruitless search paths or, more positively, providing certain instantiations for key variables that his experience suggests may be relevant or providing lemmas the prover might attempt to prove and use. Most systems provide for such interaction.

Advice -- direction from the builder or user that is superimposed upon the underlying proof scheme. The advice may be encoded in a procedure, appear in the overdirector (e.g., in an agenda design), or entered at runtime via man-machine interaction. The latter use of advice is not restricted to human-oriented provers.

Examples -- useful to help guide search, especially to suggest instantiation of key variables. Examples help direct search down possibly useful paths, which is an essential task faced in theorem proving. Unfortunately, we are not very sophisticated yet in use of examples, but some work is being done here. (See Bledsoe [Bl1] for use of examples to instantiate set variables.)

Models and counterexamples -- used to reject wrong paths of pursuit. Unlike an example, every true statement about the problem is true in the model so a statement false in the model can be abandoned in the proof search. We previously noted that the geometry theorem proving machine [GH1] provides the classic example of use of models in ATP; also see Reiter [Re1]. Counterexamples serve the same role of using semantic information to reject false statement, thus truncating erroneous search subtrees. Even if one cannot devise a suitable model for the entire problem setting, one might hope to find counterexamples for specific situations if such exist. (See Ballantyne and Bledsoe [BB1] for illustration of use of counterexamples.)

Analogy -- the use of a similar reasoning pattern in a related situation to guide search. This is a very important concept where little has been done, because progress is hard to come by. For comprehensive results here one needs good characterization of reasoning chains, including a metric over these chains. The best known work was done by Kling [Kl1] some years ago (1971) with interesting but less known work done by Munyer [Mu1]. A very simple use of analogy using syntactic symmetry occurred in the Gelernter *et al.* geometry theorem prover [GH1]. There a simple uniform replacement of points (point labels) for original points in a proof could yield a new result, because of the frequent occurrence of symmetric premises.

Learning -- the improvement of performance by automated retention of prior knowledge or processing thereof. Very little progress has been made here. Low level semi-automated learning takes place in the Boyer-Moore system summarized in this review when theorems are retained after conversion to rewrite rules (at the discretion of the user). Retention of previously established results is natural and common to most systems in some degree. The state-of-the-art in learning within a mathematics context occurs in concept formation and theorem generation in the work of Doug Lenat [Le1] (also see this volume).

We conclude the paper with some remarks regarding non-standard applications of automated theorem provers, that is, their use outside the role of proving theorems of interest to the mathematical community. An area receiving considerable attention over the last decade is formal program verification, where a program is "proved" correct by proving a series of "verification conditions" which are propositions relating a portion of code to formal specifications of the program. Although the logician Turing had anticipated the basic concept, John McCarthy [Mc1] and Robert Floyd [Fl1] are the direct creators of the methodology of program verification. The methodology of verification was advanced by the important early contributions of Naur [Na1] and Hoare [Ho1], the latter providing a specific formalization of the program semantics of an ALGOL (or PASCAL) type programming language via an axiom set. The first actual implementation of a program verifier was that of James King [Ki1] completed in 1969. Extensive hand computed formal verification of modest-sized programs by Ralph London about 1970 stimulated the development of more than one verification project. Since 1970 there have been numerous programs which had program verification as their primary or secondary goal (e.g. the Boyer-Moore System mentioned earlier). This relative intensity occurred because of the real need to find ways of locating non-trivial errors in programs, and ideally "certifying" the program correct, and because the verification condition statements are often long and complex, although the proofs are not deep in comparison with a typical theorem of interest to mathematicians.

A system that began quite early, in 1972, is the Stanford PASCAL verifier. The effort began with Shigeru Igarashi, Ralph London and David Luckham with later development by Nori Suzuki and others (see [Po1]). Its significance for us is that its development led to some notable work on decision procedures by Oppen and Nelson [NO1], work we have alluded to previously. The primary techniques used are rewrite rules and decision procedures. The key to good theorem proving of verification conditions is simplification of expressions. There is disagreement as to whether rewrite rules or decision procedures are the preferred technology for simplification, with Oppen and Nelson on the apparent minority side favoring decision procedures.

The GYPSY verification project directed by Donald Good [Go2],[GC1] at the U. of Texas has demonstrated itself as a major system by verifying a particular security property holds throughout a rather large (about 4000 lines) and complex program. The project, begun about 1974, uses a version of the Bledsoe theorem prover for its inference-making. Another verifier of importance is the AFFIRM system [GM1] developed by Dave Musser and Susan Gerhart at ISI (Santa Monica, California) beginning about 1976. Neither GYPSY nor AFFIRM need work directly with executable code but can verify algorithms at a somewhat more abstract level. AFFIRM is notable for its capability with abstract data types, which are data structures and associated operators that are governed by axioms. AFFIRM also uses a rewrite rule system, and the project has led to contributions to our understanding of such systems (see Musser [Mu2]). We mention one other verification system because of its unique application -- to verifying machine design, actually microcode within a computer. A microcode verifier was developed at IBM from about 1974 to 1980 by William Carter, Daniel Brand, William Joyner and George Leeman [CJ1]. Although (semi)automated verification is under active investigation for use by some government agencies and contractors, this use will be limited by the tremendous effort needed to prepare precise specifications of the process to be verified and then to bludgeon the proofs from the provers. There is still much manual work needed to use the semi-automated verifiers.

Program synthesis has been pursued using theorem proving techniques (e.g., see Manna and Waldinger [MW1] and Bibel [Bi2]), but this is in a very early stage of development.

We previously mentioned that theorem provers could be the heart of many artificial intelligence (AI) endeavors such as data base systems ("question-answerers") and robots. An even more surprising realization has been thrust on AI researchers, namely, that a theorem proving system can form the basis of a *general purpose* AI language. This concept was embodied in the language Prolog in 1972 by Colmerauer and Roussel ([Ck1], [RO3]), which was used for general problems on natural language processing. At this time, Prolog was mainly a theorem prover based upon SL-resolution, with draconian restrictions to shrink the search space. In fact these restrictions were equivalent to the use of Horn-clauses, discussed later by Kowalski in [Ko1] and [Ko3]. Kowalski emphasized the importance of separation of program logic and control and the value of a subset of logic called Horn clause logic, and, together with Van Emden, he determined the formal semantics of Prolog-like languages [KV1]. Prolog is now viewed by many as the most powerful existing language for implementing many designs of natural language systems, data base systems, knowledge-based expert systems and other AI tasks. It is the prototype language for the fifth generation computer systems being undertaken by the Japanese. One can regard Prolog crudely as an input resolution theorem prover with a few added control and convenience mechanisms. One may view Prolog programming as writing axioms about (or specifications for) the task to be performed and then giving the system a statement (goal) which the system must show follows from the axiom set. Traversing the correct proof (when found) forces the correct variable instantiations and side effects that solve the task. Readers intrigued by this should read Kowalski [Ko3]. It should be noted that some of the ideas used in logic programming go back to the idea of answer-extraction and related ideas of Green and Raphael ([GR1],[Gr1]).

This last quarter century has seen automated theorem proving grow from infancy to an adolescence, where its concepts have aided AI and where a few open problems in discrete mathematics have been proven. The next twenty-five years may well see open problems of major import proved where one of the coauthors of the work is a computer program.

Bibliography

[Ac1] Ackerman, W. *Solvable Cases of the Decision Problem.* North-Holland, Amsterdam, 1954.

[An1] Andrews, P. Refutations by matings. *IEEE Trans. on Computers*, C-25, 1976, 801-807.

[An2] Andrews, P. Theorem proving by general matings. *Jour. Assoc. for Comput. Mach.*, 1981, 193-214.

[An3] Andrews, P.B. Transforming matings into natural deduction proofs. *Proc. Fifth Conf. on Auto. Deduction* (Bibel and Kowalski, Eds.), Lecture Notes in Comp. Sci. 87, Springer-Verlag, Berlin, July 1981, 281-292.

[BB1] Ballantyne, A.M. and W.W. Bledsoe. On generating and using examples in proof discovery. *Mach. Intell.* 10, Ellis Harwood Ltd., Chichester, 1982, 3-39.

[BE1] Bläsius, K., N. Eininger, J. Siekmann, G. Smolka, A. Herold and C. Walther. The Markgraf Karl refutation procedure (Fall 1981). *Proc. Seventh Intern. Joint Conf. on Artif. Intell.*, Aug. 1981, 511-518.

[BF1] Breben, M., A. Ferro, E.G. Omodeo and J.T. Schwartz. Decision procedures for elementary sublanguages of set theory II: Formulas involving restricted quantifiers, together with ordinal integer, map, and domain notions. *Commun. Pure and Appl. Math.*, 1981, 177-195.

[Bi1] Bibel, W. An approach to a systematic theorem proving procedure in first-order logic. *Computing*, 1974, 43-55.

[Bi2] Bibel, W. Syntax-directed, semantics-supported program synthesis. *Artif. Intell.*, 1980, 243-262.

[Bi3] Bibel, W. Mating in matrices. *German Workshop on Artificial Intelligence 81* (Siekmann, Ed.), Informatik-Fachberichte 47, Springer-Verlag, Berlin.

[Bi4] Bibel, W. *Automated Theorem Proving*. Vieweg Verlag, Braunschweig, 1982.

[Bl1] Bledsoe, W.W. Non-resolution theorem proving. *Artif. Intell.*, 1977, 1-35. Also in *Readings in Artif. Intell.* (Webber and Nilsson, Eds.), Tioga, Palo Alto, 1981, 91-108.

[Bl2] Bledsoe, W.W. Using examples to generate instantiations for set variables. *Proc. Intern. Joint Conf. Artif. Intell.*, 1983.

[BM1] Boyer, R.S. and J.S. Moore. Proving theorems about LISP functions. *Jour. Assoc. for Comput. Mach.*, 1975, 129-144.

[BM2] Boyer, R.S. and J.S. Moore. *A Computational Logic*. Academic Press, New York, 1979.

[BM3] Boyer, R.S. and J.S. Moore. A verification condition generator for FORTRAN. *The Correctness Problem in Computer Science* (Boyer and Moore, Eds.), Academic Press, London, 1981.

[Bo1] Boyer, R.S. *Locking: a Restriction of Resolution.* Ph.D. dissertation, Univ. of Texas, Austin, 1971.

[BT1] Bledsoe, W.W. and M. Tyson. The UT interactive theorem prover. *Univ. Texas Math Dept. Memo ATP-17A*, May 1978.

[Ch1] Chang, C. The unit proof and the input proof in theorem proving. *Jour. Assoc. for Comput. Mach.*, 1970, 698-707.

[CJ1] Carter, W.C., W.H. Joyner, Jr. and D. Brand. Microprogram verification considered necessary. *AFIPS Conf. Proc. 47* (Nat. Computer Conf.), 1978, AFIPS Press, Montvale, 657-664.

[CK1] Colmerauer, A., H. Kanoui, R. Pasero and P. Roussel. Un Système de communication homme-machine en francais. Research Report CRI 72-18, Groupe Intelligence Artificielle, Université Aix-Marseille II, 1973.

[CL1] Chang, C. and R.C. Lee. *Symbolic Logic and Mechanical Theorem Proving.* Academic Press, New York, 1973.

[CM1] Clocksin, W.F. and C.S. Mellish. *Programming in Prolog.* Springer-Verlag, Berlin, 1981.

[Co1] Cook, S. The complexity of theorem-proving procedures. *Proc. Third ACM Symp. on Theory of Computing*, Shaker Hts., 1971, 151-158.

[Da1] Davis, M. Eliminating the irrelevant from mechanical proofs. *Proc. Symp. Applied Math.*, XV, 1963, 15-30.

[Da2] Davis, M. The prehistory and early history of automated deduction. *The Automation of Reasoning I* (Siekmann and Wrightson, Eds.), Springer-Verlag, 1983.

[DL1] Davis, M., G. Logemann and D. Loveland. A machine program for theorem proving. *Comm. Assoc. for Comput. Mach.*, 1962, 394-397.

[DP1] Davis, M. and H. Putnam. A computing procedure for quantification theory. *Jour. Assoc. for Computing Mach.*, 1960, 201-215.

[DS1] Downey, P.J., R. Sethi and R. Tarjan. Variations on the common subexpression problem. *Jour. Assoc. for Comput. Mach.*, to appear.

[Fi1] Fitch, F.B. *Symbolic Logic, an Introduction*. Ronald Press, New York, 1952.

[Fl1] Floyd, R. Assigning meaning to programs. *Mathematical Aspects of Computer Science* (Schwartz, Ed.), *Amer. Math. Soc.*, Providence, 1967, 19-32.

[FN1] Fikes, R.E. and N.J. Nilsson. STRIPS: a new approach to the application of theorem proving to problem solving. *Artif. Intell.*, 1971, 189-208.

[FO1] Ferro, A., E.G. Omodeo and J.T. Schwartz. Decision procedures for elementary sublanguages of set theory I: Multi-level syllogistic and some extensions. *Commun. Pure and Appl. Math.*, 1980, 599-608.

[GC1] Good, D., R.M. Cohen and J. Keeton-Williams. Principles of proving concurrent programs in Gypsy. *Proc. Sixth Conf. on Principles of Progr. Lang.*, 1979.

[Ge1] Gelernter, H. Realization of a geometry theorem-proving machine. *Proc. Intern. Conf. Information Proc.*, Paris:UNESCO House, 1959, 273-282. Also in *Computers and Thought* (Feigenbaum and Feldman, Eds.), McGraw-Hill, 1963, 134-152.

[Ge2] Gentzen, G. Untersuchungen uber das logische Schliessen. *Mathematische Zeitschrift*, 1934, 176-210, 405-431.

[GH1] Gelernter, H., J.R. Hanson and D.W. Loveland. Empirical explorations of the geometry-theorem proving machine. *Proc. West. Joint Computer Conf.*, 1960, 143-147. Also in *Computing and Thought* (Feigenbaum and Feldman, Eds.), McGraw-Hill, 1963, 153-163.

[Gi1] Gilmore, P.C. A program for the production of proofs for theorems derivable within the first order predicate calculus from axioms. *Proc. Intern. Conf. on Infor. Processing*, Paris: UNESCO House, 1959.

[Gi2] Gilmore, P.C. A proof method for quantification theory: its justification and realization. *IBM Jour. Research and Devel.*, 1960, 28-35.

[Gi3] Gilmore, P.C. An examination of the geometry theorem machine. *Artif. Intell.*, 1970, 171-187.

[GM1] Gerhart, S.L., D.R. Musser, D.H. Thompson, D.A. Baker, R.L. Bates, R.W. Erickson, R.L. London, D.G. Taylor and D.S. Wile. An overview of AFFIRM: a specification and verification system. *Proc. Intern. Federation of Infor. Proc.* 80, Australia, 1980, 343-348.

[Go1] Goldfarb, W.D. The undecidability of the second-order unification problem. *Theor. Computer Science*, 1981, 225-230.

[Go2] Good, D. The proof of a distributed system in GYPSY. Univ. of Texas, Austin, Inst. for Comp. Sci. Technical Report 30, 1982.

[Go3] Gould, W.E. A matching procedure for ω-order logic. Air Force Cambridge Research Laboratories, Report 66-781-4, 1966.

[GO1] Guard, J.R., F.C. Oglesby, J.H. Bennett and L.G. Settle. Semi-automated mathematics. *Jour. Assoc. for Comput. Mach.*, 1969, 49-62.

[Gr1] Green, C. Theorem-proving by resolution as a basis for question-answering systems. *Mach. Intell. 4* (Meltzer and Michie, Eds.), Edinburgh University Press, Edinburgh, 1969, 183-205.

[GR1] Green, C. and Raphael, B. The use of theorem-proving techniques in question-answering systems. *Proc. of the 23rd National Conf. of ACM*, Brandon Systems Press, Princeton, 1968, 169-181.

[He1] Hewitt, C. Description and theoretical analysis (using schemata) of PLANNER: a language for proving theorems and manipulating models in a robot. Ph.D. thesis, MIT, 1971. Also as MIT-AI-lab report AI-TR-258, 1972.

[Ho1] Hoare, C.A.R. An axiomatic basis for computer programming. *Commun. Assoc. for Comput. Mach.*, 1969, 576-580.

[HO1] Huet, G. and D.C. Oppen. Equations and rewrite rules: a survey. *Formal Languages - Perspectives and Open Problems*. (Book, Ed.), Academic Press, New York, 1980, 349-405.

[Hu1] Huet, G.P. A unification algorithm for typed λ-calculus. *Theor. Comput. Science*, 1975, 27-57.

[Hu2] Huet, G. Confluence reductions: Abstract properties and applications to term rewriting systems. *Jour. Assoc. for Comput. Mach.*, 1980, 797-821.

[JP1] Jensen, D.C. and T. Pietrzykowski. Mechanizing -order type theory through unification. *Theor. Computer Sci.*, 1976, 123-171.

[KB1] Knuth, D.E. and P.B. Bendix. Simple word problems in universal algebra. *Combinatorial Problems in Abstract Algebras* (Leech, Ed.), Pergamon, New York, 1970, 263-270.

[Ki1] King, J.C. *A program verifier*. Ph.D. thesis, Carnegie-Mellon Univ., 1969.

[KK1] Kowalski, R. and D. Kuehner. Linear resolution with selection function. *Artif. Intell.*, 1971, 227-260.

[Kl1] Kling, R.E. A paradigm for reasoning by analogy. *Artif. Intell.*, 1971, 147-178.

[Ko1] Kowalski, R.A. Predicate logic as a programming language. *Intern. Federation Infor. Proc. 74*, North-Holland, 1974.

[Ko2] Kowalski, R. A proof procedure based on connection graphs. *Jour. Assoc. Comput. Mach.*, 1975.

[Ko3] Kowalski, R.A. *Logic for Problem Solving*. North-Holland, 1980.

[Ko4] Kowalski, R.A. Private communication.

[Ku1] Kuehner, D.G. A note on the relation between resolution and Maslov's inverse method. *Mach. Intell. 6* (Meltzer and Michie, Eds.), American Elsevier, New York, 1971, 73-76.

[KV1] Kowalski, R. and M. Van Emden. The semantics of predicate logic as programming language. *Jour. Assoc. for Comput. Mach.*, 1976, 733-743.

[La1] Lankford, D. This volume.

[LB1] Lankford, D.S. and A.M. Ballantyne. Decision procedures for simple equational theories with commutative-associative axioms: Complete sets of commutative-associative reductions. Univ. of Texas Math Dept. Report ATP-35, Austin, 1977.

[Le1] Lenat, D. AM: Discovery in mathematics as heuristic search. *Knowledge-Based Systems in Artificial Intelligence* (Davis and Lenat, Eds.), McGraw-Hill, 1982, 3-149.

[LM1] Lusk, E.L., W.W. McCune and R.H. Overbeek. Logic machine architecture: (I) kernel functions (II) inference mechanisms. *Proc. Sixth Conf. on Auto. Deduction* (Loveland, Ed.), Lecture Notes in Comp. Sci. 138, Springer-Verlag, Berlin, June 1982, 70-108.

[Lo1] Loveland, D.W. Mechanical theorem proving by model elimination. *Jour. Assoc. for Comput. Mach.*, 1968, 236-251.

[Lo2] Loveland, D.W. A simplified format for the model elimination procedure. *Jour. Assoc. for Comput. Mach.*, 1969, 349-363.

[Lo3] Loveland, D.W. A linear format for resolution. *Symp. on Automatic Demonstration.* Lecture Notes in Math. 125, Springer-Verlag, Berlin, 1970, 147-162.

[Lo4] Loveland, D.W. A unifying view of some linear Herbrand procedures. *Jour. Assoc. for Comput. Mach.*, 1972, 366-384.

[Lo5] Loveland, D.W. *Automated Theorem Proving: a Logical Basis.* North-Holland, Amsterdam, 1978.

[LS1] Livesey, M. and J. Siekmann. Unification of A+C-terms (bags) and A+C+I-terms (sets). Universität Karlsruhe Interner Bericht Nr. 5/76, Karlsruhe, 1976.

[LS2] Loveland, D.W. and M.E. Stickel. A hole in goal trees: Some guidance from resolution theory. *Proc. Third Intern. Joint Conf. on Artif. Intell.*, Stanford, 1973, 153-161. Also in *IEEE Trans. on Computers* C-25, 1976, 335-341.

[Lu1] Luckham, D. Refinement theorems in resolution theory. *Symp. on Automatic Demonstration*, Lecture Notes in Math. 125, Springer-Verlag, Berlin, 1970, 163-190.

[Ma1] Maslov, S. Ju. An inverse method of establishing deducibility in classical predicate calculus. *Dokl. Akad. Nauk. SSR*, 1964, 17-20.

[Ma2] Maslov, S. Ju. Proof-search strategies for methods of the resolution type. *Mach. Intell. 6* (Meltzer and Michie, Eds.), American Elsevier, New York, 1971, 77-90.

[Mc1] McCarthy, J. Computer programs for checking mathematical proofs. *Proc. Amer. Math. Soc. Recursive Function Thy.*, New York, 1961, 219-227.

[MC1] Miller, D.A., E.L. Cohen and P.B. Andrews. A look at TPS. *Proc. Sixth Conf. on Auto. Deduction* (Loveland, Ed.), Lecture Notes in Comp. Sci. 138, Springer-Verlag, Berlin, June 1982, 50-69.

[MF1] Martin, W.A. and R.J. Fateman. The MACSYMA system. *Second Symp. on Symbolic Manipulation* (Petrick, Ed.), Los Angeles, 1971, 59-75.

[Mo1] Moses, J. Algebraic simplification, a guide for the perplexed. *Second Symp. on Symbolic Manipulation* (Petrick, Ed.), Los Angeles, 1971, 282-304.

[Mu1] Munyer, J.C. Analogy as a heuristic for mechanical theorem-proving. *Collected Abstracts of the Workshop on Auto. Deduction*, MIT, Aug. 1977.

[Mu2] Musser, D.R. On proving inductive properties of abstract data types. *Proc. Seventh ACM Symp. on Princ. of Progr. Lang.*, 1980.

[MW1] Manna, Z. and R. Waldinger. A deductive approach to program synthesis. *ACM Trans. on Progr. Lang.*, 1980, 90-121.

[Na1] Naur, P. Proof of algorithms by general snapshots. *BIT*, 1966.

[Ne1] Nevins, A.J. A human oriented logic for automatic theorem proving. *Jour. Assoc. for Comput. Mach.*, 1974, 606-621.

[Ne2] Nevins, A.J. Plane geometry theorem proving using forward chaining. *Artif. Intell.*, 1975, 1-23.

[Ne3] Nevins, A.J. A relaxation approach to splitting in an automatic theorem prover. *Artif. Intell.*, 1975, 25-39.

[No1] Norton, L.M. ADEPT - a heuristic program for proving theorems of group theory. Ph.D. Thesis, MIT, 1966. Also as *MIT Project MAC Report MAC-TR-33*, 1966.

[NO1] Nelson, G. and D. Oppen. Fast decision procedures based on congruence closure. *Jour. Assoc. for Comput. Mach.*, 1980, 356-364.

[NO2] Nelson, G. and D. Oppen. Simplification by cooperating decision procedures. *ACM Trans. on Progr. Lang. and Systems*, 1979.

[NS1] Newell, A., J.C. Shaw and H.A. Simon. Empirical explorations of the logic theory machine: a case study in heuristics. *Proc. Western Joint Computer Conf.*, 1956, 218-239. Also in *Computers and Thought* (Feigenbaum and Feldman, Eds.), McGraw-Hill, 1963, 134-152.

[Pl1] Plaisted, D.A. A simplified problem reduction format. *Artif. Intell.*, 1982, 227-261.

[Pl2] Plotkin, G.D. Building-in equational theories. *Mach. Intell. 7* (Meltzer and Michie, Eds.), Halsted Press, New York, 1972, 73-90.

[Po1] Polak, W. Program verification at Stanford: past, present, future. *German Workshop on Artificial Intelligence 81* (Siekmann, Ed.), Informatik-Fachberichte 47, Springer-Verlag, Berlin, 1981, 256-276.

[Pr1]   Prawitz, D. An improved proof procedure. *Theoria*, 1960, 102-139.

[Pr2]   Prawitz, D. Advances and problems in mechanical proof procedures. *Mach. Intell. 4* (Meltzer and Michie, Eds.), Edinburgh University Press, Edinburgh, 1969, 73-89.

[PS1]   Peterson, G.E. and M.E. Stickel. Complete sets of reductions for some equational theories. *Jour. Assoc. for Comput. Mach.*, 1981, 233-264.

[Re1]   Reiter, R. A semantically guided deductive system for automatic theorem proving. *Proc. Third Intern. Joint Conf. Artif. Intell.*, 1973, 41-46. Also in *IEEE Trans. on Computers*, 1976, 328-334.

[Ro1]   Robinson, J.A. A machine-oriented logic based on the resolution principle. *Jour. Assoc. for Comput. Mach.*, 1965, 23-41.

[Ro2]   Robinson, J.A. Automatic deduction with hyper-resolution. *Intern. Jour. of Computer Math.*, 1965, 227-234.

[Ro3]   Roussel, P. Prolog: manuel de réference et d'utilisation. Rapport de recherche CNRS, Groupe Intelligence Artificielle, Université Aix-Marseille II, 1975.

[RS1]   Raulefs, P., J. Siekmann, P. Szabo and E. Unvericht. A short survey on the state of the art in matching and unification problems. *SIGSAM Bulletin*, May 1979, 14-20.

[Sa1]   Sandford, D.M. *Using Sophisticated Models in Resolution Theorem Proving*. Lecture Notes in Computer Science 90, Springer-Verlag, Berlin, 1980.

[Sh1]   Shostak, R.E. Refutation graphs. *Artif. Intell.*, 1976.

[Sh2]   Shostak, R.E. A practical decision procedure for arithmetic with function symbols. *Jour. Assoc. for Comput. Mach.*, 1979, 351-360.

[Sh3]   Shostak, R.E. Deciding combinations of theories. *Proc. Sixth Conf. on Auto. Deduction* (Loveland, Ed.), Lecture Notes in Comp. Sci. 138, Springer-Verlag, Berlin, June 1982, 209-223.

[Si1] Sickel, S. Interconnectivity graphs. *IEEE Trans. on Computers C-25*, 1976.

[SJ1] Suzuki, N. and D. Jefferson. Verification decidability of Presburger array programs. *Proc. Conf. on Theor. Comput. Science*, Univ. of Waterloo, Aug. 1977.

[Sl1] Slagle, J.R. Automatic theorem proving with built-in theories of equality, partial order and sets. *Jour. Assoc. for Comput. Mach.*, 1972, 120-135.

[Sl2] Slagle, J.R. Automated theorem proving for theories with simplifiers, commutativity and associativity. *Jour. Assoc. for Comput. Mach.*, 1974, 622-642.

[St1] Stickel, M.E. A complete unification algorithm for associative-commutative functions. *Proc. Fourth Intern. Joint Conf. on Artif. Intell.*, Tbilisi, USSR, 1975. Also see *Jour. Assoc. for Comput. Mach.*, 1981, 423-434.

[SW1] Siekmann, J. and G. Wrightson (Eds.). *The Automation of Reasoning*. Vols. I and II, Springer-Verlag, 1983.

[Tu1] Turing, A.M. Computing machinery and intelligence. *Mind*, 1950, 433-460.

[Wa1] Wang, H. Toward mechanical mathematics. *IBM Jour. Research and Devel.*, 1960, 2-22. Also in *Logic, Computers and Sets*, Chelsea, New York, 1970.

[Wa2] Wang, H. Proving theorems by pattern recognition, Part I. *Commun. Assoc. for Comput. Mach.*, 1960, 220-234.

[Wa3] Wang, H. Proving theorems by pattern recognition, Part II. *Bell System Technical Jour.*, 1961, 1-41.

[WC1] Wos, L., D. Carson and G. Robinson. The unit preference strategy in theorem proving. *AFIPS Conf. Proc. 26*, Spartan Books, Wash., D.C., 1964, 615-621.

[Wo1] Wos, L. Solving open questions with an automated theorem-proving program. *Proc. Sixth Conf. on Auto. Deduction* (Loveland, Ed.), Lecture Notes in Comp. Science 138, Springer-Verlag, Berlin, June 1982, 1-31.

[WR1] Wos, L., G. Robinson, D. Carson and L. Shalla. The concept of demodulation in theorem proving. *Jour. Assoc. for Comput. Mach.*, 1967, 698-709.

[WR2] Wos, L. and G. Robinson. Paramodulation and set of support. *Symp. on Automatic Demonstration*, Lecture Notes in Math. 125, Springer-Verlag, Berlin, 1970, 276-310.

[WW1] Winker, S.K. and L. Wos. Procedure implementation through demodulation and related tricks. *Proc. Sixth Conf. on Auto. Deduction* (Loveland, Ed.), Lecture Notes in Comp. Science 138, Springer-Verlag, Berlin, June 1982, 109-131.

[WW2] Winker, S., L. Wos and E. Lusk. Semigroups, antiautomorphisms, and involutions: a computer solution to an open problem, I. *Math. of Computation*, 1981, 533-545.

[WW3] Wos, L., S. Winker and E. Lusk. An automated reasoning system. *AFIPS Conf. Proc.: Natl. Comp. Conf.*, AFIPS Press, Montvale, 1981, 697-702.

[Wu1] Wu, Wen-tsün. On the decision problem and the mechanization of theorem-proving in elementary geometry. *Scientia Sinica* XXI (2), Mar.-Apr. 1978, 159-172.

[YR1] Yates, R.A., B. Raphael and T.P. Hart. Resolution graphs. *Artif. Intell.*, 1970, 257-289.

AUTOMATED THEOREM PROVING

Some major contributions to ATP classified by philosophy of design

Logic	Human Simulation
Presburger prover Davis 1954	
	Logic Theory Machine, Newell, Shaw, Simon 1957
Beth tableau prover Gilmore 1959	
	Geometry Theorem Proving Machine, Gelernter et. al. 1959
Gentzen-Herbrand prover Hao Wang 1959	
D-P procedure Davis, Putnam 1960	
Matching Prawitz 1960	
Resolution Robinson 1964	Semi-Automated Math. Guard, Bennett, Easton 1963-1967

(continued next page)

Resolution prover
Wos et.al.1965

Resolution restrictions,
strategies et.al.
1965-1970

Knuth-Bendix theorem
Knuth, Bendix 1970

ATP project
Bledsoe et.al.1970

AURA
Overbeek, Wos, Winker,
Lusk,Smith 1972-

Human-oriented prover
Nevins 1971

Computational
Logic Theorem
Prover, Boyer,
Moore 1973 -

TPS - higher order logic
Andrews 1976

Decision procedures
for quantifier-free
theories, Shostak,
Oppen et.al.1977

Markgraf Karl
Procedure,
Siekmann
et.al.1980 -

LMA-an ATP tool kit.
ITP-a people's machine
built from LMA tools.
Lusk, McCune, Overbeek
1980-

FIGURE 1

AUTOMATED THEOREM PROVING

	Resolution Refinement	Creators [Approx. Date]	Importance
1.	Basic resolution	J.A. Robinson [1964]	Original resolution proof procedure
2.	Unit preference	Wos, G. Robinson, Carson [1964]	Used in AURA
3.	Set-of-support	Wos, G. Robinson, Carson [1964]	Used in AURA and ITP
4.	Hyperresolution	J.A. Robinson [1965]	Used in AURA and ITP
5.	Model elimination	Loveland [1965]	Completion of problem reduction format; Basis for SL-resolution
6.	Linear resolution	Loveland Luckham [1968]	Implements depth-first search in resolution
7.	SL-resolution (TOCS resolution)	Kowalski Kuehner [1970] (Loveland)	Very tightly constrained linear resolution; Used in early version of PROLOG language
8.	Locking	Boyer [1970]	Very tightly constrained ordered-clause resolution
	Incomplete refinements		
9.	Unit clause resolution Input clause resolution	Wos et al. [1964] Chang, others [1970]	These forms are complete for Horn clause logic; Input resolution is the underlying format in the present PROLOG language
	Handling equality		
10.	Paramodulation (Demodulation)	Wos, G. Robinson (+ Carson, Shalla) [1968] ([1967])	Eliminates the need for almost all equality axioms

Some resolution variants and related procedures

FIGURE 2

Citation for Hao Wang as Winner of The Milestone Award in Automated Theorem-Proving

The first "milestone" prize for research in automated theorem proving is hereby awarded to Professor Hao Wang of Rockefeller University for his fundamental contributions to the founding of the field. Among these, the following may be listed:

1. He emphasized that what was at issue was the development of a new intellectual endeavor (which he proposed to call "inferential analysis") which would lean on mathematical logic much as numerical analysis leans on mathematical analysis.
2. He insisted on the fundamental role of predicate calculus and of the "cut-free" formalisms of Herbrand and Gentzen.
3. He implemented a proof-procedure which efficiently proved all of the over 350 theorems of Russell and Whitehead's "Principia Mathematica" which are part of the predicate calculus with equality.
4. He was the first to emphasize the importance of algorithms which "eliminate in advance useless terms" in a Herbrand expansion.
5. He provided a well-thought out list of theorems of the predicate calculus which could serve as challenge problems for helping to judge the effectiveness of new theorem-proving programs.

Martin Davis, Chairman
David Luckham
John McCarthy

Articles by Hao Wang on Automated Theorem Proving.

1. Towards Mechanical Mathematics. *IBM Journal of Research and Development*, Vol. 4 (1960) pp. 2-22.

2. Proving Theorems by Pattern Recognition - I. *Communications of the ACM*. Vol. 3 (1960) pp. 220-234.

3. Proving Theorems by Pattern Recognition II. *Bell System Technical Journal*, Vol. 40 (1961), pp. 1-41.

4. Mechanical Mathematics and Inferential Analysis. Braffort and Hirschberg (Editors), *Computer Programming and Formal Systems*, North-Holland, 1963, pp. 1-20.

5. The Mechanization of Mechanical Arguments. *Proceedings of the Symposium in Applied Mathematics*, Vol. 15 (1963), pp. 31-40.

6. Formalization and Automatic Theorem-Proving. *Proceedings of the IFIP Congress*, 1965, pp. 51-58.

7. On the Long Range Prospects of Automatic Theorem Proving. *Lecture Notes in Mathematics*, 125, Springer-Verlag, 1970, pp. 101-111.

COMPUTER THEOREM PROVING AND ARTIFICIAL INTELLIGENCE

Hao Wang[1]

It gives me a completely unexpected pleasure to be chosen as the first recipient of the milestone prize for ATP, sponsored by the International Joint Conference on Artificial Intelligence. I have worked in a diversity of fields; I am correspondingly limited in my capacity to appreciate, or express my appreciation of, a large range of efforts in each of these fields; and I tend to shun positions of power. Undoubtedly to a considerable extent as a result of these innocent shortcomings, honors have a way of passing me by. I have indeed slowly grown used to this. Hence, the present reward has surprised me.

I had hoped to arrive at an understanding of current work in the field in order to relate it to my early aspirations and expectations. For this purpose, I had asked Professor Woody Bledsoe for help, who has kindly sent me some papers and references. I soon realized that with my present preoccupation with philosophy I shall not be able to bring myself to concentrate enough to secure a judicious evaluation of these writings which at first sight appear largely alien to my own way of thought. Indeed several years ago Joshua Lederberg thoughtfully sent me a copy of Dr. D. Lenat's

---
[1] Mathematics Dept., Rockefeller University, 1230 York Ave., New York, NY 10021

large dissertation *(AM: an artificial intelligence approach to discovery in mathematics as heuristic search)* which I found thoroughly unwieldy and could not see how one might further build on such a baffling foundation. This previous experience strengthens my reluctance to plunge into a study of current work. Hence, I have decided to limit myself to a summary of my own views with some comments only on one line of current research with which I happen to have some familiarity. Actually I have just noticed, in the announcement of the prize, an explicit statement that "the recipients will present one-hour lectures on their work." Hence, restricting to work familiar to me is quite proper. It is possible that certain aspects of my thought have been bypassed so far and they may be of use to future research. My work in computer theorem proving and related papers are given in the bibliography of this paper.

Around the beginning of 1953 I became dissatisfied with philosophy (as seen at Harvard) and, for other reasons, I also wanted to do something somewhat more obviously useful. Computers struck me as conceptually elegant and closely related to my training. The first idea was to see in computers a home for the obsessive formal precision of (the older parts of) mathematical logic which mathematicians tend to find irrelevant and worse, pedestrian and perhaps a hindrance to creativity. With little conception of the industrial world, I accidentally got an appointment as 'research engineer' at Burroughs. Unfortunately I was not permitted to use the only local computer and was even discouraged from taking a course for electronics technicians. I was reduced to speculations which led to a programming version of Turing machines and some idle thoughts on 'theorem-proving machines'. The material was written up only quite a bit later (probably in 1955) and published in January 1957 (see reference [7]; section 6 is devoted to theorem-proving).

The section begins with the observation that logicians and computer scientists share a common concern. 'Both groups are interested in making thoughts articulate, in formalization and mechanization of more or less vague ideas.' Decidable theories are mentioned in passing. But the main emphasis is on 'the possibility of using machines to aid theoretical mathematical research on a large scale'. A few modest research goals are mentioned: the independence of axioms in the propositional calculus; 'we often wish to test whether an alleged proof is correct'; this is related to the matter of filling in gaps when 'the alleged proof is only presented in sketch'; to confirm or disprove our hunches; 'to disentangle a few exceedingly confusing steps' in a (proposed) proof. Herbrand's theorem on cut free proofs is mentioned as possibly helpful. 'We can instruct a proving machine to select and print out theorems which are short but require long proofs (the 'deep' theorems).' And a general observation: 'The important point is that we are trading qualitative difficulty for quantitative complexity'.

It was not until the summer of 1958 that I began to work actually with a computer (an IBM 704, using SAP, the Share Assembly Programming Language), thanks to Bradford Dunham's arrangement. It was an enjoyable long summer (owing to the fact that Oxford begins school only late in October) when I learned programming, designed algorithms, and wrote three surprisingly successful programs. 'The first program provides a proof-decision procedure for the propositional calculus' and proves all such theorems (about 220) in *Principia mathematica* (briefly *PM)* in less than three (or five?) minutes. 'The second program instructs the machine to form itself propositions of the propositional calculus from basic symbols and select nontrivial theorems.' 'Too few theorems were excluded as being trivial, because the principles of triviality actually included in the program were too crude.' The third program dealt with the predicate calculus with equality and was able to prove 85% of the over 150 theorems in *PM* in a few minutes. This was improved in the summer of 1959 (see below) to prove all these over 150 theorems. The improved program embodies a proof-decision procedure for the 'AE predicate calculus with equality' '(i.e., those propositions which can be transformed into a form in which no existential quantifier governs any universal quantifier) which includes the monadic predicate calculus as a subdomain.'

What is amazing is the discovery that the long list of all the theorems of *PM* in the undecidable predicate calculus (a total of ten long chapters) falls under the exceedingly restrictive domain of the AE predicate calculus. This seems to suggest that of the vast body of mathematical truths, most of the discovered theorems are among the easier to prove even in mechanical terms. Such a suggestion would be encouraging for the project of computer theorem proving. A more definite general implication of the situation of *PM* with regard to the special case of the predicate calculus is the desirability of selecting suitable subdomains of mathematical disciplines which are, like the AE calculus, quickly decidable and yet contain difficult theorems in the human sense. The recent work of Wu (see below) in elementary and differential geometry would seem to be an excellent illustration of such a possibility.

The work of the summer was written up and submitted to the *IBM Journal* in December 1958. But for some curious reason it appeared only in January 1960. The paper (see reference 8) begins with a comparison between calculation and proof, arguing that the differences present no serious obstacles to fruitful work in mechanical mathematics. In particular, 'much of our basic strategies in searching for proofs is mechanizable but not realized to be so because we had had little reason to be articulate on such matters until large, fast machines became available. We are in fact faced with a challenge to devise methods of buying originality with plodding, now that we are in possession of slaves which are such persistent plodders.' 'It is, therefore, thought that the general domain of algorithmic analysis can now begin to be enriched by the inclusion of inferential analysis as a younger companion to the fairly well-established but still rapidly developing leg of numerical analysis.'

The paper concludes with a long list of remarks such as the relation between quantifiers and functions, quantifier-free number theory, the central place of induction in number theory, the *Entscheidungsproblem*, "formalizing and checking outlines of proofs, say, from textbooks to detailed formulations more rigorous than *PM*, from technical papers to textbooks, or from abstracts to technical papers," etc. (By the way, I now feel that the reverse process of condensing is of interest both for its own sake and for instructiveness to the original project.) 'It seems as though logicians had worked with the fiction of man as a persistent and unimaginative beast who can follow rules blindly, and then the fiction found its incarnation in the machine. Hence, the striving for inhuman exactness is not pointless, senseless, but gets direction and justification.'

At that time, several reports were in circulation by Newell-Shaw-Simon which deal with the propositional calculus in an inefficient manner and contain claims against 'automatic' procedures for producing proofs. Their claims were refuted conclusively by my work. Hence, I included some discussion of their work. 'There is no need to kill a chicken with a butcher's knife. Yet the net impression is that Newell-Shaw-Simon failed even to kill the chicken with their butcher's knife.' A larger issue is the tendency at that time to exclude what are now called 'expert systems' from artificial intelligence. Relative to the project of theorem proving, I put forward at that time a plea for 'expert systems':

> 'Even though one could illustrate how much more effective partial strategies can be if we had only a very dreadful general algorithm, it would appear desirable to postpone such considerations till we encounter a more realistic case where there is no general algorithm or no efficient general algorithm, e.g., in the whole predicate calculus or in number theory. As the interest is presumably in seeing how well a particular procedure can enable us to prove theorems on a machine, it would seem preferable to spend more effort on choosing the more efficient methods rather than on enunciating more or less familiar generalities. And it is felt that an emphasis on mathematical logic is unavoidable because it is just as essential in this area as numerical analysis is for solving large sets of simultaneous numerical equations.'

In the summer of 1959 I went to Murray Hill to continue with the work of the preceding summer, and before long a program adequate to handling the predicate calculus with equality in *PM* was completed. It contains about 3,200 instructions. All the over 350 theorems of this part of *PM* are proved in 8.4 minutes with an output of about 110 pages of 60 lines each. The result was written up before the end of 1959, published in April 1960 (see reference 9) and then presented in May at the ACM Conference on Symbol Manipulation. Most of the other papers at the conference were devoted to introducing programming languages. I recall that several of the speakers illustrated the power of their language by giving a succinct presentation of 'the Wang algorithm' viz. the simple program for the propositional calculus described in reference 8. With a special emphasis on the role of logic (which I now consider an unnecessary limitation), I suggested that 'computers ought to produce in the long run some fundamental change in the nature of all mathematical activity.' The recent extensive use of computers in the study of finite groups and in the solution of the four color problem would seem to confirm this expectation and point to an effective general methodological principle of paying special attention to and taking advantage of potentially algorithmic dimensions of problems in pure mathematics.

Toward the end of 1959, I was struck by the interesting patterns of the Herbrand expansions of formulas in the predicate calculus. My thought at that time was that if we can handle the interlinks within each pattern in an effective way, we would not only decide many subdomains of the predicate calculus but also supply a generally efficient procedure to search for proofs as well. At that time Paul Bernays and Kurt Scütte happened to be at the Princeton Institute. Schütte impressed on me the importance of the challenging and long-outstanding problem of whether the apparently simple AEA case is decidable. The attraction of such a clean problem lured me away from the computer enterprise which seemed to suggest indecisively a large number of different directions to continue.

While working on this theoretical problem, since I enjoyed discussions with my colleagues who had little training in logic, I came to introduce a class of tiling problems (the 'domino problems') that captures the heart of the logical problem. I was surprised to discover that the 'origin-constrained domino problem' can simulate Turing machines and is therefore undecidable. These results were written up in reference 10 but unfortunately I neglected to include the easy result on the origin-constrained problem, which was made into a technical report only a year later (in the summer of 1961). Soon after that, with the assistance of my student A.S. Kahr, I was able to prove, using the domino problems, that the AEA case is undecidable, contrary to my previous conjecture.

Reference 10 also includes a few concrete examples. An analysis is given of how $x \neq x + 1$ and $x + y = y + x$ would be proved in quantifier-free number theory from the Peano axioms. A research paper by J. Hintikka is reformulated into a relatively simple example of deciding a formula in the predicate calculus. This last example was further elaborated in reference 12 in which the vague concept of 'stock-in-trade systems' was introduced and two examples from number theory (the infinitude of primes and the irrationality of $\sqrt{2}$) were mentioned. It was only in the 1965 IFIP address (reference 14) that I offer more detailed discussions of four examples from number theory and three examples from the predicate calculus. I am not aware that anybody has actually solved any of these examples on a computer, either by working out my suggestions or by other means. I believe that these examples remain useful today because particularity often sharpens the challenge to provoke sustained thinking. Some general points underlying the comments on these examples are made more explicit in reference 15, which also goes into more general speculation such as the following.

'We are invited to deal with mathematical activity in a systematic way .... one does expect and look for pleasant surprises in this requirement of a novel combination of psychology, logic, mathematics and technology.' There is the matter of contrasting the formal and the intuitive: for example, Poincaré compares Weierstrass and Riemann, Hadamard finds Hermite's discoveries more mysterious than those of Poincaré. The four stages of an intellectual discovery are said to be: (1) preparation, (2) incubation, (3) illumination, (4) verification. To mechanize the first and the last 'stages appear formidable enough, but incubation leading to illumination would seem in principle a different kind of process from the operations of existing computers.'

A chapter of reference 16 is devoted to a general discussion of 'minds and machines,' with a long section on the relation to Gödel's incompleteness theorems, as well as a section contributed by Gödel. An account of some aspects of computer theorem proving is included in reference 17. I should like to take this opportunity to publicize an idea of using some simple mathematics to bridge the gap between computer proofs and human proofs. I have in mind proofs which include more formal details than usual which are more likely candidates for comparatively easier computerization. In particular, around 1965 I wrote a number of set-theoretical exercises, some by myself and some with my students: see references 18 to 23. In all these cases I am confident that I can without a good deal of effort turn the fairly detailed outlines into completely formal proofs. My suggestion is to use these papers as initial data to help the designing of algorithms for discovery and verification of sketches of proofs.

For some time I have assumed that it is desirable to have accomplished mathematicians involved in the project of using computers in their special areas. This has certainly taken place in several regions as auxiliaries to broader projects of settling certain open problems (say on finite groups and map coloring). A less frequent phenomenon is to design efficient algorithms wherever possible and then examine their ranges of application. It used to be the case that pure mathematicians tended to shun computers as something ugly and vulgar. In recent years, with the rapid spreading of computers, more and more younger mathematicians are growing up with computers. Hence, a basic obstacle to their use has been removed and the remaining division is less sharp: some would only use computers as a hobby while others would more or less consciously think of taking advantage of computers to extend their own capacity for doing more and better mathematics.

Along the line of searching for algorithms in branches of mathematics, I happen to have some familiarity with the work of Wu Wen-tsün, an accomplished geometer. (At one time Wu taught secondary schools and, for lack of contact with modern mathematics, became an exceptional expert of elementary geometry. This misguided training has since not only helped him in his research in advanced mathematics but also turned out to be very useful for his recent project of mechanizing certain parts of geometry.) In 1977 Wu discovered a feasible method which can prove mechanically most of the theorems in elementary geometry involving only axioms of incidence, congruence and parallelism but not axioms of order and continuity (in Hilbert's famous axiomatization). The method was later extended to elementary differential geometry. These are reported in 1978 and 1979 (references 26 and 27).

Among the theorems proved on an HP computer (9835A and then 1000) are the gou-gu or Pythagorean theorem, some trigonometric identities, the Simson-line theorem, the Pappus theorem, the Pascal theorem, and the Feuerbach theorem. Among the 'new theorems' discovered and proved are: (1) the 'anticenter theorem and the anticenter line theorem'; (2) the 'Pappus-point theorem'; (3) the 'Pascal-conic theorem'. In differential geometry the results proved include Dupin's theorem on triple orthogonal families of surfaces, the affine analogue of the Bertrand curve-pair theorems, and the Bäckland theorems on the transformation of surfaces. For more details, please consult Wu's recent paper, reference 28.

In 1978 Douglas McIlroy and I experimented with Wu's algorithm for elementary geometry and found that without proper caution the requirement for storage tended to be large. For example, a crucial step is to test whether a polynomial g vanishes modulo given irreducible polynomials $p_1,...,p_n$. If we work with each coefficient separately rather than deal with the whole expansion of g, the size of calculation is reduced considerably. McIlroy has since continued to toy with matters surrounding the algorithm and accumulated an amount of interesting data.

I should now like to turn to a few general remarks on the larger topic of artificial intelligence (briefly AI) as well as on its relation to theorem proving.

The area of AI is rather indefinite. In a broad sense it includes 'expert systems' (or 'knowledge engineering') which, according to Michie ([4], p. 195) 'transfer human expertise in given domains into effective machine forms, so as to enable computing systems to perform convincingly as advisory consultants. Expert systems development is becoming cost effective.' Michie goes on to list more than ten examples. One might argue about the cost effectiveness but at least we get here a tangible criterion of success. Most of the 'basic' or 'theoretical' work in AI does not admit any such transparent standards of evaluation: this is undoubtedly one of the several reasons why the field is so controversial. But let me delay over expert systems a bit.

DENDRAL takes the pattern generated by subjecting an unknown organic chemical to a mass spectrometer, and infers the molecular structure. This program is commonly regarded as a success. Indeed I was full of enthusiasm about it when I first heard of the project in the 1960s. But the outcome fell short of my high expectations. One problem is that experienced chemists are often unable to state explicitly the rules they know how to use when confronting actual particular situations. An unquestionable authority told me that the direct outcome of DENDRAL was not worth all the investment put into it but that when one took into consideration its more general influence in starting the trend to develop expert systems, the evaluation was different. The more interesting meta-DENDRAL project has apparently been left in abeyance for complex sociopsychological reasons.

I can no longer be sure whether my programs for proving theorems in the predicate calculus qualify as an expert system. They did use some 'expert' knowledge of logic. And I was told that people did use my programs for other purposes. In this case it is clear that they were cost effective since it involved only a few months' labor of one person plus less than ten hours on a 704 machine. At any rate, my getting a prize from an international AI organization shows that I am finally being accepted into the AI community, twenty years after I was commended as a 'cybernetist' in the Soviet Union. This example illustrates why I get confused over the range of AI which, like any other subject, undoubtedly evolves with time.

Critics of AI quote the less interesting part of my work only in order to berate the unprofessional job of Newell-Shaw-Simon (e.g., Dreyfus [1], first edition, p. 8 and Weizenbaum [25], p. 166). (By the way, Dreyfus reduces all the 220 theorems to only the 53 selected by Newell-Shaw-Simon.) They miss the central point that although the predicate calculus is undecidable, all the theorems of *PM* in this area were so easily proved. Enthusiasts for AI simply leave me alone. The professional writer McCorduck [3] at least makes a bow to theorem proving: 'and I rationalize such neglect by telling myself that theorems would only scare away the nonspecialist reader this book is intended for.' This incidentally leads to some other factors which render the field of AI controversial. It incites popular interest yet, unlike physics, it is a more mixed field in which natural scientists and social scientists (not to mention philosophers) with quite different standards meet. More, since it is near big technology, it is close to industrial and government money. As a result of all these factors, public relations tend to play a larger role than in more mature disciplines. And that tends to put some people off even though they find the intellectual core of many problems in the area appealing and challenging. Exaggerated and irresponsible claims and predictions, instead of being chastised, appear to be a central ingredient of the glory of many of the 'giants' in this field.

The controversies over AI are a mixed bag. At one end there are relatively solid accomplishments which may be evaluated in terms of 'cost effectiveness' and cleancut 'milestones.' There is a middle region which contains results which are not particularly decisive in themselves but seem to promise more exciting future developments. But in most cases the promises are fairly subjective and precarious. When we move to 'basic' AI which in many ways is the most exciting for the widest audience, we seem to be arguing over conclusions for or against which our universal ignorance of crucial aspects offers so little evidence that individual psychology over ethical, aesthetic, political, and philosophical matters generally appears to play a dominant role. The fascinating question seems to shift from who is right (science?) to the motley of sources of the vehement disagreements (sociopsychology of convictions). Let me try to sort out some of the larger issues as I darkly see them.

Is the brain a computer? By now it seems generally agreed that the brain is certainly not a computer of the kind we are familiar with today. This point was argued, for example, by John von Neumann, who used to say that the brain is a professional job (which the computer isn't), in *The computer and the brain*, 1958. There remains the position that whatever the brain can do, some computer of the current type can do it as well, some day. This seems implausible and in any case so indefinite that we don't know how to begin assembling evidence for such a position. Recent proponents of mechanism take rather different forms. For example, according to Hofstadter ([2], p. 579), the AI thesis is: 'As the intelligence of machines evolves, its underlying mechanisms will gradually converge to the mechanisms underlying human intelligence.' I presume that 'the evolution of machines' will require quite a bit of human intervention. I can't help wondering that if we postulate the possibility of such convergence, why not include also a divergence? Since human intelligence is not perfect, if the machine is believed to be on the way of converging to the complex human intelligence, surely it can be expected to acquire other superiorities (apart from the obvious one of speed in certain situations). There is no indication how long the evolution will take.

Webb [24] undertakes to argue that Gödel's incompleteness theorems are for rather than against mechanism. (By the way, he also argues the stronger thesis that these theorems are for rather than against Hilbert's program. Here he is certainly wrong since Hilbert's requirement of 'finitist viewpoint' is sufficiently definite to admit a conclusive refutation of his program by Gödel's results.) Webb is certainly only arguing for a matter of principle. 'But the ultimate test would be to simulate Hilbert's metatheorem about Thales' theorem, undertaken to analyze his implicit commitment to spatial intuition in plane geometry. Existing machines are presumably light years away from results like this, much less the philosophical motivation for them, but I see no obstacle of principle' (p. 111). Is he merely saying that nobody has mathematically refuted mechanism (or, for that matter, the existence of God)?

The more familiar battleground over larger claims of AI involves the issues whether machines can handle natural language, acquire commonsense understanding, and especially situated understanding. In this area we of course easily come upon perennial problems in philosophy and the methods of psychology. Before getting into this more nebulous area, it may be better to discuss a somewhat neutral and modest thesis: Every 'precisely described' human behavior (in particular, any algorithm) can be simulated (realized) by a suitably programmed computer. (In the 1950s I heard something like this attributed to John von Neumann who believed that not all human behavior can be precisely described.) There is a slight ambiguity in this thesis which relates to the by now familiar distinction between theoretical and feasible computability. We can precisely describe a procedure for finding a nonlosing strategy for playing chess or deciding theorems in elementary geometry. But, as we know, these computations are not feasible. It could be argued that carrying out such computations is not normal human behavior. But, for example, in deciding tautologies in the propositional calculus or dealing with cases of the travelling salesman problem, we do naturally try the standard simple algorithms which are in the general case not feasible. One would have thought at first that this type of behavior is exactly what the computer can profitably simulate and get beyond our limited capacity in managing complexity. The point is that there are many different levels of precise description which, even with the computer's increased speed and storage, are often inseparable from commonsense understanding or sophisticated reasoning, if the descriptions are to lead to feasible computation. For example, if we increase a thousandfold the matrix for pattern recognition, the computers could do a much better job, except for the fact that the procedures are no longer feasible for the computers.

There is a tendency to view theorem proving as of little central interest to the large goals of AI. The idea is that it is too pure and clean to run into the tougher problems. But I disagree. For example, the surprising human ability to make connections may be illustrated by the discovery of hundreds of problems equivalent to P = NP. I mentioned earlier the stage of incubation leading to illumination which is typically applicable to mathematical discoveries. Poincaré reports on how dreams played a crucial role in the process of some of his discoveries. If we use the distinction of conscious, preconscious and unconscious, it is said that phenomenology places its main emphasis on the line between the conscious and the preconscious, while psychoanalysis is preoccupied with the line between the unconscious and the preconscious. As we know, dreams are one of the main tools for getting at the unconscious. Hence, the experience of Poincaré points to the involvement of the unconscious in mathematical discoveries. That is certainly not something 'pure and clean' nor irrelevant to the larger contrast of minds and machines.

I recently observed to a colleague, 'How are we to introduce the unconscious into the machines?' After a pause, he answered. 'I would say that machines are entirely unconscious.' In other words, we haven't yet made machines 'conscious' or able to simulate our conscious behaviors, it seems very remote to get at the unconscious and then endow the machines with a reservoir of the unconscious. This sort of consideration may be what has led to the talk of evolution, child development, and the 'Society of Minds.' Minsky speaks of trying 'to combine methods from developmental, dynamic, and cognitive psychological theories with ideas from AI and computational theories. Freud and Piaget play important roles. In this theory, mental abilities, both "intellectual" and "affective" (and we ultimately reject the distinction) emerge from interactions between "agents" organized into local, quasi-political hierarchies.' ([5], p. 447). This is certainly an ambitious and awe-inspiring project which would seem to be remote from existing computers. I am much in sympathy with the project, but more as one in philosophy rather than one in AI. Perhaps the idea is to simulate the development of mankind and the universe but to speed up evolution and individual as well as societal developments to such a high degree that the mental states of living human beings all get mirrored into a gigantic computer program? In one sense, since the project is interesting in itself, a good theory is welcome more or less independently of its connections with AI. The problem is rather whether a preoccupation with AI helps or hinders the development of a good theory of such a dimension.

Minsky and Seymour Papert draw on Freud and Piaget but say nothing about the sources of their political theory. Dreyfus rather appeals to Heiddeger, to Wittgenstein, to Merleau-Ponty, and, more recently, also to John Dewey. Dreyfus sees a convergence of Minsky's earlier 'frame theory' to Husserl's phenomenology which has, for Dreyfus, been superseded by the work of the four philosophers just mentioned ([1], p. 36). In fact, Dreyfus explicitly follows Heiddegger in his view of the history of western philosophy and it is refreshing to see him relate such a view to a critique of AI. While I find the debates stimulating for my philosophical pursuit, I am unable to render clear to myself direct connections between these interesting observations and my mundane reflections on AI.

One problem I find tangibly enticing is how one chooses an appropriate research project. Of course we are often ignorant of the real causes behind our choices and we all make more or less serious mistakes in our choosing. (Closely related to the individual's choice, there is a problem of the society's choice.) For instance, given the present state of AI, what sort of young people would be making a good choice to enter the field? That means, among other things, they would be happier or do better work or both, than in some other field. 'Happier' and 'better work' are tough and elusive concepts. Yet I feel that even using such considerations as a guide would be more realistic than trying to find out whether either the AI enthusiasts or their opponents are more in the right. For one thing, they seem to argue over a motley of diverse issues. More, it is particularly difficult to decide on remote possibilities and impossibilities, even harder to have them guide one's action and short-term goals. -- Certainly the problem for a popularizer is quite different from that for a prospective research worker, and will not be discussed here.

Once Freud argued against mysticism in the following words (letter of 1917 to Groddeck):

'Let us grant to nature her infinite variety which rises from the inanimate to the organically animated, from the just physically alive to the spiritual. No doubt the Unconscious is the right mediator between the physical and the mental ... . But just because we have recognized this at last, is that any reason for refusing to see anything else? ... the monistic tendency to disparage all the beautiful differences of nature in favour of tempting unity. But does this help to eliminate the differences?'

That the unconscious should play such a key role certainly throws a long shadow over the attempts to understand better the relation between mind and body. It does in a way tempt people to some form of mysticism. But the larger question raised is the dilemma between the cleanliness of universality and the richness of particularity. It is so difficult to take full advantage of the proper particularity while striving successfully for universality. Applied to problems in AI, for instance, somebody's proving a theorem or recognizing a pattern is a particular event which generally resists the universalizing process of a 'precise description' of what is involved in it.

A particular act of recognition or reasoning depends on one's genes, the history of one's mind and body, some essential relation to desire, even race, class, sex and many other factors. Of most of these factors we have at best an imperfect knowledge and understanding. In AI we can only take into consideration some small residues of these factors which happen to be noticeable by perception or introspection. This is not to deny that we have often unexpected surprises in technology for one reason or another. But the courses of advance are hard to predict and certainly to not coincide with the history of human development. For example, I once thought of getting a gadget which would do the work of a dog. I was surprised to get a box which sounds the alarm when bodily temperature is suddenly introduced. It is surprising that the computer can play a good game of chess. Yet it can't yet drive a car well in traffic. But for the human intelligence situated in a technological society the latter task is easier than the former.

In short, I am not able to find the right mix of participation in and distancing from the AI project to offer any clear and distinct ideas on the loudest current controversies in this area. To rationalize my own failure in this regard, I venture to give my impression that the strong voices come usually from committed believers who begin with conclusions, perhaps reached through incommunicable private sources. Hence, it is easier to be more decisive since merely assembling arguments to support predetermined positions suffers from less distractions. Those who wish to examine the available evidence with an open mind, as is to be expected, are at a disadvantage.

Bibliography

1. Dreyfus, H.L. *What Computers Can't Do: The Limits of Artificial Reason*, revised edition, Harper and Row, New York, 1979 (original edition 1972).

2. Hofstadter, D.K. *Gödel, Escher, Bach: An Eternal Golden Braid*, Basic Books, New York, 1979.

3. McCorduck, P. *Machines Who Think*, W.H. Freeman, San Francisco, 1979.

4. Michie, D. *Machine Intelligence and Related Topics: An Information Scientist's Weekend Book*, Gordon and Breach, New York, 1982.

5. Minsky, M. "The Society Theory of Thinking", *Artificial Intelligence: An MIT Perspective*, Vol. 1, 1979, pp. 421-450.

6. Turing, A.M. "Computing Machinery and Intelligence", *Mind*, Vol. 59 (1950), pp. 433-460.

7. Wang, H. "A Variant to Turing's Theory of Computing Machines", *Journal of the Association for Computing Machinery*, Vol. 4 (1957), pp. 63-92; reprinted in reference 11 below.

8. Wang, H. "Toward Mechanical Mathematics", *IBM Journal*, Vol. 4 (1960), pp. 2-22; reprinted in reference 11 below.

9. Wang, H. "Proving Theorems by Pattern Recognition", Part I, *Communications of the Association for Computing Machinery*, Vol. 3 (1960), pp. 220-234.

10. Wang, H. "Proving Theorems by Pattern Recognition", Part II, *Bell System Technical Journal*, Vol. 40 (1961), pp. 1-41. These two parts also appeared as Bell Technical Monograph 3745.

11. Wang, H. *A Survey of Mathematical Logic*, Science Press, Peking, 1962, 652 pp. + x; also distributed by North-Holland Publishing Company, Amsterdam, 1963. Reprinted by Chelsea, New York, 1970 under the title *Logic, Computers and Sets*.

12. Wang, H. "Mechanical Mathematics and Inferential Analysis", *Seminar on the Relationship Between Nonnumerical Programming and the Theory of Formal Systems*, P. Braffort and D. Hirschberg (eds.) (1963), pp. 1-20.

13. Wang, H. The Mechanization of Mathematical Arguments", *Proceedings of Symposia in Applied Mathematics*, Vol. 15, American Mathematical Society (1963), *Experimental Arithmetic, High Speed Computing and Mathematics*, pp. 31-40.

14. Wang, H. "Formalization and Automatic Theorem Proving", *Proceedings of IFIP Congress 65*, 1965, Washington, D.C., pp. 51-58. Russian translation in *Problems of Cybernetics*, Vol. 7 (1970), pp. 180-193.

15. Wang, H. "On the Long-range Prospects of Automatic Theorem-Proving", *Symposium on Automatic Demonstration*, Springer-Verlag, 1970, pp. 101-111.

16. Wang, H. *From Mathematics to Philosophy*, Routledge & Kegan Paul, 1974, 431 pp. + xiv.

17. Wang H. *Popular Lectures on Mathematical Logic*, Science Press, Beijing; Van Nostrand Reinhold Company, New York, 1981.

18. Wang, H. (with S.A. Cook) "Characterizations of Ordinal Numbers in Set Theory", *Mathematische Annalen*, Vol. 164 (1966), pp. 1-25.

19. Wang, H. (with K.R. Brown) "Finite Set Theory, Number Theory and Axioms of Limitation", *ibid.*, pp. 26-29.

20. Wang, H. (with K.R. Brown) "Short Definitions of Ordinals", *Journal of Symbolic Logic*, Vol. 31 (1966), pp. 409-414.

21. Wang, H. "On Axioms of Conditional Set Existence", *Zeitschr. f. Math. Logik und Grundlagen d. Math.*, Vol. 13 (1967), pp. 183-188.

22. Wang, H. "Natural Hulls and Set Existence", *ibid.*, pp. 175-182.

23. Wang, H. "A Theorem on Definitions of the Zermelo-Neumann Ordinals", *ibid.*, pp. 241-250.

24. Webb, J.C. *Mechanism, Mentalism, and Metamathematics*, Reidel, Dordrecht, 1980.

25. Weizenbaum, J. *Computer Power and Human Reason: From Judgement to Calculation*, W.H. Freeman, San Francisco, 1976.

26. Wu Wen-tsün. "On the Decision Problem and the Mechanization of Theorem Proving in Elementary Geometry", *Scientia Sinica*, Vol. 21 (1978), pp. 159-172. Also in this volume.

27. Wu Wen-tsün. "Mechanical Theorem Proving in Elementary Differential Geometry", *Scientia Sinica, Mathematics Supplement*, I (1979), pp. 94-102.

28. Wu Wen-tsün. "Mechanical Theorem Proving in Elementary Geometry and Differential Geometry", *Proc. 1980 Beijing DD-symposium*, Vol. 2 (1982), pp. 1073-1092.

Citation For Lawrence Wos And Steven Winker As Winners Of The Current Research Award In Automated Theorem Proving

Lawrence Wos and Steven Winker are awarded the automated theorem proving current research prize for their leadership in the systematic pursuit of developing and applying automated, general-purpose theorem provers as tools in solving new and open problems. Wos and Winker have used automatic theorem provers to attach problems in ternary Boolean algebra, finite semigroups and equivalential calculus. Innovative work in finding interesting digital circuits and assessing the presence of hazards (faults) also has been undertaken. In making our nomination we also acknowledge the role of colleagues involved in some of their investigations: L. Henschen, E. Lusk, R, Overbeek, B. Smith, and R. Veroff.

The techniques developed by Wos and Winker involve interaction with human reasoning and employ a resolution-paramodulation type automated theorem prover. A key technique was to have the theorem prover help determine and complete a model of an appropriate structure -- e.g., a ternary Boolean algebra or a finite semigroup. In the equivalential calculus the substitution rule and the detachment rule were combined in a condensed detachment rule. The theorem prover was used to examine indirectly all theorems generable from a given statement and to help classify the theorems so created. Statements that were not sole axioms could be discovered by observing that the inferences all had a specific characteristic not true of all theorems of the equivalential calculus.

Nils J. Nilsson, Chairman
Robert Boyer
Donald Loveland
R. Daniel Mauldin

Selected Articles by S. Winker and L. Wos.

1. S. Winker and L. Wos. Procedure Implementation Through Demodulation and Related Tricks. *Sixth Conference on Automated Deduction*, Courant Institute of Mathematical Sciences, June 1982, pp. 109-131.

2. S. Winker and L. Wos. Automated Generation of Models and Counterexamples and its Application to Open Questions in Ternary Boolean Algebra, *Proceedings of the Eighth Internatonal Symposium on Multiple-valued Logic*, Rosemont, Illinois, IEEE and ACM Publication, (1978), pp. 251-256.

3. S. Winker, L. Wos, and E. Lusk. Semigroups, Antiautomorphisms, and Involutions: A Computer Solution to an Open Problem, I, *Mathematics of Computation*, Vol. 37, (October 1981), pp. 533-545.

4. S. Winker. Generation and Verification of Finite Models and Counter-examples Using an Automated Theorem Prover Answering Two Open Questions, *Journal of the ACM*, Vol. 29 (2), (April 1982), pp. 273-284.

5. L. Wos, S. Winker, R. Veroff, B. Smith, and L. Henschen. Questions Concerning Possible Shortest Single Axioms in Equivalential Calculus: An Application of Automated Theorem Proving to Infinite Domains, *Notre Dame Journal of Formal Logic*, Vol. 24 (2), (April 1983), pp. 205-223.

6. L. Wos. Solving Open Questions with an Automated Theorem-Proving Program, (June 1982). *Sixth Conference on Automated Deduction*, Courant Institute of Mathematical Sciences, June 1982, pp. 1-31.

Contemporary Mathematics
Volume 29, 1984

OPEN QUESTIONS SOLVED WITH THE ASSISTANCE OF AURA

by

L. Wos and S. Winker[1]

ABSTRACT. The computer program AURA (AUtomated Reasoning Assistant) has provided invaluable assistance in solving open questions in mathematics and in formal logic. Historically, the questions were taken from studies of ternary Boolean algebra, finite semigroups, Robbins algebra, equivalential calculus, the R-calculus, and the L-calculus. In answering the questions, AURA was used both to generate proofs and to find models and counterexamples. In particular, the program assisted in suggesting and verifying conjectures, testing for the presence and absence of various mappings, and characterizing certain infinite domains of theorems. No modifications to the existing program AURA were required to solve the open questions. Research in mathematics and in formal logic may be sharply enhanced by employing an automated assistant such as AURA. Open questions requiring either the generation of a proof or the finding of a small model or counterexample are being sought.

1. INTRODUCTION

The activity that occupies much of a mathematician's or a logician's time is that of proving new theorems. Obviously, always present is the possibility that the new proof will be faulty. Often a colleague is shown the proof with the intention of catching such errors, especially before publication. With the advent of the computer, interest was expressed in attempting to automate the activity of proving theorems. Perhaps understandably, some originally thought that the entire activity eventually could be automated--thought that a computer program could be written to accept a purported theorem, and return a proof. After all, the computer is very fast and dispatches certain tasks

---

[1]Authors' address: Mathematics and Computer Science Division, Argonne National Laboratory, 9700 South Cass Avenue, Argonne, IL 60439

This work was supported in part by the Applied Mathematical Sciences Research Program (KC-04-02) of the Office of Energy Research of the U.S. Department of Energy under Contract W-31-109-ENG-38 (Argonne National Laboratory) and in part by NSF grant MCS79-03870 (Northern Illinois University).

© 1984 American Mathematical Society
0271-4132/84 $1.00 + $.25 per page

simply by being assigned them. With a computer program--a theorem-proving program--of the desired type, the burden of proof finding would be shifted from the mathematician and logician to the computer. In addition, such a theorem-proving program would have the fortuitous property that faulty proofs would never occur. As history proved, this all-powerful program would not be found. Proving theorems in mathematics and in logic is too complex a task for total automation, for it requires insight, deep thought, and much knowledge and experience. Even with this disappointment, the game was far from over.

Despite the complexity, a computer program [7,8,9,10,11,14,26] has been written that, although not fulfilling the original expectation of some, can be and is used to assist in finding proofs of theorems. In fact, the program under discussion has assisted in solving previously open questions from various fields of mathematics and formal logic [17,19,24,27]. This theorem-proving program--AURA (AUtomated Reasoning Assistant)--like other theorem-proving programs, does have the desirable property of yielding proofs that rely on inference rules that are sound. By a sound inference rule, we mean that a conclusion yielded from its use follows logically from the premisses. Although we have found no error in any proof yielded by our theorem-proving program in the last four years, one must always note that a (large) computer program cannot be pronounced free of flaws. However, since the program AURA makes explicit every step of any proof it finds, one can, if desired, check all the details of such a proof.

With AURA's assistance, much of the burden of solving the previously open questions was shifted to the computer. What will become evident is the fact that the researchers involved in the effort knew and know almost nothing of the fields from which the questions were selected. Rather than indicating the triviality of the questions (some of which are far from easy to answer), this fact shows the potential value of having access to an automated assistant--a colleague in the form of a theorem-proving program.

The questions are taken from the fields of ternary Boolean algebra, finite semigroups, equivalential calculus, the R-calculus, and the L-calculus. We shall discuss these questions from two viewpoints. First, we shall demonstrate that a theorem-proving program can provide invaluable assistance in research. Second, we shall focus on aspects that, if possessed by a problem, make it more amenable to the joint attack of a researcher and a theorem-proving program.

By providing some criteria for partitioning problems into those suitable and those unsuitable for treatment with an automated theorem-proving program, we may make possible the attainment of two goals. Our intention is to obtain additional open questions to attack with a theorem-proving program. Further, we wish to interest others in conducting research with the assistance of such a program.

2. FINITE SEMIGROUPS

The following anecdote is enlightening and perhaps amusing. The anecdote concerns solving a previously open question in finite semigroups. Although this question was not the first open question that was answered with the assistance of the theorem-proving program AURA, the story of its solution best illustrates various points that must be made. So let us tell the story exactly as it occurred.

Elated with having solved certain previously open questions in ternary Boolean algebra, we began the search for new territory to conquer. Since ternary Boolean algebra has at best a small following, and since the questions we had answered resulted in the formulation of techniques for finding small models or counterexamples, we sought some problem that was rather more mainstream. The sought-after problem could address either side of the question— require finding a proof or require generating a small model or counterexample —and still be of interest.

One phone call to I. Kaplansky was sufficient. He suggested the following problem for finite semigroups concerning the possibility of various mappings occurring in combination. Specifically, does there exist a finite semigroup admitting a nontrivial antiautomorphism but admitting no nontrivial involutions? Two obvious choices existed. We could attempt to prove that every finite semigroup admitting a nontrivial antiautomorphism must also admit a nontrivial involution, or we could seek a counterexample to the theorem. Since the theorem would be difficult to state in the language required by an automated theorem-proving program, and since we had the new toy for generating small models and/or counterexamples, we chose the latter approach. We must point out that neither of us knew or now know very much about finite semigroups. The importance of this point will shortly become obvious, and the accuracy of it will be transparent. However, despite our lack of knowledge, we were also encouraged to take the approach of seeking a counterexample since that was our intuitive inclination.

After formulating an approach that permitted us to rely heavily on an automated theorem-proving program, we made a series of computer runs. We were rewarded. We found the sought-after counterexample. We found various semigroups of order 4 that admit a nontrivial antiautomorphism but fail to admit a nontrivial involution. What we failed to do was write the corresponding paper and submit it for possible publication. Instead, we announced the result to various colleagues in the field of automated theorem proving. We were, of course, congratulated profusely.

Approximately 18 months later, in another conversation with Kaplansky, a remark was passed that caused him to point out that we had not solved the problem he had suggested. Yes, it was no doubt true that we had a counterexample to something, but the something was not particularly interesting. How fortunate that we had not written the paper, much less submitted it for publication! How puzzled, though polite, were our colleagues! What had we done wrong? Where was the misunderstanding?

The small detail that was misunderstood was a definition. An involution is a one-to-one onto mapping whose square is the identity and that preserves products, isn't it? At least, this was the definition of involution we employed in solving the problem. But it is not the definition that should have been employed. Rather, an involution is an antiautomorphism whose square is the identity. An antiautomorphism h (for noncommutative semigroups) does not preserve products, but rather is such that $h(xy) = h(y)h(x)$, and not a mapping with $h(xy) = h(x)h(y)$.

What had we demonstrated? First and foremost, we had proved that we knew very little about finite semigroups. Otherwise we would not have employed the wrong definition. We had shown that, with the assistance of a theorem-proving program, we could answer some question whose answer we did not know--even if the question were not interesting mathematically. Might there be something else we had accomplished? If the method for doing the wrong problem was general, then it should be applicable to the right problem. All we need do is replace the incorrect definition of an involution and, as it turned out, make a few other relevant changes and then attempt to solve the much harder and more interesting problem.

Three hours after discovering that the many references to having solved another open problem were at best misleading, we had in fact solved the interesting problem. The solution was obtained by making a series of computer runs that took 40 seconds of time on a large-scale computer. That we were able to solve the correct problem was very gratifying. Perhaps of greater importance, we had demonstrated the flexibility and power available to one using an automated theorem-proving program. We had not been forced to modify the program itself in any way. We were able to recover from a mistake in very short time. With much assistance from AURA, we were, despite knowing very little about the field, able to solve an open question of a somewhat interesting nature.

Finally, we state the precise problem with its correct definitions, and give the solution [19]. Does there exist a finite semigroup admitting a nontrivial antiautomorphism but admitting no nontrivial involution? An antiautomorphism is a one-to-one onto mapping h with h(xy) = h(y)h(x). An involution is an antiautomorphism whose square is the identity. To be nontrivial, the mapping must not be the identity map. Commutative semigroups are of course not of interest for, in such a semigroup, an antiautomorphism is simply an automorphism. Yes, there does exist such a semigroup. The first found has order 83, and is generated by b, c, d, and e. The defining relations are: bbc = dde; dcc = bee; and all products of four or more elements are equal. The antiautomorphism is such that h(b) = c, h(c) = d, h(d) = e, and h(e) = b. The smallest such semigroup has order 7, and there are four such semigroups [20].

3. TERNARY BOOLEAN ALGEBRA

The technique for generating small models and counterexamples [16] with an automated theorem-proving program did not originate with the study of finite semigroups. Rather, it originated with an earlier study, namely, the study of open questions in ternary Boolean algebra [17]. The attempt to answer those open questions by relying heavily on AURA in fact led directly to the model generation technique that proved so valuable in the study of finite semigroups.

A ternary Boolean algebra is a nonempty set satisfying the following axioms.

1) $f(f(v,w,x),y,f(v,w,z)) = f(v,w,f(x,y,z))$;
2) $f(y,x,x) = x$;
3) $f(x,y,g(y)) = x$;
4) $f(x,x,y) = x$;
5) $f(g(y),y,x) = x$

The function f acts as product, while the function g acts as inverse.

The question is: Which, if any, of the first three axioms is independent of the remaining four? It was known [2] that axiom 4 and axiom 5 are dependent axioms. In fact, proofs to that effect were easily obtained with AURA, as well as obtained with other automated theorem-proving programs. As in the semigroup problem discussed in Section 2, two choices exist. A proof of dependence can be sought, or a model can be sought that satisfies, for example, axioms 1, 3, 4, and 5 while violating axiom 2. We first tried to find proofs of the dependence of axioms 2 and 3 by submitting the corresponding purported theorems to the theorem-proving program. The program found no proofs of dependence.

Unfortunately, because of a fundamental theorem of logic, we could not conclude that dependence cannot be established. We could merely conclude that no proof had yet been found. AURA is often very successful at finding short proofs, especially of simple theorems. We therefore suspected that dependence might not be the case, and turned to the consideration of finding appropriate models.

At the time, we knew of no means to have a theorem-proving program attempt to generate models, even small finite models. Thus, we were forced to formulate a method for using a theorem-proving program to do so. Our effort resulted in a general approach to finding small models and counterexamples [16]. The approach yielded models establishing the independence of each of the first three axioms of a ternary Boolean algebra [17]. As mathematicians and logicians so well know, the capacity to find models and counterexamples is a necessary complement to finding proofs. Indeed, as was seen in the previous section, this capacity was vital to solving the semigroup problems. The possible uses for an automated theorem-proving program had finally been extended to fill an obvious gap.

4. EQUIVALENTIAL CALCULUS

Here again we have recourse to an anecdote before giving the precise statement of the problem. The anecdote begins with our colleague, Brian Smith, running a set of experiments with AURA. In automated theorem proving, experiments of various kinds are absolutely necessary. Through them, one tests the additions and modifications to a theorem-proving program for possible flaws. As a result of experimentation, new inference rules and strategies and notation are found.

Smith's set of experiments were designed to test certain additions to the basic program. (No such additions are necessary for the mathematician and/or logician to utilize a theorem-proving program to assist in research. We add to the program to give it more power and effectiveness. But, as it stands, we have shown that AURA is quite useful now.) One of Smith's experiments [28] was with a known theorem [6] of the equivalential calculus [5], a branch of formal logic that is concerned with the abstraction of "equivalence". Smith was attempting to find a proof using AURA. To this point, we had never obtained a proof of this theorem with our program. By using a mechanism within AURA known as weighting [10,14] and ordinarily used to impose one's intuition on the search for a proof or for a model, Smith forced the program to traverse the known proof, closely but not (necessarily) identically. Rather than duplicating the known proof as Smith intended, AURA instead found a proof half as long [28].

His success caused the authors of this paper to consider the possibility of answering certain questions still open [12] in the calculus. Yet another member of the group, at the time not directly involved in the study, suggested a formula for investigation. The problem was to prove that the selected formula is a single axiom for the equivalential calculus, or prove that it is too weak to be such. (At the time, 11 formulas [6,12] were known to be shortest single axioms for the calculus.) A single computer run elicited much excitement, for the theorems deducible from the formula appeared to have a predictable pattern of behavior--appeared to be ever-increasing in length. If this proved true, then the formula would quickly be established as too weak to be a single axiom, for there are theorems of the calculus that are shorter than the formula under discussion.

The excitement proved well founded. Another colleague, Bob Veroff, and one of the authors of this paper easily proved that the conjecture was true-- that the theorems are ever-increasing in length--and a celebration occurred. More important, the success with this formula led immediately to the investigation of the corresponding question for the remaining, as yet unclassified formulas. The authors of this paper wholeheartedly attacked the set of at-the-time open questions. Two of the formulas yielded immediately--proved too weak to be single axioms. Again the key was AURA.

But perhaps the sole credit should go to the researchers, and not to a theorem-proving program. Did the researchers know much about the calculus? As indicated earlier, apparently not, for a startling discovery then occurred. After solving two open questions--attacked because of the ease with which the first question was answered--the researchers discovered that the first and easy question had not been an open question after all. In fact, its answer was well known to those with any knowledge of the field. Only a transcription error had led to the consideration of that easy-to-solve question. Thus, as with the semigroup study, again good fortune was present. Not only had two previously open questions been answered, but a method had been developed that would prove sufficient to answer the remaining open questions of the type under discussion.

We view this and the earlier anecdote with amusement. Each supports the contention that we were far from knowledgeable in the fields of study from which the open questions were selected. Thus, the results strongly indicate what can be done with the assistance of an automated theorem-proving program. Were we experts in some field of mathematics or logic, and were we to conduct a concentrated investigation with the assistance of AURA, how far might we get? Equally, how far might a mathematician or logician get using AURA as an assistant or colleague? We would clearly like the answer to this question, and encourage some mathematician and/or logician to consider a joint effort of the type that parallels our studies.

The precise statement of the problem is: Are any of the following seven formulas [12] shortest single axioms for the equivalential calculus [5]?

1) XJL = E(x,E(y,E(E(E(z,y),x),z)))
2) XKE = E(x,E(y,E(E(x,E(z,y)),z)))
3) XAK = E(x,E(E(E(y,z),x),z),y))
4) BXO = E(E(E(E(x,E(y,z)),z),y),x)
5) XCB = E(x,E(E(E(x,y),E(z,y)),z))
6) XHK = E(x,E(E(y,z),E(E(x,z),y)))
7) XHN = E(x,E(E(y,z),E(E(z,x),y)))

The first five are too weak to be a single axiom [18,27], while the last two are in fact each single axioms [18]. The conjecture was that no additional shortest single axioms existed, and thus the results (as yet unpublished) concerning the last two of the seven formulas are of particular interest.

The weakness of each of the first five formulas was demonstrated by means of a complete characterization of the deducible theorems therefrom. With that characterization, we were able to show that, for each of the five formulas, certain known theorems are absent from those deducible from the formula. The approach of generating models, which was used for the previous two classes of problem, is not adequate for obtaining the desired characterizations. The obstacle rests with the fact that each of them is sufficiently powerful that an infinite number of theorems can be deduced from each. Thus, a new technique was required to obtain the desired characterization. A complete characterization is a goal sought by mathematicians and logicians, but not always obtained. Our result was, therefore, unusually rewarding.

The proofs that XHK and XHN are each a single axiom consist of 87 and 162 steps, respectively. Since there are steps in the first proof that contain 71 symbols exclusive of commas and grouping symbols, and in the second proof steps containing 103, the proofs might be considered rather complex. The methodology formulated to obtain these two proofs, as well as obtain the characterization of the theorems deducible from each of the other five, is sufficiently general to apply to related calculi.

5. THE R-CALCULUS AND THE L-CALCULUS

We give here the briefest of notes. In the R-calculus [5], a field of formal logic, there are formulas whose status with respect to being a single axiom is unknown [12]. The technique employed to answer corresponding questions in the equivalential calculus was successfully applied to certain formulas with the following results.

In the R-calculus, two previously unclassified formulas were examined.

XEH = E(x,E(E(y,E(E(y,z),x)),z))
XGJ = E(x,E(E(y,E(z,x)),E(y,z)))

One of these, XEH, was proved too weak to be a single axiom, while the other, XGJ, was proved to be a "new" single axiom for the R-calculus [18].

In the L-calculus [5], the following previously unclassified formula was proved too weak to be a single axiom [18].

XCK = E(x,E(E(E(y,z),E(x,z)),y))

Thus, the study of the equivalential calculus led to similar results for related calculi.

6. PROBLEMS SUITABLE FOR AN ATTACK WITH AN AUTOMATED THEOREM-PROVING PROGRAM

We come now to the discussion of which types of problem are suitable for attack with the assistance of an automated theorem-proving program. We focus first on the open questions that were solved, listing by example some of the properties that enabled us to consider them. Then we give various examples that indicate what concepts can and what concepts cannot be easily mapped into the language required by an automated theorem-proving program. We are of course promoting the use of theorem-proving programs by others, but we are also attempting to interest mathematicians and logicians in finding open questions suitable for attack with such a program.

An automated theorem-proving program is used primarily to find proofs of purported theorems. But, as demonstrated, it can also be used to find small models and small counterexamples. With its assistance, valuable lemmas occasionally are found--found by examining the results of a computer run. Such was the case in the study of ternary Boolean algebra. This algebra is particularly suitable for study with an automated theorem-proving program since its axioms take the form of simple equalities, and can thus be written as unit clauses in the predicate EQUAL. For example, axiom 1 is given to the program as

EQUAL(f(f(v,w,x),y,f(v,w,z)),f(v,w,f(x,y,z))),

with the remaining axioms treated similarly. A lemma that was found in one of the computer runs proved valuable in the search for models establishing the independence of various axioms. Although axiomatic studies are currently not in vogue, nevertheless, occasionally such a question might arise and might be answered with an automated theorem-proving program.

The axioms for a semigroup, like those for a ternary Boolean algebra, can also be written as simple equalities. For example, the associative law is written

EQUAL(f(f(x,y),z),f(x,f(y,z))).

Although no means exists for stating that a structure under study is finite, the use of generators and relations is extremely compatible with the language requirements of a theorem-proving program. In answering the initial question--that concerning the existence of a semigroup of the desired type--the use of relations appeared to be of greater value than the use of the entire multiplication table, although the latter can also be done. For example, one relation that was used in defining the first semigroup possessing the desired properties is written

EQUAL(f(b,f(b,c)),f(d,f(d,e))).

The relation that forces products of four or more elements to be identical is written

EQUAL(f(x,f(y,f(z,w))),f(b,f(b,f(b,b)))).

When the minimality question was considered, the multiplication tables were used, as well as the defining relations.

For mappings such as antiautomorphisms, their general action can be represented with

EQUAL(h(f(x,y)),f(h(y),h(x))).

The specific action of the chosen antiautomorphism h is represented with

EQUAL(h(b),c)

and similar equalities. For an involution j, one simply gives the clause that corresponds to that given for the general action of an antiautomorphism, and also writes

EQUAL(j(j(x)),x).

In this study, the program derived any and all nontrivial relations implied by those that were given--derived all relations that were needed to characterize the multiplication table. Various tests were made by the program--tests that associativity held throughout, that the proposed antiautomorphism held, and that no involutions were present. Not only is the automation of various tests convenient, but such automation often immediately shows that the assumptions given to the program must be modified. For example, the first attempt at finding the required semigroup failed--failed because the proposed semigroup was discovered to admit a number of involutions. We were forced to modify the relations defining the semigroup, with the result that the next attempt succeeded. In the study of the corresponding minimality question, both proof finding and model generation were required. Needless to say, the program was forced to cope with questions of isomorphism and even with "without loss of generality" arguments.

The study of the equivalential calculus was amenable to attack with AURA because of the structure of its formulas and because of an inference rule used to generate formulas of the calculus. The formulas can be represented merely by prefixing each with a predicate symbol. For example, the formula XJL was studied by writing

P(E(x,E(y,E(E(z,y),x),z)))).

The inference rule, condensed detachment [12], was encodable by writing

-P(E(x,y)) -P(x) P(y)

and using either of two well-known inference rules extant in automated theorem-proving, namely, hyperresolution [13] and UR-resolution [9]. However, even with these properties, the study presented problems that we had not previously encountered. The most serious of the problems was that of examining the entire set of theorems deducible from a given formula. This examination was deemed necessary because we conjectured that certain of the formulas under investigation might be too weak to be a single axiom. A typical approach to establishing such weakness is that of presenting an appropriate model. The desired model is intended to demonstrate that certain known single axioms of the calculus are not among the theorems so deducible. But the set of deducible theorems, at least for the formulas under study, is infinite. Furthermore, while seeking such a model for certain of the formulas under investigation, Peterson [12] had already examined without success a number of order 8. Since the number of models of order 8 that might be examined is 8 to the 64th, the model generation technique applied to the studies of ternary Boolean algebra and of finite semigroups was classed as unpromising.

A new approach [27] was needed and was therefore developed. The approach is based on the use of schemata. With schemata, the fact that the set to be characterized is infinite was no longer an obstacle. Thus, in those cases where schemata suffice to circumvent the difficulties presented by examination of an infinite class, a theorem-proving program can also be used as an assistant. In the study, AURA was both used to find proofs and to generate counterexamples, the latter by means of obtaining a complete characterization of the set of deducible theorems. The conjecture that led eventually to the solutions of all seven problems was made directly as a result of examining the output from a single computer run. With the assistance of AURA, the needed schemata were found, conjectures were proved, and the heart of the key induction argument was obtained.

What becomes evident is an overall description of the activity. A theorem-proving program is written and tested on easy theorems. Then it is tested on harder theorems, and often found lacking. So new inference rules are formulated, and new strategies are found to enable the program to be more effective in its search for a solution. Finally, an open question is considered. The question forces a new technique to be invented--for example, one for finding models, or for studying mappings, or for manipulating schemata. The new techniques often require no additional programming, but are phraseable instead strictly within the language employed to present problems and questions to an automated theorem-proving program. Then the new methodology is tried on similar questions and modified and extended as necessary. New open questions are sought on which to test the new methodology. It is often found lacking, and more methodology is developed. Thus, more open questions are needed. What would be of great value and is missing is the use by other mathematicians and logicians of an automated theorem-proving program. Use of an automated theorem-proving program by an expert in some field might lead to startling results.

But precisely what are the properties of the "new" open questions to be considered? Succinctly, the best case occurs when the entire problem can be simply phrased in first-order predicate calculus. Equality presents no intrinsic problem, for there exist inference rules available to a theorem-proving program that "build in" equality--that treat equality as if its meaning is understood. Because of the existence of such inference rules, a theorem from universal algebra becomes relevant. If a class of algebras is a variety, and if the variety is finitely based, then an algebra of the class can be characterized with a finite set of equations [3,15]. From the viewpoint of automated theorem proving, such a variety is said to be representable with a finite set of equality units. Such varieties--groups, rings, semigroups, for example--are most amenable to consideration by a theorem-proving program.

In contrast to such varieties, many concepts cannot simply be phrased in first-order predicate calculus. No way exists to simply state that the domain is finite, and the obvious way of informing a theorem-proving program that the domain is infinite is not particularly useful. Between these extremes is, for example, the statement that "the group has order 120". This statement can be conveyed, but conveyed in a very cumbersome fashion. Statements such as "the group has exponent 3" are trivial to phrase. But statements of the form "all mappings from G to H are onto" are very difficult to phrase, if not impossible. We can, obviously, have recourse to certain methods employed in mathematics and in logic. For example, if told "there exists a function", we can simply name the function. No such maneuver, however, yet exists for transfinite ordinals, for partial differential equations, and for very many other concepts. Nevertheless, the study of some fields and of parts of other fields is somewhat amenable to relying on the assistance of an automated theorem-proving program.

At this point we give, for those who are curious, some of the actual figures (in seconds) required to obtain the results quoted in this paper. These figures do not account for the time spent by the researchers in developing the methodology, but are only the computer time that was required.

To find the consequences that can be deduced from a given set of relations requires approximately 15 seconds of computer time on an IBM 3033, if several hundred relations are derivable. For the semigroup of order 83, 2 seconds sufficed to find the derived relations, and 10 seconds to check that no involutions are present. A number of such runs were required to find the desired semigroup. Several additional runs of the same order of time were required to find a semigroup of order 7 with the desired properties. To find, among the over 800,000 semigroups of order 7 [4], the remaining three of the desired type required some 80 runs, each of which was approximately 30 seconds in duration.

In the equivalential calculus study, 15 to 30-second runs produced many formulas that, when examined, led to the key conjecture. Runs of 15 seconds sufficed to establish that each of three formulas is too weak to be a single axiom, while a fourth was proved too weak in a 30-second run. To prove that the fifth formula is too weak, a series of runs was required with a total computer time of about 4 minutes. The establishment of each of XHK and XHN as new shortest single axioms required roughly 50 runs, each of which was from 5 to 45 seconds in duration.

7. OTHER APPLICATIONS

Although this paper focuses chiefly on the use of an automated theorem-proving program for research in mathematics and logic, other applications [24,25,26] might be of interest. AURA has been used in both logical design [21,22] and design validation [23]. In the first of these two uses, the program is given the descriptions of various components, and is asked to design a circuit with chosen properties. In the second, the program is given the design of a circuit purported to have certain properties, and is asked to prove that the circuit in fact has the claimed properties. The evidence to this point indicates that both applications are quite promising. Some of the circuits designed with the aid of AURA proved superior [22] to others of the type previously known to those in the field of logic design. As for the second application, the validations of various designs of adders required surprisingly little computer time. For example, given the design of a 16-bit adder, AURA proved [23] that the design has the desired properties in 297 steps and in less than 20 seconds on an IBM 3033.

A rather different application is that of chemical synthesis. The problem is to determine which combination of compounds selected from a large database of compounds will, when combined subject to given reaction rules, produce the desired compound. This study has just commenced.

A potential application that interests some engineers is that of using a theorem-proving program to validate the design of a nuclear plant. Such an effort would, of course, require far more than a program like AURA, for many computation questions would likely occur. But questions of a logical nature would also exist and could be submitted to such a program.

Finally, the application that occupies so much of the attention of both R. Boyer and J Moore [1] deserves mention. Their interest is in using an automated theorem-proving program to prove claims made for other computer programs. With no exaggeration, their work can be evaluated as extremely promising.

8. ACKNOWLEDGMENTS

The results on which this paper is based would not have been possible were it not for the existence of a team of researchers of which we are but two. The team--which exists only on a very informal level--consists of L. J. Henschen, E. L. Lusk, W. McCune, R. A. Overbeek, B. T. Smith, R. L. Stevens, R. L. Veroff, S. K. Winker, and L. Wos. These individuals have collaborated on a number of research topics in automated theorem proving. Although the authors of this paper and some members of the team are from Argonne National Laboratory, other members are from Northern Illinois University, Northwestern University, the University of New Mexico, and Western Michigan University.

The questions were answerable because of access to AURA and its array of inference rules, strategies, data structures, and various other implementation and design features. The value of a powerful theorem-proving program that simultaneously makes available the basic procedures of the field cannot be overestimated.

9. CONCLUSIONS

The evidence given here indicates that an automated theorem-proving program can and does provide invaluable assistance in research. The program AURA clearly played a vital role in answering the various open questions. This claim is substantiated by the fact that the principal investigators were able to answer the open questions despite having little knowledge about any of the three fields involved. Since the results presented here were obtained without recourse to program modification, we have demonstrated that an important obstacle to conducting research with an automated theorem-proving program has been removed.

Additional open questions would be of great value. Those answered here led to important advances in automated theorem proving itself and in the use of the corresponding program. Of even greater value would be the use by mathematicians and logicians of such a program as an assistant and colleague. Perhaps the union of mathematics and/or logic with automated theorem proving is finally at hand.

BIBLIOGRAPHY

[1] Boyer, R. and Moore, J, A Computational Logic, ACM Monograph Series, Academic Press, 1979.

[2] Chinthayamma, "Sets of independent axioms for ternary Boolean algebra," Notices of the American Mathematical Society, Vol. 16(1969), p. 654.

[3] Graetzer, G., Universal Algebra, 1st ed., van Nostrand, 1968, pp. 152-171.

[4] Jurgensen, H. and Wick, P., "Die Halbgruppen der Ordnungen $\leq$ 7," Semigroup Forum, Vol. 14 (1977), pp. 69-79.

[5] Kalman, J., "Axiomatizations of logics with values in groups," J. London Math. Soc. (2), 14 (1976), pp. 193-199.

[6] Kalman, J., "A shortest single axiom for the classical equivalential calculus," Notre Dame Journal of Formal Logic, Vol. 19, No. 1, January 1978, pp. 141-144.

[7] Lusk, E., "Input translator for the environmental theorem prover - user's guide," to be published as an Argonne National Laboratory technical report.

[8] Lusk, E. and Overbeek, R., "Experiments with resolution-based theorem-proving algorithms," Computers and Mathematics with Applications 8 (1982), pp. 141-152.

[9] McCharen, J., Overbeek, R. and Wos, L., "Problems and experiments for and with automated theorem proving programs," IEEE Transactions on Computers, Vol. C-25(1976), pp. 773-782.

[10] McCharen, J., Overbeek, R. and Wos, L., "Complexity and related enhancements for automated theorem-proving programs," Computers and Mathematics with Applications, Vol. 2(1976), pp. 1-16.

[11] Overbeek, R., "An implementation of hyper-resolution," Computers and Mathematics with Applications, Vol. 1(1975), pp. 201-214.

[12] Peterson, J., "The possible shortest single axioms for EC-tautologies," Auckland University Department of Mathematics Report Series No. 105, 1977.

[13] Robinson, J., "Automatic deduction with hyper-resolution," International Journal of Computer Mathematics, Vol. 1(1965), pp. 227-234.

[14] Smith, B., "Reference manual for the environmental theorem prover, An Incarnation of AURA," to be published as an Argonne National Laboratory technical report.

[15] Taylor, W., "Equational Logic," Houston J. Math. 5 (Survey 1979), pp. 1-83.

[16] Winker, S., "Generation and verification of finite models and counterexamples using an automated theorem prover answering two open questions," J. ACM, Vol. 29 (1982), pp. 273-284.

[17] Winker, S. and Wos, L., "Automated generation of models and counterexamples and its application to open questions in ternary Boolean algebra," Proc. of the Eighth International Symposium on Multiple-valued Logic, Rosemont, Illinois, 1978, IEEE and ACM Publ., pp. 251-256.

[18] Winker, S. and Wos, L., "New shortest single axioms for the equivalential calculus," in preparation.

[19] Winker, S., Wos, L. and Lusk, E., "Semigroups, antiautomorphisms, and involutions: a computer solution to an open problem, I," Mathematics of Computation, Vol. 37 (1981), pp. 533-545.

[20] Winker, S., Wos, L. and Lusk, E., "Semigroups, antiautomorphisms, and involutions: a computer solution to an open problem, II," in preparation.

[21] Wojciechowski, W. and Wojcik, A., "Multiple-valued logic design by theorem proving," Proc. of the Ninth International Symposium on Multiple-valued Logic, Bath, England, May 1979, IEEE, pp. 196-199.

[22] Wojciechowski, W. and Wojcik, A., "Automated design of multiple-valued logic circuits by automatic theorem proving techniques," to appear in IEEE Transactions on Computers.

[23] Wojcik, A., "Formal design verification of digital systems," Proceedings of the Twentieth Design Automation Conference, Miami Beach, June 1983, pp. 228-234.

[24] Wos, L., "Solving open questions with an automated theorem-proving program," 6th Conference on Automated Deduction, Vol. 138, Lecture Notes in Computer Science, ed. D. W. Loveland, Springer-Verlag, Berlin, Heidelberg, New York, 1982, pp. 1-31.

[25] Wos, L., Overbeek, R., Lusk, E., and Boyle, J., "Automated reasoning: introduction and Applications," Prentice-Hall, Englewood Cliffs, New Jersey, 1984.

[26] Wos, L., Winker, S. and Lusk, E., "An automated reasoning system," AFIPS Conference Proceedings, Vol. 50 (1981), National Computer Conference (Chicago, Ill., 1981), AFIPS Press, pp. 697-702.

[27] Wos, L., Winker, S., Veroff, R., Smith, B. and Henschen, L., "Questions concerning possible shortest single axioms for the equivalential calculus: an application of automated theorem proving to infinite domains," Notre Dame Journal of Formal Logic, Vol. 24, No. 2, April 1983, pp. 205-223.

[28] Wos, L., Winker, S., Veroff, R., Smith, B. and Henschen, L., "A new use of an automated reasoning assistant: open questions in equivalential calculus and the study of infinite domains," submitted for publication.

ABSTRACT. We list here a number of theorems in introductory analysis that have been proven automatically by theorem provers, and describe in a general way some of the techniques used in their proofs.

SOME AUTOMATIC PROOFS IN ANALYSIS

W. W. Bledsoe[1]

1. Introduction

The work described in this talk is mainly that of our own group, or that of others that we know best.

During the last few years a limited number of theorems in introductory analysis and related areas, have been proved by automatic theorem provers. These fall in the following subareas:

    Set Theory

        Elementary Set Theory

        Theorems in set theory requiring Induction

    The Limit Theorems of Calculus

    Intermediate Analysis

    Elementary Topology

We list below a number of these theorems, and describe in a general way some techniques used in their proofs. All of these have been proved by "stand-alone" provers, with no human help.

---

[1] The University of Texas, Austin, Texas.

This work was supported by NSF Grant MCS-801-1417.

LIST OF THEOREMS PROVED

1. Set Theory
    1.1 Elementary Set Theory
        .1 $A \subseteq A$
        .2 $A \subseteq A \cup B$
        .3 $A \cup (B \cup C) = (A \cup B) \cup C$
        .4 $A \cap (B \cap C) = (A \cap B) \cap C$
        .5 $A \cup (B \cap C) = (A \cup B) \cap (A \cup C)$
        .6 $A \cap (B \cup C) = (A \cap B) \cup (A \cap C)$
        .7 $Sb(A \cap B) = Sb(A) \cap Sb(B)$[1]
        etc.

    1.2 Set Theory Theorems Requiring Induction[2]
        .1 $\omega = \sigma \omega$
        .2 $\omega = \omega \cap Sb\,\omega$

    1.3 Higher Order Set Theory Theorems
        .1 $\{x\} = \{y\} \rightarrow x = y$
        .2 (Cantor's Theorem) N^N is not denumerable. (N here is the set of integers).

2. The Limit Theorems of Calculus
    2.1 (Limit of a sum and product)
        If $\lim_{x \to a} f(x) = L$ and $\lim_{x \to a} g(x) = K$, then

        (+) $\lim_{x \to a} (f + g)(x) = L + K$

    and (.) $\lim_{x \to a} (f \cdot g)(x) = L \cdot K$.

    2.2 The sum of two continuous functions is continuous.

    2.3 The product of two continuous functions is continuous.

---

[1] $Sb(x)$ means the family of subsets of X, i.e., $Sb(X) = \{Y: Y \subseteq X\}$.

[2] We use here the definitions: 0 = empty set, $1 = \{0\}$, $2 = \{0,1\}$, etc., $\omega = \{0,1,2,3...\}$. And σF is the union of the members of F, i.e., $\sigma F = \bigcup_{A \in F} A$

SOME AUTOMATIC PROOFS IN ANALYSIS

2.4 The composition of two continuous functions is continuous.

2.5 Differentiable functions are continuous.

2.6 A uniformly continuous function is continuous.

3. Intermediate Analysis (On the Real Numbers).

   3.1 If the function f is continuous on the compact set S then f is uniformly continuous on S.

   3.2 If the function f is continuous on the compact set S then $f[S]$ is compact.

   3.3 If a sequence s_n converges to some limit p, then s_n is a Cauchy sequence.

   3.4 If s_n is a Cauchy sequence, then s_n converges to some limit p.

   3.5 If a sequence s_n converges simultaneously to two numbers a and b, then $a = b$.

   3.6 If the function f is continuous at the point a, and
   $$\lim_{n \to \infty} s_n = a, \quad \text{then} \quad \lim_{n \to \infty} f(s_n) = f(a).$$

   3.7 (Balzano-Weierstrass Theorem). If S is an infinite, bounded set, then there exists an accumulation point p of S.

   3.8 Let f and g be two continuous functions. Then the set of points on which f and g agree is a closed set.

   3.9 If, for the sequences x_n and y_n, we have
   $$\lim_{n \to \infty} x_n = a \quad \text{and} \quad \lim_{n \to \infty} y_n = b,$$
   then $\lim_{n \to \infty} (x_n + y_n) = a + b$, and $\lim_{n \to \infty} (x_n \cdot y_n) = a \cdot b$.

3.10 If x_n is a sequence and $\lim_{n \to \infty} x_n = a \neq 0$, then $\lim_{n \to \infty} 1/x_n = 1/a$.

3.11 (Explicit examples).

   .1 $\lim_{n \to \infty} n/(n + n^2) = 0$.

   .2 If $f(x) = 2x^2 + 3x + 1$, then $f'(a) = 4a + 3$.

3.12 (Intermediate Value Theorem). If f is continuous on $[a,b]$, $a \leq b$, $f(a) \leq 0$, and $0 \leq f(b)$, then $f(x) = 0$ for some x in $[a,b]$.

4. Elementary Topology

4.1 If a set B contains an open neighborhood of each of its points, then B is open.

4.2 If F is a family of open sets covering the regular topological space X, then there exists a family G of open sets which covers X and for which $\bar{G} \subseteq \subseteq F$.[3]

It should be noted that this list does not represent all the theorems (in analysis) that can be proved automatically, but rather more or less all that have been. A particular research group might spend most of its time developing new methods rather than trying to extend the number of theorems that have been proved automatically.

This list of theorems needs explanation, because a number of methods (sometimes special heuristics) were used in their proofs. And a casual reader cannot properly understand what has been done until he has a feeling for these methods and their applicability to other theorems.

For example, the method of REDUCTIONS (Rewrite rules or "demodulators") was used to "build-in" elementary set theory into the prover [6], thereby making it easy to prove automatically the theorems of 1.1, and making it possible to prove those of 1.2 by induction. This and algebraic simplification and other methods will be discussed in later sections.

----

[3] $\bar{G}$ denoted the family of the closures of members of G, i.e., $\bar{G} = \{\bar{A}: A \in G\}$, and ($H \subseteq \subseteq F$) means that H is a refinement of F, i.e., each member of H is a subset of some member of F.

Also a special limit heuristics [10] was used to prove the limit theorems of calculus, 2. Later some of these were proved without this special heuristic but the limit heuristic made all such proofs easier. A method of variable restriction was also used for these limit theorems and other theorems.

A technique based on Nonstandard Analysis [3] was used to prove the intermediate analysis theorems 3.1-3.11, and the limit theorems 2.1. This interesting method greatly simplifies proofs but has limited applicability.

A method for automatically instantiating set variables [8], along with techniques for handling general inequalities [12], were used to prove 3.12, 4.1-4.2, and a number of others.

Much of real analysis uses inequalities in a fundamental way. Special provers have been developed to handle ground inequalities theorems (those without variables) and general inequality theorems (those with variables to be instantiated). Also special methods have been developed for handling equality.

A prover for a particular theory (such as the first order logic) is said to be <u>complete</u> for that theory if it can prove all theorems within the theory. It is a <u>decision procedure</u> if it can also detect non-theorems.

Completeness is important in ATP because a prover is not worth much if it exhibits very little generality. On the other hand, experience so far has shown that complete procedures have tended to be weak, in the sense that they take too long to prove easy theorems (or cannot prove them at all). So various attempts have been made to obtain speed without sacrificing (much) completeness.

There are two basic types of automatic provers in use, <u>Resolution</u> based provers [33,16,27] and <u>Natural Deduction</u> type provers [7,10,27 Ch. 6,9]. These are described briefly below.

Resolution based provers tend to be complete whereas Natural Deduction provers tend not to be, especially when various auxillary procedures are added (for speed).

For both Resolution and Natural Deduction a theorem is first converted to quantifier-free form by a Skolemization process and then proved. (See Section 2.2 below).

The theorems listed above were all proved by Natural Deduction or Resolution (or both), using the various procedures and heuristics listed in the following table.

LIST OF METHODS USED

Theorems	Type of Prover	Special Heuristics and Procedures	Complete?
1. Set Theory			
1.1 Elem. set Th.	Nat. Deduction Resolution	None	No
1.2 Set Th. Thms using induction	Nat. Deduction Resolution	Induction	No
1.3 Higher order set theory theorems	Resolution	Higher Order Resolution	No
2. Limit Theorems			
2.1-2.6	Nat. Deduction	Limit Heuristic, Variable Restrictions	No
3. Intermediate Analysis			
3.1-3.11, 2.1-2.6	Nat. Deduction	Non-standard Analysis Variable Restrictions	No
3.12 (Part)[4] 2.1.1(+)	Nat. Deduction Resolution	Variable elimination and Shielding Term	Yes[5]
4. Elementary Topology			
4.1-4.2	Nat. Deduction	Set Variable Instantiation	No

All of the above, except 1.3, used reductions and simplifications.

----

[4] This is an inequality theorem derived from the Intermediate Value theorem. See [12] and Section 5.2 below.

[5] This prover [12] is complete for the first order logic, but the actual implementation which proved this theorem used <u>mandatory</u> variable elimination which is not known to be complete.

SOME AUTOMATIC PROOFS IN ANALYSIS

The limit heuristic used in the proofs of 2.1-2.6 and the non-standard analysis technique used in the proofs of 3.1-3.11, are very powerful agents for theorems in the domain for which they apply, but unfortunately do not seem to have much applicability in other areas.

On the other hand the Reduction and Simplification routines have general applicability throughout all mathematics, and seem to be an essential part of any good prover, especially in analysis. And the Variable restriction and variable elimination techniques seem to have general applicability wherever real inequalities (with variables) are encountered. It is not clear at this time what generality the set variable methods, used in 4.1-4.2, and 3.12, will have.

2. REDUCTIONS, Skolemization, and Induction

2.1 Building-in Elementary Set Theory.

A number of researchers have exercised their provers on theorems like those of 1.1-1.6 from elementary set theory, with varying degrees of success. One difficulty with such experiments has been that the (human) user has had to give to the prover, as additional hypotheses, a number of axioms and definitions from which these simple theorem are derived. This does not appear to be a difficulty until one realizes that when a prover is given too many hypotheses the proof time is drastically increased (because there are many more ways for variables to match). Also when the prover is working on harder theorems, which use these elementary theorems as lemmas, one is obliged to add these theorems themselves as additional hypotheses or forever carry along the needed axioms and definitions. Not surprisingly, such procedures cannot prove difficult theorems.

A similar situation applies to the ordered field axioms, for the real numbers.

In both cases we decided to "build-in" these theories in such a way that such elementary theorems, when encountered in the proof of other theorems, are proved automatically without the need for any additional axioms and definitions.

The heart of the built-in procedure used by our group, is a set of REDUCTIONS or rewrite rules, whereby certain terms are always rewritten as other equivalent terms.

For example, one of our reductions is

$$x \in (A \cap B) \longrightarrow x \in A \wedge x \in B$$

Thus whenever an expression of the form $x \in (A \cap B)$ appears within any formula being processed by the prover, it is immediately replaced by $(x \in A \wedge x \in B)$. This is a one way thing, $(x \in A \wedge x \in B)$ is never replaced by $x \in (A \cap B)$.

Wos's group at Argonne National Laboratory, also uses such rewrite rules (they call them "demodulators") in their prover, which has achieved a great deal of success.

The prover in [6] employed a REDUCTION table of about thirty entries to prove most theorems of elementary set theory. Table 1 lists some of these, and Table 2 lists some of the definitions used in those proofs.

Table 1
Some Reductions

	INPUT	OUTPUT
1.	$x \in (A \cap B)$	$x \in A \wedge x \in B$
2.	$x \in (A \cup B)$	$x \in A \vee x \in B$
3.	$A \in Sb(B)$	$A \subseteq B$ [6]
4.	$A \subseteq (B \cap C)$	$A \subseteq B \wedge A \subseteq C$
5.	$t \in \{x: P(x)\}$	$P(t)$ [6]
...		
10.	$\emptyset \subseteq A$	TRUE
11.	$A \subseteq A$	TRUE
12.	$A \cap A$	A
13.	$P \wedge \text{TRUE}$	P
...		

Table 2
Some DEFINITIONS

$A = B$ (set equality)	$A \subseteq B \wedge B \subseteq A$
$A \subseteq B$	$\forall t(t \in A \rightarrow t \in B)$
...	

----

[6] Rule 3 was actually used with the output $(A \subseteq B \wedge A \in U)$, where U is the universal set. Similarly for Rule 5; the additional conjunct $t \in U$ is added to the output. Such a restriction is needed when t is a non-set (i.e., is not a member of any set).

Let us consider two examples to see the effectiveness of these reductions.

$\underline{EX1}$. $\forall A \; \forall B \; (A \subseteq A \cup B)$

(1) $A \subseteq A \cup B$ The Goal

       $t_0 \in A \rightarrow t_0 \in (A \cup B)$ Defn. of $=$[7]

       $t_0 \in A \rightarrow t_0 \in A \lor t_0 \in B$ REDUCTION Rule 2

           TRUE[8]

$\underline{EX2}$. $\forall A \; \forall B [Sb(A \cap B) = Sb(A) \cap Sb(B)]$

(1) $Sb(A \cap B) = Sb(A) \cap Sb(B)$ The Goal

       $[Sb(A \cap B) \subseteq Sb(A) \cap Sb(B)]$

       $\land \; [Sb(A) \cap Sb(B) \subseteq Sb(A \cap B)]$ Defn. of $=$[7]

(1 1) $Sb(A \cap B) \subseteq Sb(A) \cap Sb(B)$ Subgoal 1

       $t \in Sb(A \cap B) \rightarrow t \in [Sb(A) \cap Sb(B)]$ Defn. of $\subseteq$

       $t \subseteq A \cap B \rightarrow t \in Sb(A) \land t \in Sb(B)$ REDUCTION Rules 3,1

       $t \subseteq A \land t \subseteq B \rightarrow t \subseteq A \land t \subseteq B$ REDUCTION Rules 4,3

          TRUE

(1 2) $Sb(A) \cap Sb(B) \subseteq Sb(A \cap B)$ Subgoal 2

       Proved similarly.

Pastre [30] has generalized the notion of Reductions in a prover that has proved most of theorems 1.1 and 1.2, and others.

Notice that the REDUCTION table is a convenient place to store unit facts such as ($\emptyset \subseteq A$ = TRUE), ($A \cap A = A$), ($P \land$ TRUE $= P$), etc. Such rules can be powerful simplifiers when used within a large proof.

----

[7] Here the prover instantiates the definition of $\subseteq$. This prover [6] (see Section 3.2 below), does not automatically instantiate every definition where possible, but does so only under limited control. In this case, when it had nothing more that it could do, it instantiated (only) the main connective of the conclusion.

[8] This prover, as are most, is able to easily detect that such expressions as ($P \rightarrow P \lor Q$), ($P \land Q \rightarrow P \land Q$) etc., are true.

Actually the use of reductions rules is just a way of "substituting equals for equals", because each entry in the reduction table is an equality or an equivalence. The real savings in efficiency comes from our insistence that it substitutes only one way. (E.g., it always replaces $x \in (A \cap B)$ by $(x \in A \land x \in B)$ but never the other way.) This greatly reduces the combinatorics of the search, whenever such a one-way substitution can be done. When can it be done? When is it possible to replace all equality axioms of a given theory by an equivalent set of reductions without losing completeness? This question is addressed by the field of <u>Complete Sets of Reductions</u> [24,25, 26]. For example, ordinary group theory lends itself to such a treatment [25]. See also Lankford's paper in this volume.

Reduce tables, such as the one above, have been generalized to include conditional reductions, whereby a rule is invoked only if some condition is satisfied. For example, the rule

INPUT	CONDITION	OUTPUT
IF P G H	$\sim$P	H

would rewrite the formula (IF $(0 \leq x)$ 2x+1 5) as 5, if it has a hypothesis, the entry $x \leq -1$.

2.2 Skolemization.

As mentioned earlier, most automatic provers require that a formula be quantifier-free when it is presented for proof. For example, the theorem

$$\forall x \; [P(x) \to Q(x)] \land P(a) \to Q(a)$$

is converted to the form

$$[P(x) \to Q(x)] \land P(a) \to Q(a),$$

where a is a constant and x is a variable that can be instantiated during the proof process. (In this example x is bound to the value a.) This process, which is called <u>skolemization</u>, can be automated for any formula in first order logic [16 Ch. 4; 27 Sect. 1.5; 9 Appendix]. The more complicated example,

$$(\forall x \, \exists y \, P(x,y) \to Q)$$

becomes

$$(P(x,g(x)) \to Q),$$

when skolemized. The 'g' is a new function symbol, called a <u>Skolem function</u> symbol. The hypothesis asserts that for each x there is a (corresponding) y for which $P(x,y)$. Since the y depends on x, we write it as $g(x)$.

When instantiating a definition the computer must take account of the <u>position</u> in the theorem of the formula being replaced, in order to insert the proper skolemization. Thus if $(A \subseteq B)$ is replaced by $\forall x(x \in A \to x \in B)$ in the theorem

$$(H \to A \subseteq B)$$

(where H is some hypothesis). We get

$$(H \to (x_0 \in A \to x_0 \in B)),$$

where x_0 is a new skolem constant, but in

$$(A \subseteq B \to C),$$

we get

$$((x \in A \to x \in B) \to C),$$

where x is a variable which can be instantiated during the proof.

2.3 Using Induction.

A number of researchers have used induction in automatic proofs [17, 6, 15, 14]. The main difficulties in using induction automatically, is in determining

      (1) when to use induction,

      (2) what variable to induct upon, and

      (3) what to use for the induction hypothesis.

For example, in the proof of 1.2.1, $\omega = \sigma\omega$, when the subgoal

(*) $x \in \omega \to (t \in x \to t \in \omega)$

is encountered, the prover decided to try induction on x with the formula (*) itself as the induction hypothesis. Thus it was required to prove the two subgoals

$$(t \in 0 \to t \in \omega)$$

and

$$x \in \omega \land \forall s (s \in x \to s \in \omega) \implies (t \in \text{scsr } x \to t \in \omega),$$

which it could do.

In this example it was able to use the subgoal (*) itself as the induction hypothesis. But of course, this sometimes will not work, we often must use a <u>generalized</u> induction hypothesis. To discover such a needed generalized induction hypothesis remains an essentially unsolved problem for automatic provers. Except for the work of Boyer and Moore [14], no automatic prover, that we know of, uses any induction hypothesis other than the subgoal itself.

2.4 Simplification of Algebraic Expressions.

As mentioned earlier, automatic provers have difficulty coping with the ordered field axioms for the real numbers:

$$x + y = y + x$$
$$x + (y + z) = (x + y) + z$$
$$x \cdot y = y \cdot x$$
$$x \cdot (y \cdot z) = (x \cdot y) \cdot z$$
$$x \cdot (y + z) = x \cdot y + x \cdot z$$
$$\ldots$$

when these are given as additional hypotheses of the theorem being proved, because there are so many ways in which the variables x,y,z in the formulas can match the terms of the theorem being proved. (We believe this difficulty is the reason that most efforts in Automated Theorem Proving have been in areas other than real analysis.)

We have been able to avoid adding such axioms by again using REDUCTIONS, to simplify each algebraic expression to a <u>canonical form</u>. Such simplifications are widely used in computer mathematics, especially in such systems as MACYMA, which performs automatic differentiation, integration, equation solving, limit taking, etc., [2]. Such automatic simplification has found increased use in ATP since about 1970.

3. Types of Provers

3.1 Resolution.

As mentioned earlier there are two basic types of automatic provers in use, Resolution and Natural Deduction. Resolution [33, 16, 27] is a refutation method whereby a quantifier-free theorem is negated and converted to conjunctive normal form (CNF) and then shown to be unsatisfiable (inconsistent) by performing a series of simple inference steps (resolutions). For example, the theorem

$$\forall x\, (P(x) \rightarrow Q(x)) \wedge P(a) \rightarrow Q(a)$$

is skolemized (i.e., has its quantifier removed),

$$(P(x) \rightarrow Q(x)) \wedge P(a) \rightarrow Q(a),$$

(x is now a variable which can be instantiated), and is then negated

$$(P(x) \rightarrow Q(x)) \wedge P(a) \wedge {\sim}Q(a),$$

and placed in CNF,

$$(\overset{1}{{\sim}P(x) \vee Q(x)}) \wedge \overset{2}{P(a)} \wedge \overset{3}{{\sim}Q(a)}.$$

The disjuncts 1, 2, 3, are called <u>clauses</u>

1. ${\sim}P(x) \vee Q(x)$
2. $P(a)$
3. ${\sim}Q(a)$

Clause 1 is resolved with Clause 2 (with substitution a/x) to obtain

4. $Q(a)$,

which is then resolved with Clause 3 to obtain

5. □ ,

a contradiction. See the references quoted above for details, and also

Reference [2, 448-462] for a brief elementary introduction to Resolution and ATP. All theorems in first order logic can be proved in this way by Resolution.

3.2 Natural Deduction.

Natural Deduction is a procedure whereby the theorem being proved is manipulated by a set of production rules, in which a goal is converted to one or more subgoals [7, 10, 27 Ch. 6, 9]. The initial goal is first converted to quantifier-free form by skolemization, but <u>not negated</u>.

The prover in [9] is given a goal G and a hypothesis H, and is required to find and return a (most general) substitution θ for which $(H\theta \rightarrow G\theta)$ is a tautology. Some of its rules are:

I4. $(H \Longrightarrow A \wedge B)$ "SPLIT"

   If $(H \Longrightarrow A)$ returns θ
   and $(H \Longrightarrow B\theta)$ returns λ
   then return $\theta \circ \lambda$.

   ($\theta \circ \lambda$ is the composition of the substitutions θ and λ. See [16, p. 76].)

H2. $(H \wedge P \Longrightarrow C)$ "MATCH"

   If $P\theta \equiv C\theta$, return θ

H7. $H \wedge (A \rightarrow D) \Longrightarrow C$ "BACKCHAINING"

   If $(D \Longrightarrow C)$ returns θ
   and $(H \Longrightarrow A\theta)$ returns λ,
   then return $\theta \circ \lambda$.

   ...

To prove the example

$$\forall x(P(x) \rightarrow Q(x)) \wedge P(a) \rightarrow Q(a),$$

backchaining (Rule H7) is applied to the original goal

GOAL: $(P(x) \rightarrow Q(x)) \wedge P(a) \Longrightarrow Q(a)$

to obtain Subgoals 1 and 2.

SOME AUTOMATIC PROOFS IN ANALYSIS

SUBGOAL 1: $(Q(x) \Longrightarrow Q(a))$

returns $\theta = a/x$, by matching (Rule H2),

SUBGOAL 2: $(P(x) \to Q(x)) \wedge P(a) \Longrightarrow P(x)(a/x)$

returns TRUE (because $P(x)(a/x) = P(a)$).

So a/x is returned for the original GOAL, to complete the proof.

The procedure in [9] is not complete but seems to be adequate to prove a wide variety of theorems. A similar but extended system by Loveland and Stickle [27] is complete (can prove all theorems of first order logic). See [7] for a list of other papers on Natural Deduction type provers. The theorems listed in Section 1 were all proved by Natural Deduction or Resolution type provers. Also some used various heuristics and procedures as indicated in Section 1.

3.3 Provers for Higher Order Logic.

Andrews' prover which is described elsewhere in this volume has successfully proved Cantor's Theorem 1.3.2, and a number of others. See also [1, 23, 22, 18, 31].

4. Limit Heuristic

The Natural Deduction prover [10] was used to prove the limit theorems of Calculus 2.1-2.6, employed a <u>limit heuristic</u> and a <u>variable restriction</u> mechanism, as well as reduction and simplification routines.

LIMIT HEURISTIC: when proving a goal of the form

$$|B| < E$$

in the presence of a hypothesis of the form

$$|A| < E'$$

(and other hypotheses H), first try to find a substitution σ, for which $B\sigma$ can be expanded as a non-trivial linear combination of $A\sigma$, i.e.,

$$[B = kA + L]_\sigma$$

where σ is a most general substitution, and, if this is possible, then try to prove the following three subgoals:

SG1: $[|K| < M]_\sigma$ for some M, $0 < M < \infty$,

SG2: $[|A| < E' \rightarrow |A| < E/(2 \cdot M)]_{\sigma \circ \sigma_1}$

SG3: $[|L| < E/2]_{\sigma \circ \sigma_1 \circ \sigma_2}$

where σ, σ_1, σ_2, are returned from subgoals SG1, SG2, and SG3, respectively.

So, for example, in proving that the limit of a product of two functions is the product of their limits (if the two limits exist), the prover encounters the goal

$$|f(x) \cdot g(x) - L_1 \cdot L_2| < E$$

in the presence of the hypothesis

$$|f(x_1) - L_1| < E_1$$

(and other hypotheses, H) where E, x, and x_1 are variables. Using the limit heuristic, it expresses $(f(x) \cdot g(x) - L_1 \cdot L_2)$ as a linear combination of $(f(x_1) - L_1)$, as follows:

$$\underbrace{f(x) \cdot g(x) - L_1 \cdot L_2}_{B} = \underbrace{g(x)}_{K} \cdot \underbrace{[f(x) - L_1]}_{A} + \underbrace{g(x) \cdot L_1 - L_1 \cdot L_2}_{L}$$

with substitution $\sigma = (x/x_1)$, and then proceeds to prove the subgoals

Sb1: $|g(x)| < M$

Sb2: $|f(x) - L_1| < E_1 \rightarrow |f(x) - L_1| < E/(2 \cdot M)$

Sb3: $|g(x) \cdot L_1 - L_1 \cdot L_2| < E/2$,

where x, M, and E, are variables. (See Section 5, in regard to proving these inequalities).

5. Inequality Provers

One of the most urgent needs for automatic proofs in analysis is an efficient technique for handling general inequalities. Ground inequalities are relatively easy, but those with variables to instantiate are not.

5.1 Variable Restrictions.

One such technique is due to Slagle and Norton [35]. Another is the variable restrictions of Bledsoe [11]. When faced with a subgoal of the form $(1 < x < 3)$ this prover will not instantiate x with a particular value such as 2.5, but instead will give x the "restriction <1 3>", (which simply means that x is between 1 and 3) and store this fact in the data base, leaving x as a variable to be instantiated by later subgoals. Of course any value x_0 of x obtained subsequently must satisfy the restriction $(1 < x_0 < 3)$.

For example in proving the theorem,

$$P(2.5) \to \exists x (1 < x < 3 \land P(x))$$

the prover first encounters the subgoal

(1) $\qquad P(2.5) \to (1 < x < 3)$

which it satisfies by giving x the restriction <1 3>, but leaving x as a variable; and then encounters the subgoal

(2) $\qquad P(2.5) \to P(x)$

which it satisfies by giving x the value 2.5. It then finishes the proof by verifying that $1 < 2.5 < 3$, which is immediate.

This concept was used by the prover described in [10] in proving the limit-of-a-product theorem, mentioned in Section 4. In proving the three subgoals Sb1, Sb2, and Sb3, (see above) the prover encounters the three subgoals

- (1) $(0 < D)$,
- (2) $(|x - a| < D \to |x - a| < D_2)$,
- (3) $(|x - a| < D \to |x - a| < D_1)$,

(as well as other subgoals) where D is a variable and D_1 and D_2 are constants.

The prover satisfies Subgoal (1) by giving D the restriction $<0\ \infty>$, and then updates this restriction to $<0\ D_2>$ to satisfy Subgoal (2). (It contains in its data base the facts that D_1 and D_2 have restrictions $<0\ \infty>$.) Finally, it satisfies Subgoal (3) by further updating the D restriction to $<0, \min(D_1,D_2)>$.

Several other subgoals in this proof are also satisfied by variable restrictions. This technique along with the algebraic simplification that accompanies it, helps avoid the explicit use of the inequalities axioms and the real field axioms, which tend to clutter and degrade an automatic prover.

Other methods for handling inequalities with variables are given in Section 5.2. Methods similar to these and others like Slagle and Norton's [35] have been very useful in proofs that arise in program verification [21, 36].

5.2 Variable Elimination and Shielding Term Removal.

Another approach to general inequalities (on the reals with variables to be instantiated), is found in [12]. This is a resolution based prover which uses

$$\text{Inequality Chaining}$$
$$\text{Variable Elimination}$$
$$\text{Shielding Term Removal}$$

These will be described shortly. The resulting prover is complete for the first order logic [13].

First a theorem is negated and converted to clausal form (see Section 3.1). If the resulting clauses are ground, i.e., have no variables to be instantiated, the proof is usually easy. There are a number of fast decision procedures for ground inequalities [29, 34].

We will first describe variable elimination (VE) which plays a central role in the prover. Consider the clause

(1) $\qquad x < a \ \lor \ b < x, \qquad$ (x is a variable)

which was derived by negation from the formula

(2) $\qquad \exists x (a \leq x \leq b)$

to be proved. Since (2) is equivalent (on the real numbers) to

(2´) $\qquad a \leq b$

it follows that (1) is equivalent to

(1´) $\qquad b < a$.

So we can replace (1) by (1´), thereby eliminating the variable x. Similarly, the clauses,

$$x \leq a \quad \lor \quad b < a,$$
$$x < a,$$
$$x \leq a \quad \lor \quad x \leq b \quad \lor \quad c < x,$$

can be replaced by $(b \leq a)$, $\square$, and $(c \leq a \ \lor \ c \leq b)$, respectively.

Thus in proving the simple theorem

$$a \leq b \quad \longrightarrow \quad \exists x (a \leq x \leq b),$$

it is first converted to clausal form,

1. $\quad a \leq b$
2. $\quad x < a \ \lor \ b < x,$

and then VE is used to obtain

3. $\quad b < a \qquad\qquad$ 2, VE x

and then Clauses 1 and 3 are resolved to obtain

4. $\quad \square , \qquad\qquad$ 1, 3

to complete the proof.

A similar proof does <u>not</u> work for the three clauses

1. $g(y) \leq b \lor 1 \leq y$
2. $y' < g(y') \lor 1 \leq y'$
3. $b < 1$,

because the variable y, (and also the variable y'), cannot be eliminated from either 1 or 2. But if we first "chain" 1 and 2 on $g(y)$ we get,

4. $y < b \lor 1 \leq y$ 1, 2, removing $g(y)$

and then VE can be applied to 4 getting,

5. $1 \leq b$,   4, VE y
6. □ .   3, 5

The term $g(y)$ in the above is called a "shielding term", because it shields the variable y; once all shielding terms (of a particular variable) have been removed that variable becomes eligible for elimination.

Inequality chaining [35] is simply the concept of applying the transitivity axioms

$$x \leq y \land y \leq z \longrightarrow x \leq z$$
$$x \leq y \land y < z \longrightarrow x < z$$
$$\text{etc.}$$

to two literals, as was done in the above examples to the literals, $g(y) \leq b$ and $y < g(y)$ to get $y < b$. Also unification is permitted, e.g., $y < g(y)$ and $g(c) \leq b$ result in $c < b$.

It is also possible to chain on the <u>variable</u> y' in Clause 2, but note that it can match in three ways. In fact there are 12 different chain resolvents from Clauses 1, 2, 3, if we allow chaining on variables, and only three if we do not. We avoid this proliferation of clauses by forbidding <u>chaining on variables</u>. Without chaining on variables the prover is still complete, and much more powerful.

Random chaining on terms tends to greatly enlarge the search space, even without chaining on variables, so we have attempted to devise an overall <u>strategy</u> which will guide the search. Three facts motivate this strategy

SOME AUTOMATIC PROOFS IN ANALYSIS

- Ground proofs are easy
- VE removes variables from clauses (makes them "more ground")
- Shielding term removal tends to make variables eligible for VE

So our strategy is as follows:

- If a ground proof is possible, do it
- If not, try to eliminate a variable by VE
- If no variable is eligible for VE then try to remove a shielding term by chaining (but don't chain on variables)

This can be depicted this way

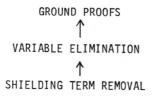

$$\begin{array}{c} \text{GROUND PROOFS} \\ \uparrow \\ \text{VARIABLE ELIMINATION} \\ \uparrow \\ \text{SHIELDING TERM REMOVAL} \end{array}$$

Algebraic simplification and reductions are also used in this prover.

Such a system has been shown to be complete for first order logic [13]. (Recall that any first order literal $P(x_1, x_2, \ldots, x_n)$ can be converted to an equivalent inequality literal $g(x_1, x_2, \ldots, x_n) \leq 0$, so any first order theorem can be converted into an inequality theorem.)

Our implementation described in [12] uses <u>mandatory</u> VE, whereby VE must be applied to any clause with a variable which is eligible for VE, and the parent clause discarded (a highly desirable act since we seek to remove all variables). The completeness proof given in [13] does not allow mandatory VE: it is still an open question whether the prover is complete with mandatory VE.

We now give a few other examples theorems, followed by the corresponding clauses, and proofs (or parts of proofs). See [12] for further details.

@ $\quad (\forall \varepsilon (0 < \varepsilon \rightarrow A \leq B + \varepsilon) \longrightarrow A \leq B)$

Clauses

1. $\varepsilon \leq 0 \ \lor \ A \leq B + \varepsilon$
2. $B < A$

---

3. $A \leq B$ 1, VE ε
4. $\square$ 2,3

@ $\forall \varepsilon [(0 < \varepsilon \rightarrow A \leq B(\varepsilon) + \varepsilon) \land B(\varepsilon) \leq C] \longrightarrow A \leq C$

(ε is a variable, $B(\varepsilon)$ is a function of ε.)

Clauses

1. $\varepsilon \leq 0 \ \lor \ A \leq B(\varepsilon) + \varepsilon$
2. $B(\varepsilon) \leq C$
3. $C < A$

---

4. $\varepsilon \leq 0 \ \lor \ A \leq C + \varepsilon$ 1, 2
5. $A \leq C$ 4, VE ε
6. $\square$ 3,5

@ The sum of two continuous functions is continuous.

Clauses

1. $f(x_\delta) + g(x_\delta) + \varepsilon_0 < f(x_0) + g(x_0)$
   $\lor \ f(x_0) + g(x_0) + \varepsilon_0 < f(x_\delta) + g(x_\delta) \ \lor \ \delta \leq 0$
2. $0 < \varepsilon_0$
3. $0 < \delta_\varepsilon \ \lor \ \varepsilon \leq 0$
   ...
5. $f(x_0) \leq f(y) + \varepsilon \ \lor \ \delta_\varepsilon + x_0 < y \ \lor \ \delta_\varepsilon + y < x_0 \ \lor \ \varepsilon \leq 0$
   ...
10. $x_\delta \leq x_0 + \delta \ \lor \ \delta \leq 0$

where $\delta, \varepsilon, \varepsilon'$, and y are variables, x_δ is a skolem function of δ, δ_ε is a skolem function of ε, etc.

11. $g(x_\delta) + \varepsilon_0 < g(x_0) + \varepsilon$
    $\vee\ f(x_0) + g(x_0) + \varepsilon_0 < f(x_\delta) + g(x_\delta)$
    $\vee\ \delta \leq 0\ \vee\ \delta_\varepsilon + x_0 < x_\delta$
    $\vee\ \delta_\varepsilon + x_\delta < x_0\ \vee\ \varepsilon \leq 0$ 1, 5, x_δ/y, removing $f(x_\delta)$

    ...

18. $\varepsilon_0 < \varepsilon + \varepsilon'\ \vee\ \delta \leq 0\ \vee\ \delta_\varepsilon < \delta$
    $\vee\ \varepsilon \leq 0\ \vee\ \delta'_{\varepsilon'} < \delta\ \vee\ \varepsilon' \leq 0$

19. $\varepsilon_0 < \varepsilon + \varepsilon'\ \vee\ \delta_\varepsilon \leq 0\ \vee\ \delta'_{\varepsilon'} \leq 0$
    $\vee\ \varepsilon \leq 0\ \vee\ \varepsilon' \leq 0$ 18, VE δ

    ...

24. □

@ 3.12´ (From the intermediate value theorem - See Section 7).

<u>Clauses</u> (We will omit the $\vee$ symbol between literals of clauses.)

1. $x < a$ $b < x$ $f(x) \leq 0$ $t_x < x$
2.   "   "   " $s \leq t_x$ $x < s$ $0 < f(s)$
3.   "   " $0 \leq f(x)$ $x < t_x$
4.   "   "   " $t_x \leq s$ $s < x$ $f(s) < 0$
5. $b < x$ $0 < f(x)$ $x \leq 1$
6. $1 \leq y$ $z_y \leq b$
7.   " $f(z_y) \leq 0$
8.   " $y < z_y$
9. $a \leq b$
10. $f(a) \leq 0$
11. $0 \leq f(b)$
12. $0 < f(x)$ $f(x) < 0$

where x, s, and y are variables, and t_x, z_y are functions of x and y, respectively.

The prover described in [12] was unable to prove this theorem. There are too many ways to match even with our strategy. But later versions by Hines and another by Tie-Cheng Wang (not yet published) using multi-step planning, have been able to prove this and other harder theorems.

6. Proofs Using Non-standard Analysis

It is well known that a number of proofs of theorems in real analysis are made easier by the use of non-standard analysis [32]. This is also true for automatic proofs: the theorems 3.1-3.11, 2.1-2.6, were proved by the prover described in [3] by first converting them to "non-standard form" and then finishing the proofs using various properties of the non-standard concepts.

Some Concepts, Properties, and Definitions in Non-standard Analysis

$x \approx y$	means that x and y belong to the same monad [32] (i.e., "x and y are infinitely close together")
$\approx$	is an equivalence relation (in particular, it is transitive)
$st(x)$	means the standard part of x
$st(x) \approx x$	
Standard x	means that x is an ordinary real number
Standard $st(x)$	

$x \approx y \rightarrow st(x) \approx st(y)$

S is compact iff $\forall x(x \in S \rightarrow st(x) \in S)$

f is continuous at r iff standard r
$\wedge \forall y(r \approx y \rightarrow f(r) \approx f(y))$

f is uniformly continuous on S iff
$\quad \forall x \in S \; \forall y \in S(x \approx y \rightarrow f(x) \approx f(y))$

The theorem,

> If f is continuous on the compact set S, then f is uniformly continuous on S,

is proved by first converting it to non-standard form,

(1) $\quad (x \in S \rightarrow st(x) \in S)$ \hfill Compact

(2) $\quad (r \in S \wedge y \in S \wedge$ Standard r
$\quad \wedge r \approx y \rightarrow f(r) \approx f(y))$ \hfill Continuity

(3) $\quad x_0 \in S \;\land\; y_0 \in S \;\land\; x_0 \approx y_0$
$\quad\quad\longrightarrow\quad f(x_0) \approx f(y_0)$

and then establishing the following facts:

$x_0 \in S,\; y_0 \in S,\; x_0 \approx y_0,$	Given (Hypothesis (3))
$st(x_0) \in S,\; st(y_0) \in S,$	Hypothesis (1)
$x_0 \approx st(x_0),\; y_0 \approx st(y_0),$	Property of $\approx$
Standard $st(x_0)$, Standard $st(y_0)$,	Property of $\approx$
$st(x_0) \approx st(y_0),$	Property of $\approx$, since $x_0 \approx y_0$
$f(x_0) \approx f(st(x_0))$	Hypothesis (2), with $st(x_0)/r,\; x_0/y$
$\approx f(st(y_0))$	Hypothesis (2), with $st(x_0)/r,\; st(y_0)/y$
$\approx f(y_0)$	Hypothesis (2), with $st(y_0)/r,\; y_0/y$
$f(x_0) \approx f(y_0)$	Transitivity of $\approx$.

The prover described in [3] uses a typing mechanism (whereby entities are typed as "real", "infinitesimal", etc.), and a data base of relevant facts, reductions, simplification, etc., to facilitate its proof, and others like it.

As was mentioned earlier, the prover has great success on those theorems for which it applies but is severely limited in its applicability.

7. Set Variable Instantiation

Theorems in higher order logic are, in general, harder to prove than first order theorems: higher order logic is incomplete and (proper) higher order instantiations are more difficult to find. When the higher order variables involved are all universally quantified, then the thoerem is essentially first order, but when one or more higher order variables are existentially quantified, the proof is often much harder. Examples are: the set variable A in

$$\exists A \subseteq R \ (A \text{ is dense in } R \ \wedge \ (R \sim A) \text{ is dense in } R);$$

the function variable f in

$$\exists f \ \forall x \in R \ (\text{Continuous } f \ x \ \wedge \ \sim \text{differentiable } f \ x);$$

the family variable G in

> If F is a family of open sets covering a regular topological space X, then there exists a family G of open sets which covers X and for which the family of closures of members of G is a refinement of F;

and the set variable A in

3.12 (Intermediate Value Theorem)

$$\forall A \ (A \neq \emptyset \ \wedge \ \text{bounded } A \ \rightarrow \ \exists L(L = \sup A))$$
$$\wedge a \leq b \ \wedge \ f \text{ is continuous on } [a,b] \ \wedge \ f(a) \leq 0 \leq f(b)$$
$$\longrightarrow \ \exists x(a \leq x \leq b \ \wedge \ f(x) = 0)$$

Andrews' prover, described elsewhere in this volume, is ideally suited for such theorems. See also [1, 23, 22, 18, 31]. These provers, though powerful in concept, have yet to be developed to the point where they can prove moderately hard theorems in higher order logic. (Though there are a number of interesting exceptions.) Therefore we have looked to special ad hoc methods to handle a subset of higher order logic, those theorems which require the instantiation of a set variable.

So given a theorem of the form

$$\exists A \ P(A)$$

we desire to give a "value" to A of the form

$$A = \{x: \ Q(x)\} \ ,$$

where Q(x) is described in terms of the symbols of P(A). The central concept in this work is to determine this value for A (i.e., the description of Q) in a series of steps, by keying on subformulas of the form

(1) $\quad\quad\quad (x \in A \rightarrow P(x))$

and

(2) $\quad\quad\quad t \in A$

and others, within the theorem. Each of these triggers the building of a partial description of Q (e.g., $\{x: P(x)\}$ for (1)), and these partial descriptions are combined to make up the complete description.

The basic notion for this is due to Darlington [19] and Bledsoe [8], and is also closely related to some earlier work of Behmann [5] on a decision procedure for monatic logic. See also [28, 20].

Once a description, $\{x: Q(x)\}$ has been obtained, the symbol "A" in the theorem is replaced by $\{x: Q(x)\}$, and the resulting first order theorem is proved. Examples 3.12, 4.1-4.2, and others were proved using these methods by the prover described in [8].

When this prover was applied to the intermediate value theorem 3.12 (see above), the value

$$A = \{x: x \leq b \land f(x) \leq 0\}$$

was obtained, which, when substituted for A in the theorem, resulted in the new (first order) theorem

3.12´ $\quad$ (LUB $\land$ L1 $\land$ L2 $\land$ $a \leq b$ $\land$ $f(a) \leq 0 \leq f(b)$
$\quad\quad\quad \longrightarrow \exists x(a \leq x \leq b \land f(x) \leq 0 \leq f(x)))$

where LUB, L1, and L2 are as follows

LUB: $([\exists r(r \leq b \land f(r) \leq 0)$
$\quad\quad \land \exists u \, \forall t(t \leq b \land f(t) \leq 0 \rightarrow t \leq u)]$
$\quad\quad \longrightarrow \exists L[\forall x(x \leq b \land f(x) \leq 0 \rightarrow x \leq L)$
$\quad\quad\quad\quad \land \forall y \{\forall z(z \leq b \land f(z) \leq 0 \rightarrow z \leq y) \longrightarrow L \leq y\}])$.

L1: $\quad \forall x(a \leq x \leq b \land 0 < f(x)$
$\quad\quad \longrightarrow \exists t(t < x \land \forall s(t < s \leq x \rightarrow 0 < f(s))))$.

L2: $\quad \forall x(a \leq x \leq b) \land f(x) < 0$
$\quad\quad \longrightarrow \exists t(x < t \land \forall s(x \leq s < t \rightarrow f(s) < 0)))$.

(In this proof we have used the continuity lemmas L1 and L2 instead of the full definition of continuity of f.)

While 3.12´ is of first order, it too is a difficult theorem for automatic provers (and fairly hard for humans).

During the first pass when the value $\{x: x \leq b \wedge f(x) \leq 0\}$ was obtained for A, the prover also obtained the binding L/x. Of course the proof of 3.12´ becomes much easier when this value, L, is substituted for x. In fact the general-inequality prover [12] described in Section 5 can easily prove 3.12´ when x is replaced by L and cannot when x is left as a variable. (More recent versions by Hines and by Tie-Cheng Wang (unpublished) have proved 3.12´.)

It is interesting to note that the original theorem, 3.12, though of higher order, is nevertheless, easier for the automatic prover to prove than its first order derivate, 3.12´, provided that both the instantiation for A and that for x are saved and used in the second pass. This is consistent with human behavior; most mathematicians find 3.12 easier to prove directly than 3.12´.

8. Remarks

Much remains to be done before we can have automatic provers which will compete successfully with their human counterparts on difficult theorems. We feel that such power will not become available until we begin to incorporate yet other concepts, such as

>Analogy
>
>The use of examples (as counterexamples, and as
>   aids in discovering the proof) [4]
>
>Use of special cases
>
>Conjecturing (see Lenat's paper in this volume)

and much better overall planning (agenda mechanisms, etc.). See [7].

We expect sizable advances to be made in this field during the next several years, especially if and when a number of new researchers are added, who have a strong background in mathematics and some knowledge of computers.

Bibliography

1. Andrews, P.B., [1971], Resolution in Type Theory, *J. Sym. Logic 36*, 414-432.
2. Arden, B.W. (ed.), [1980], Automatic Theorem Proving, in *What Can Be Automated*, MIT Press, 448-462 and 513-526.
3. Ballantyne, A.M., Bledsoe, W.W., [1977], Automatic Proofs of Theorems in Analysis Using Non-standard Techniques, *JACM 24*, 353-374.
4. Ballantyne, A.M., Bledsoe, W.W., [1982], On Generating and Using Examples in Proof Discovery, *Machine Intelligence 10*, 3-39.
5. Behmann, H., Beitrage Zur Algebra Der Logik: Insbesondere Zum Enischeidungsproblem, Mathematische Annalen, 86, 163-229.
6. Bledsoe, W.W., [1971], Splitting and Reduction Heuristics in Automatic Theorem Proving, *Artificial Intelligence 2*, No. 1, 55-77.
7. Bledsoe, W.W., [1977], Non-Resolution Theorem Proving, *Artificial Intelligence* 9, 1-35.
8. Bledsoe, W.W., [1982], A Maximal Method for Set Variables in Automatic Theorem Proving, *Machine Intelligence 9*, (J.E. Hayes, D. Michie, L.I. Mikulich, eds.), Ellis Harwood Lim., Chichester, 53-100.
9. Bledsoe, W.W., [1983], The UT Natural Deduction Prover, University of Texas Mathematics Dept. Memo ATP-17B, April.
10. Bledsoe, W.W., Boyer, R.S., Henneman, W.H., [1971], Computer Proofs of Limit Theorems, *Artificial Intelligence 3*, 27-60.
11. Bledsoe, W.W., Bruell, P., Shostak, R., [1978], A Prover for General Inequalities, University of Texas Mathematics Dept. Memo ATP40; also IJCAI-79.
12. Bledsoe, W.W., Hines. L.M., [1980], Variable Elimination and Chaining in a Resolution-based Prover for Inequalities, Proc. 5th Conference on Automated Deduction, Les Arcs, France, July 8-11, (W. Bibel, R. Kowalski, eds.) Springer-Verlag, 70-87.
13. Bledsoe, W.W., Kunen, K., Shostak, R., [1982], Completeness Proofs for Inequality Provers, University of Texas Mathematics Dept. Memo ATP-65, (submitted).
14. Boyer, R.S., Moore, J S., [1979], A Computational Logic, Academic Press.
15. Burstall, R.M., [1969], Properties of Programs by Structural Inducation, *Comput. J. 12*, 41-48.
16. Chang, C.L., Lee, R.C.T., [1973], Symbolic Logic and Mechanical Theorem Proving, Academic Press.
17. Darlington, J.L., [1968], Automatic Theorem Proving with Equality Substitutions and Mathematical Induction, *Machine Intelligence 3*, 113-127.
18. Darlington, J.L., [1971], A Partial Mechanization of Second-order Logic, *Machine Intelligence 6*, 91-100.
19. Darlington, J.L., [1972], Deductive Plan Formation in Higher Order Logic, *Machine Intelligence 7*, 129-137.
20. Ferro, A., Omodeo, E.G., Schwartz, J.T., Decision Procedures for Elementary Sublanguages of Set Theory, (NYU).
21. Good, D.I., London, R.L., Bledsoe, W.W., [1975], An Interactive Verification System, Proc. 1975 Intl. Conf. on Reliable Software, Los Angeles, CA., 482-492.

22. Huet, G.P., [1973], A Mechanization of Type Theory, IJCAI-73, Stanford, 139-146.
23. Huet, G.P., [1975], A Unification Algorithm for Typed Lambda-Calculus, Theoretical Computer Science 1, 27-57.
24. Huet, G.P., Oppen, D., [1980], Equations and Rewrite Rules: A Survey, Tech. Report CSL-111, SRI-International, January.
25. Knuth, D., Bendix, P., [1970], Simple Word Problems in Universal Algebras, in Computational Problems in Abstract Algebra, (J. Leech, ed.), Pergamon Press, 263-297.
26. Lankford, D., Ballantyne, A.M., [1977], University of Texas Mathematics Dept. Memos 35, 37-39.
27. Loveland, D.W., [1978], Automated Theorem Proving: A Logical Basis, North-Holland.
28. Minor, J.T., [1979], Proving a Subset of Second-order Logic with First-order Proof Procedures, Ph.D. Dissertation, University of Texas, Computer Science Dept.
29. Nelson, G., Oppen, D., [1978], A Simplifier Based on Efficient Decision Algorithms, Proc. 5th ACM Sym. on Principles on Programming Languages.
30. Pastre, D., [1976], Demonstration Automatique de Theoremes en Theorie des Ensenbles, Ph.D. Thesis, University of Paris 6.
31. Pietrzykowski, T., [1973], A Complete Mechanization of Second-order Logic, JACM 20, 333-364.
32. Robinson, Abraham, [1966], Nonstandard Analysis, North Holland.
33. Robinson, J.A., [1965], A Machine Oriented Logic Based on the Resolution Principle, JACM 12, 23-41.
34. Shostak, R., [1977], On the SUP-INF Method for Proving Presburger Formulas JACM 24, No. 4, 529-543.
35. Slagle, J.R., Norton, L., [1973], Experiments with an Automatic Theorem Prover Having Partial Ordering Rules, CACM 16, 682-688.
36. Suzuki, N., [1975], Verifying Programs by Algebraic and Logical Reduction, Proc. Intl. Conf. on Reliable Software, IEEE, 473-481.
37. Tyson, M., [1981], A Priority-ordered Agenda Prover, Ph.D. Dissertation, University of Texas, Computer Science Dept.

Proof-Checking, Theorem-Proving, and Program Verification

ROBERT S. BOYER AND J STROTHER MOORE[1]

This article consists of three parts: a tutorial introduction to a computer program that proves theorems by induction; a brief description of recent applications of that theorem-prover; and a discussion of several nontechnical aspects of the problem of building automatic theorem-provers. The theorem-prover described has proved theorems such as the uniqueness of prime factorizations, Fermat's theorem, and the recursive unsolvability of the halting problem.

The article is addressed to those who know nothing about automatic theorem-proving but would like a glimpse of one such system. This article definitely does not provide a balanced view of all automatic theorem-proving, the literature of which is already rather large and technical.[2] Nor do we describe the details of our theorem-proving system, but they can be found in the books, articles, and technical reports that we reference.

In our opinion, progress in automatic theorem-proving is largely a function of the mathematical ability of those attempting to build such systems. We encourage good mathematicians to work in the field.

---

[1] Institute for Computing Science and Computer Applications, University of Texas at Austin, Austin, Texas 78712.

[2] Good places to start on the technical literature are [18] and [3].

The work reported here was supported in part by NSF Grant MCS-8202943 and ONR Contract N00014-81-K-0634.

© 1984 American Mathematical Society
0271-4132/84 $1.00 + $.25 per page

1. Tutorial.

1.1. *From Proof-Checking to Fully Automatic Proof.* A *formal proof* is a finite sequence of formulas, each member of which is either an *axiom* or the result of applying a *rule of inference* to previous members of the sequence. Typical rules of inference are *modus ponens* and the substitution of equals for equals. A grammar for formulas, a collection of axioms, and a collection of rules of inference together define a logical *theory*.

For the usual theories of mathematics, e.g. set theory or number theory, it is a relatively modest exercise to write a program called a *proof checker* that will check, in a reasonable amount of time, whether a given sequence of formulas is a proof. For some theories, e.g. the propositional calculus, it is possible to write a computer program called a *decision procedure* that will determine whether any given formula has a proof in the theory. But for the usual theories of mathematics, it is theoretically impossible to write a decision procedure. On the other hand, it is theoretically possible to write a *semi-decision procedure*, that is, a program which will find a proof for any given formula if there is one, but which may run forever if there is no proof. For example, one can write the practically useless program that will systematically generate all the proofs in a theory. The challenge of automatic theorem-proving is to write computer programs that find proofs in a reasonable or practical amount of time.

One of the main techniques used to meet this challenge is the invention of *heuristic* proof techniques -- algorithms that analyze the problem at hand and pursue sound and plausible lines of reasoning. Unlike decision procedures, heuristics are not guaranteed to find a proof if one exists. Because the ordinary kinds of mathematical theories, such as number theory and set theory, are undecidable, heuristics will inevitably be a part of any completely automatic theorem-proving system.

While automatic theorem-provers have occasionally contributed to the proofs of new results in mathematics,[3] the kinds of proofs discovered by today's programs would be considered trivial by most mathematicians. But an automatic theorem-prover need not be a first rate mathematician to be useful. It would be a major accomplishment with far-reaching practical consequences to produce an automatic theorem-proving program that could follow and detect errors in mathematical proofs described at the level of graduate mathematics textbooks. We will explain this remark when we discuss the applications of our theorem-prover.

---

[3]Cf. the papers of Wos and Winker and of Chou in this volume.

1.2. *Our Automatic Theorem-Prover.* Our work on automatic theorem-proving can be regarded as an attempt to construct a *high-level proof checker* for elementary number theory and recursive function theory. The basic axioms of our theory are those of Peano arithmetic, plus similar axioms defining ordered pairs. The Peano axioms characterize the natural numbers as follows. 0 is a natural number; if i is a natural number, so is the "successor" of i, usually written s(i); s(i) is never 0; s(i) is s(j) iff i is j. In addition, we are provided with the principle of mathematical induction as a rule of inference: To prove that any formula P(n) is a theorem for all natural numbers n, it is sufficient to prove P(0) and to prove that for all natural numbers i, P(i) implies P(s(i)).

Starting from such concepts the user of our system introduces such recursive definitions as "sum," "product," and "remainder" and uses the theorem-prover to prove theorems about them. Among the theorems proved by our machine are:

- the existence and uniqueness of prime factorizations [4]
- Fermat's theorem: $M^{(p-1)} = 1$ (mod p) if p is prime and does not divide M [7]
- Wilson's theorem: $p-1! = -1$ (mod p) if p is prime [22]
- Gauss' law of quadratic reciprocity
- the existence of nonprimitive recursive functions
- the soundness and completeness of a propositional calculus decision procedure similar to the Wang algorithm [4]
- the Turing completeness of the Pure LISP programming language [10]
- the recursive unsolvability of the halting problem [8]

To guide our system towards finding proofs for these theorems, the user of the system suggested intermediate steps for the proofs. But the total amount of assistance given by the user was approximately the text that one might find in a graduate level exposition of these theorems.

1.3. *Three Heuristics used by Our Program.* When, in searching for a proof, does a mathematician decide to make an argument by induction, and how does he decide on which formula "P" to do induction? Answers to this question are likely to involve a heuristic guess. Furthermore, there are many variants on the induction principle some of which may be more appropriate for a given problem than others, e.g., course-of-values, simultaneous induction on several variables, induction up to certain bounds. The major emphasis of our work has been the development of a heuristic for mechanizing mathematical induction.

The heuristic, which takes several pages to describe fully [4], considers the definitions of the recursive functions that are mentioned in the conjecture and selects an induction argument that is the "dual" of some combination of the definitions. We illustrate our induction heuristic in the next section of this paper.

In addition, our program contains many other heuristics. The most complex select instances of axioms, definitions, and previously proved lemmas to use in a proof. For example, suppose some function symbol f is defined recursively. When does one decide to replace a "call" of f by the definition of f? The main heuristic our program uses is to check whether the recursive calls of f that would be introduced are already in the conjecture at hand. As will be illustrated below, hypotheses about such recursive calls are frequently provided by the induction heuristic.

A third important heuristic in our theorem-proving system is generalization, i.e. considering a harder problem than the one at hand. Generalization seems to be *necessary* in order to get certain conjectures proved inductively, but generalization is a very dangerous business. The most common form of generalization that our system uses is to "throw away" an induction hypothesis once it has been "used."

1.4. *Two Examples.* We will now illustrate our induction, expansion, and generalization heuristics by describing our system's proofs of two theorems from elementary number theory: the associativity of multiplication and Fermat's theorem.

The addition function may be defined recursively as follows:
$$0+y = y$$
$$s(x)+y = s(x+y).$$

Our program admits such equations as new axioms only after proving that one and only one function satisfies them.

We define multiplication in terms of addition:
$$0*y = 0$$
$$s(x)*y = y+(x*y).$$

Suppose the user of our program now submits to the system the conjecture: $(x*y)*z = x*(y*z)$. How does the theorem-prover proceed?

Our program decides to prove this by induction, after ruling out such possible "moves" as considering the cases and expanding some of the function definitions. How does it choose which induction to try? Consider for a moment a simple induction on x. Let p be the conjecture we are trying to prove. The *base case* is formed by replacing all the x's in p by 0. The *induction step* is an implication from the *induction hypothesis* to the *induction conclusion*. The induction hypothesis is p. The induction conclusion is formed from p by replacing all the x's by s(x). Thus, the x∗y in the induction hypothesis becomes s(x)∗y in the induction conclusion. But by the recursive definition of "∗," s(x)∗y is equal to y+x∗y. We say that x∗y has "stepped through" the induction on x because, after simplification, it appears in both the induction hypothesis and the induction conclusion.

Given the recursive definition of "∗," the occurrence of x∗y in the conjecture we are trying to prove suggests a simple induction on x. The occurrence of the term y∗z in the conjecture suggests a simple induction on y, but another term in the conjecture, namely x∗y, does not step through an induction on y. By such considerations our induction heuristic elects to induct on x.

The base case is trivial: both sides reduce to 0 by the definition of "∗." The induction step is more interesting. The hypothesis is

  hyp: (x∗y)∗z = x∗(y∗z)

and the conclusion is

  conc1: (s(x)∗y)∗z = s(x)∗(y∗z).

Our program attacks this problem first by simplifying the terms appearing in it. Consider the term s(x)∗y in the conclusion. By definition, this term is equal to y+(x∗y). Since x∗y already occurs in the conjecture at hand, namely, in the hypothesis, our program elects to replace s(x)∗y by y+(x∗y). By such expansions the program reduces the induction conclusion to

  conc2: (y+(x∗y))∗z = (y∗z)+(x∗(y∗z)).

Since further expansion of any term produces terms not already in the conjecture, our program stops expanding definitions at this point.

Next, the program tries to use its induction hypothesis, hyp. The right hand side of hyp, (x∗(y∗z)), has stepped through the induction and emerged inside the right hand side of the simplified conclusion, conc2. This permits the program to use its induction hypothesis by substituting the left hand side for the right in conc2. The result is:

  conc3: (y+(x∗y))∗z = (y∗z)+((x∗y)∗z).

However, although its goal is to prove that hyp implies conc3, our program adopts the stronger goal of proving conc3, without any hypothesis, on the grounds that the induction hypothesis has been used and should not contaminate future goals. In addition, because the term x*y also stepped through the induction and now appears on both sides of the equality, the program decides to adopt an even stronger generalization, obtained by replacing x*y in conc3 by the new variable w:

conc4: (y+w)*z = (y*z)+(w*z).

Observe that this sequence of heuristics has led the program to "guess" that multiplication distributes over addition. The program proves this by induction and further expansion. Thus, the system proves the associativity of multiplication without any guidance from the user. It takes our program about 10 seconds on a DEC 2060 in Interlisp to produce the proof described above.

Once the theorem-prover proves a lemma it remembers it for future use. For example, the associativity of multiplication would be used to reassociate any instance of (x*y)*z to the corresponding instance of x*(y*z). By having the theorem-prover prove key lemmas the user can lead it to the proofs of complicated theorems.

Below we exhibit the proof of Fermat's theorem. Concepts used in the theorem and proof are introduced with recursive definitions, just as we introduced "+" and "*" above. Each English sentence below corresponds to one formula (lemma) typed by the user and proved by the system. Several of the lines require induction to prove. The proof below was constructed after the system had proved many of the theorems in Chapter V of Hardy and Wright's *An Introduction to the Theory of Numbers* [16].

Fermat's Theorem: If p is prime and does not divide M, M^{p-1} mod p = 1.

Proof. Let S(n,M,p) be the sequence (n*M mod p, (n-1)*M mod p, ..., 1*M mod p).

In the text below we adopt the convention that p is a prime that does not divide M.

The product of the elements of S(n,M,p) mod p is equal to $n!*M^n$ mod p.

Observe that if i<j<p, then j*M mod p is not a member of S(i,M,p) (Hint: induct on i). Hence, if n<p, then no element of S(n,M,p) occurs twice. Furthermore, each element of S(n,M,p) is positive, each is less than or equal to p-1, and there are n elements.

Thus, from the Pigeon Hole Principle we have that the product of the elements of S(p-1,M,p) is (p-1)!. But we also have that the product of the elements of S(p-1,M,p) mod p is $(p-1)! * M^{p-1}$ mod p. Hence, Fermat's theorem. Q.E.D.

2. Applications. In this section we describe some of the applications of our theorem-prover. To do so we must first elaborate our remark above that the production of a good "high level proof checker" would have far-reaching practical significance.

Computer programs may be regarded as formal mathematical objects whose correctness can be proved in exactly the sense that theorems are proved. A "bug" in a computer program represents either (a) the failure of the programmer to prove that the program does what it is supposed to do or (b) the failure of someone, be it the programmer or his employer, to specify clearly what the program was supposed to do. In principle, bugs of the first variety can be eliminated by requiring that program proofs be mechanically checked. Nor is this a mere theoretical possibility. Widespread research into "program verification" suggests that the cost of mechanically checking the proofs of programs is currently somewhere between 2 and 30 times as great as the normal development cost. To our knowledge, the largest program mechanically verified to date consists of 4,211 lines of executable high level code [23]. The major, perhaps the only serious, difficulty in further reducing the cost is the development of better high-level proof-checkers.

There are two traditional types of program verification: Floyd-style [14] and McCarthy-style [19]. Our theorem-prover is used in both types of program verification.

The Floyd-style, which has its roots in the classic Goldstine and von Neumann reports [25], handles the usual kind of programming language, of which FORTRAN is perhaps the best example. In this style of verification, certain points in the flowchart representation of a program are annotated with mathematical assertions about what is "always true" about the program variables and the input whenever "control" reaches such points. By exploring all possible paths from one assertion to the next and analyzing the effects of intervening program statements it is possible to reduce the correctness of the program to the problem of proving certain derived formulas called *verification conditions*. Furthermore, this reduction can be done mechanically once the program has been properly annotated with assertions. The computer program that produces the theorems to be proved from the annotated program is called a *verification condition generator*.

We have written a verification condition generator for a subset of ANSI FORTRAN 66 and 77 and we use our theorem-prover to prove the resulting verification conditions. We make the following claim about our verifier:

> If a FORTRAN subprogram is accepted and proved by our system and the program can be loaded onto a FORTRAN processor that meets the ANSI specification of FORTRAN [24, 1] and certain parameterized constraints on the accuracy of arithmetic, then any invocation of the program in an environment satisfying the input condition of the program will terminate without run-time errors and will produce an environment satisfying the output condition of the program.

Among the FORTRAN programs we have proved correct mechanically are a fast string searching algorithm [5], an integer square root algorithm based on Newton's method, and a linear time majority vote algorithm [9]. However, merely browsing through our description of the verification condition generator for FORTRAN [5] -- in which we describe how to handle COMMON statements, second level definition, aliasing, undefined variables, and other arcane features -- is enough to convince most people that it is at best awkward to verify programs written in programming languages of the von Neumann style.

The McCarthy-style of program verification eschews programming languages such as FORTRAN and instead takes as the programming language a mathematical language, i.e. one in which axioms and conjectures can be stated. For example, McCarthy's language LISP [20] defines programs using lambda abstraction and recursion equations. A more recent language by Backus, the author of FORTRAN [2], is based upon combinators rather than lambda abstraction. The increasingly popular logic programming languages [17] are based on the first order predicate calculus.

It is our experience that most programs are much easier to verify if they are written in such programming languages, for several reasons:

- It is not necessary for the user of the verification system to shift constantly from one language to the another, i.e. from the programming language to the logical language.

- The tedious problems of storage allocation and deallocation are handled transparently by logical languages, but must be managed explicitly by FORTRAN style languages.

Among the McCarthy-style program verification problems that our automatic theorem-proving system has solved are:

- The correctness of a simple compiler [4] and parser [15]

- The soundness of an arithmetic simplifier [6], which is actually part of our theorem-prover

- The invertibility of the RSA public key encryption algorithm [7], which requires proving that if p and q are distinct primes, n is p*q, M<n, and e and d are multiplicative inverses in the ring of integers modulo (p-1)*(q-1) then $(M^e \bmod n)^d \bmod n = M$.

- The termination, over the integers, of the Takeuchi function [21]:

$$\text{Tak}(x,y,z) = \text{if } x \leq y \text{ then } y$$
$$\text{else Tak}(\text{Tak}(x\text{-}1, y, z),$$
$$\text{Tak}(y\text{-}1, z, x),$$
$$\text{Tak}(z\text{-}1, x, y)).$$

The later is a nontrivial theorem that we think would tax any mathematician for more than a few minutes.

Beyond these two traditional kinds of program verification, there are several new kinds of program verification that are emerging and to which our theorem-prover has been applied.

- The mechanical verification of concurrent, or parallel programs, has received much less attention than it deserves. A major reason, perhaps, is that new, improved methods for specifying and proving such programs by hand are being developed almost daily. Included here is the verification of networks, as opposed to systems resident on single computers. One mechanization of network verification has been based upon our theorem-prover [12].

- The mechanical or even hand verification of real-time programs has been almost ignored. We have made a minor investigation [11] in which we use our theorem-prover to prove that a simple program keeps a vehicle "on course" in a varying cross wind. A major problem in real time control verification is the specification of the real world with which such programs are supposed to interact. In addition, timing and interrupt handling are major problems.

- The verification of specifications, i.e., proving properties about program specifications rather than about the programs themselves, has received a surprising amount of attention. The major property checked is a certain type of "security." The federal government has issued RFQs for major systems with a requirement that the specifications be mechanically checked for security. One such checker, which uses our theorem-prover, is Feiertag's [13].

3. Nontechnical Issues.

3.1. *Developing Heuristics*. Our experience with developing heuristics has convinced us of three doctrines.

First, it is easy to "wire" any particular proof into a theorem-prover. Changes to a theorem-prover that lead to no proofs besides the examples the author had in mind when he made the changes are to be eschewed. To help us avoid this pitfall, we do not permit ourselves to use in the code for our theorem-prover the name of any logical function except the primitives of our theory. Thus, in a certain sense our program behaves the same way whether the user names his factorial function "!" or "FACT." Nevertheless, it is possible to cheat and build in subroutines that recognize when the user has defined certain functions, without ever mentioning them by name in the code. In the end, one is forced to evaluate an automatic theorem-prover by how good it is when applied to "new" problems. To this end we make it a habit not to change our theorem-prover's heuristics to solve a new problem, but rather to solve the problem with the old version of the system (thereby getting valuable information about the current arrangement of heuristics). Once we have successfully tackled a new problem we consider how we might have changed the system to have made that problem easier.

Second, it is much easier to invent heuristics than to evaluate them. Generally, heuristics are motivated by a few examples. What is not so easy to see is the effect a candidate heuristic will have on other examples. In the development of our system, we have adopted the discipline of making approximately sure that our system can do whatever it used to be able to do when we add or improve a heuristic by the brute force, and expensive, technique of running the new system on all the old problems. With this filter, we have thrown out far more heuristics than we have retained.

Third, combining heuristics with other heuristics or decision procedures cannot be usefully accomplished by merely pasting them together. On numerous occasions, we have been asked, "Why don't you incorporate into your system the decision procedures of so and so?" The answer is that adding new proof techniques is unlikely to be profitable unless the new techniques are tightly interwoven with the old. For example, we have recently added decision procedures for both "linear arithmetic" and "complete equality." In both cases, we first tried adding "black boxes" to our system that contained the code for these decision procedures, with the idea that we would periodically pass the current conjecture to those boxes. We found this approach practically useless because it almost never happened that our current conjecture was merely a consequence of linear arithmetic or pure equality. Instead, we found it necessary to interweave code for the decision procedures with the already very complicated code that heuristically selects instances of previously proved theorems because equality and arithmetic reasoning are so often necessary to relieve the hypotheses of lemmas.

3.2. *System Engineering.* There is a surprisingly large amount of work to building an automatic theorem-proving system besides developing and coding the basic mathematical techniques for finding proofs. This extra work is largely due to the fact that humans will be using the system. Even if the number of serious users of the system is small (in our case it is about 15), we have found it cost-effective to devote a lot of time to the following issues.

OUTPUT. Understanding what an automatic theorem-prover has to say can be taxing, especially if heuristics are involved, because one not only wants to know what the system has done but "why" it has done it. Perhaps twenty percent of our system is devoted to describing what is going on and why. We have found it worth the time to make the output appear in literate English prose and good typographic style. The output routine is sufficiently complex to merit a special programming language. The code for reporting what the system does is necessarily intertwined with the code for deciding what to do. Changing the heuristics can force major changes to the output routines.

ERROR-RECOVERY. Because to err is human, we allow the user to recover from "mistakes." The development of a complex proof with perhaps hundreds of lemmas seems inevitably to result in false starts. It is amazingly difficult to type perfectly accurate definitions and theorems. But implementing techniques for "backing up" and editing takes more work that it might seem.

STATE SAVING. The development of a large proof frequently requires many working days. It is necessary to be able to save the logical state of the system -- e.g., the axioms, definitions, and proved lemmas -- so that the work of one day can be continued another. Such a data base constitutes a "library" and much time can be spent designing and implementing such a library facility and (especially) building up libraries of useful lemmas.

RELIABILITY. In writing a one-off experimental automatic theorem-proving system, there is a great temptation to cut corners. For example, one can avoid checking that conjectures are well-formed formulas or he can fail to define exactly the mathematical theory in which the proofs are to be found. We have found it desirable but expensive to do our best to make our system "impenetrable." Because our system has not been mechanically verified, it probably has errors. But there are no errors or holes of which we are aware. We used to offer to jump off the Golden Gate Bridge if someone found an error in our system that would cause it to "prove" a non-theorem. However, when the first such bug was found by Topher Cooper of Digital Equipment Corporation, we merely awarded him the first Golden Gate Bridge award and moved to Texas.[4] He is the only recipient to date.

TOOLS. It is perhaps an occupational hazard of researchers in artificial intelligence that they become involved in "tool building." That is, instead of merely getting on with the job of writing programs, they spend a lot of time writing programs to help them write programs. We have suffered from this hazard. Among the "tools" implemented have been several text editors, an elaborate syntax checker for our own code that catches our common programming errors, and devices for overcoming the 1 megabyte memory address space limitation of Interlisp-10.

COMPUTERS. During the last 15 years, obtaining a decent computing environment for doing research on automatic theorem-proving has usually meant having access to an expensive machine with a large address space like a Digital Equipment Corporation 2060, costing around $1,000,000. This major problem is fortunately disappearing rapidly, due to the emergence of LISP machines sold by LMI, Symbolics, and Xerox at well under $100,000.

4. Acknowledgments. Our joint work began in 1971 in Edinburgh, Scotland, under Science Research Council support to Bernard Meltzer of the Metamathematics Unit of the University of Edinburgh. At SRI and at the University of Texas our continuing benefactors have been Thomas Keenan of NSF and Robert Grafton and Marvin Denicoff of ONR, to whom we are deeply grateful.

BIBLIOGRAPHY

1. American National Standards Institute, Inc. American National Standard Programming Language FORTRAN. Tech. Rept. ANSI X3.9-1978, American National Standards Institute, Inc., 1430 Broadway, N.Y. 10018, April, 1978.

2. J. Backus. "Can Programming Be Liberated from the von Neumann Style? A Functional Style and Its Algebra of Programs." *Comm. ACM 21* (August 1978), 616-641.

---

[4] A subroutine for calculating the value of a primitive function on constants contained a bug that caused it to deliver the wrong answer on certain constants axiomatized by the user.

3. W. W. Bledsoe. "Non-resolution Theorem Proving." *Artificial Intelligence 9* (1977), 1-36.

4. R. S. Boyer and J S. Moore. *A Computational Logic.* Academic Press, New York, 1979.

5. R. S. Boyer and J S. Moore. A Verification Condition Generator for FORTRAN. In *The Correctness Problem in Computer Science*, R. S. Boyer and J S. Moore, Eds., Academic Press, London, 1981.

6. R. S. Boyer and J S. Moore. Metafunctions: Proving Them Correct and Using Them Efficiently as New Proof Procedures. In *The Correctness Problem in Computer Science*, R. S. Boyer and J S. Moore, Eds., Academic Press, London, 1981.

7. R. S. Boyer and J S. Moore. Proof Checking the RSA Public Key Encryption Algorithm. Technical Report ICSCA-CMP-33, Institute for Computing Science and Computer Applications, University of Texas at Austin, 1982. To appear in the *American Mathematical Monthly*.

8. R. S. Boyer and J S. Moore. A Mechanical Proof of the Unsolvability of the Halting Problem. Technical Report ICSCA-CMP-28, University of Texas at Austin, 1982. To appear in the *Journal of the Association for Computing Machinery*.

9. R. S. Boyer and J S. Moore. MJRTY - A Fast Majority Vote Algorithm. Technical Report ICSCA-CMP-32, Institute for Computing Science and Computer Applications, University of Texas at Austin, 1982.

10. R. S. Boyer and J S. Moore. A Mechanical Proof of the Turing Completeness of Pure Lisp. Technical Report ICSCA-CMP-37, Institute for Computing Science and Computer Appplications, University of Texas at Austin, 1983. To appear in the *Automated Theorem Proving* volume of the Contemporary Mathematics Series of the American Mathematical Society.

11. R. S. Boyer, M. W. Green and J S. Moore. The Use of a Formal Simulator to Verify a Simple Real Time Control Program. Technical Report ICSA-CMP-29, University of Texas at Austin, 1982.

12. Benedetto Lorenzo Di Vito. Verification of Communications Protcols and Abstract Process Models. PhD Thesis ICSCA-CMP-25, Institute for Computing Science and Computer Applications, University of Texas at Austin, 1982.

13. Richard J. Feiertag. A Technique for Proving Specifications are Multilevel Secure. Technical Report CSL-109, SRI International, 1981.

14. R. Floyd. Assigning Meanings to Programs. In *Mathematical Aspects of Computer Science, Proceedings of Symposia in Applied Mathematics*, American Mathematical Society, Providence, Rhode Island, 1967, pp. 19-32.

15. P. Y. Gloess. An Experiment with the Boyer-Moore Theorem Prover: A Proof of the Correctness of a Simple Parser of Expressions. In *5th Conference on Automated Deduction, Lecture Notes in Computer Science*, Springer Verlag, 1980, pp. 154-169.

16. G. H. Hardy and E. M. Wright. *An Introduction to the Theory of Numbers.* Oxford University Press, 1979.

17. R. Kowalski. *Logic for Problem Solving.* Elsevier North Holland, Inc., New York, 1979.

18. D. Loveland. *Automated Theorem Proving: A Logical Basis.* North Holland, Amsterdam, 1978.

19. J. McCarthy. A Basis for a Mathematical Theory of Computation. In *Computer Programming and Formal Systems*, P. Braffort and D. Hershberg, Eds., North-Holland Publishing Company, Amsterdam, The Netherlands, 1963.

20. J. McCarthy, et al. *LISP 1.5 Programmer's Manual.* The MIT Press, Cambridge, Massachusetts, 1965.

21. J S. Moore. "A Mechanical Proof of the Termination of Takeuchi's Function." *Information Processing Letters 9*, 4 (1979), 176-181.

22. David M. Russinoff. A Mechanical Proof of Wilson's Theorem. Department of Computer Sciences, University of Texas at Austin, 1983.

23. M. Smith, A. Siebert, B. DiVitto, and D. Good. "A Verified Encrypted Packet Interface." *SIGSOFT 6*, 3 (1981).

24. United States of America Standards Institute. USA Standard FORTRAN. Tech. Rept. USAS X3.9-1966, United States of America Standards Institute, 10 East 40th Street, New York, New York 10016, 1966.

25. J. von Neumann. *John von Neumann, Collected Works, Volume V.* Pergamon Press, Oxford, 1961.

A Mechanical Proof of
the Turing Completeness of Pure Lisp

ROBERT S. BOYER AND J STROTHER MOORE[1]

ABSTRACT. We describe a proof by a computer program of the Turing completeness of a computational paradigm akin to Pure LISP. That is, we define formally the notions of a Turing machine and of a version of Pure LISP and prove that anything that can be computed by a Turing machine can be computed by LISP. While this result is straightforward, we believe this is the first instance of a machine proving the Turing completeness of another computational paradigm.

1. Introduction. In our paper [4] we present a definition of a function EVAL that serves as an interpreter for a language akin to Pure LISP, and we describe a mechanical proof of the unsolvability of the halting problem for this version of LISP. We claim in that paper that we have proved the recursive unsolvability of the halting problem. It has been pointed out by a reviewer that we cannot claim to have mechanically proved the recursive unsolvability of the halting problem unless we also mechanically prove that the computational paradigm used is as powerful as partial recursive functions. To this end we here describe a mechanically checked proof that EVAL is at least as powerful as the class of Turing machines. The proof was constructed by a descendant of the systems described in [2] and [3]. For an informal introduction to our system see [5], in this volume.

One of our motivations in writing this paper is to add further support to our

---

[1]Institute for Computing Science and Computer Applications, University of Texas at Austin, Austin, Texas 78712.

The work reported here was supported in part by NSF Grant MCS-8202943 and ONR Contract N00014-81-K-0634.

conjecture that, with the aid of automatic theorem-provers, it is now often practical to do mathematics formally. Our proof of the completeness theorem is presented as a sequence of lemmas, stitched together by some English sentences motivating or justifying the lemmas. Our description of the proof is at about the same level it would be in a careful informal presentation of Turing completeness. However, every formula exhibited as a theorem in this document has actually been proved mechanically. In essence, our machine has read the informal proof and assented to each step (without bothering with the motivations). Thus, while the user of the machine was describing an informal proof at a fairly high level, the machine was filling in the often large gaps.

In the next section we briefly describe the Pure LISP programming language, our formal logic, and our use of the logic to formalize Pure LISP. These topics are covered more thoroughly in [4] where we prove the unsolvability of the halting problem. We then sketch our formalization of Turing machines and we give our formal statement of the Turing completeness of Pure LISP. The remainder of the document is an extensively annotated list of definitions and lemmas leading our theorem-prover through the formalization of Turing machines and the proof of the completeness result.

2. A Sketch of Our Formalization of Pure LISP.

2.1. *A Quick Introduction to LISP*

LISP is a programming language originally defined by McCarthy in [7]. Pure LISP is a subset of LISP akin to the lambda calculus [6]. Pure LISP has no operations that have "side-effects." For example, there is no way in Pure LISP to "change" the ordered pair <1,2> into the ordered pair <2,1>. The version of Pure LISP we describe here differs from McCarthy's version primarily by its omission of "functional arguments" and "global variables." We are here concerned entirely with our version of Pure LISP, which henceforth we will simply refer to as LISP.

The syntax of LISP is very similar to that of the lambda calculus. Program names and variable symbols are simply words, such as plus, x, and y. LISP expressions (sometimes called s-expressions) are formed from symbols grouped together with parentheses. A variable symbol is an s-expression. If f is the name of a LISP program and x and y are s-expressions then $(f\ x\ y)$ is an s-expression denoting the application of the program named f to the values of x and y. Thus, (plus x y) is an s-expression. Because plus is the name of the addition operation, (plus x y) is the LISP notation for x+y. The syntax of LISP also provides a notation for writing down certain constants.

LISP provides a variety of data structures. In addition to the natural numbers, the two most commonly used data types are "atoms" and ordered pairs. Atoms are words. The properties of atoms are largely unimportant here, except that two atoms are equal if and only if they have the same spelling. Atomic constants are usually written in "quote notation." For example, (quote abc) is a LISP expression whose value is the word abc. Observe that quote is not like other LISP programs. If f is the name of a LISP program of one argument, and f is not quote, the value of the LISP expression $(f$ abc) is obtained by running the program named f on the value of the variable named abc; if f is quote, the value of $(f$ abc) is the atom abc.

Given two LISP objects, x and y, it is possible to construct the ordered pair $<x,y>$. quote notation may be used to denote ordered pairs. For example, (quote (1 . 2)) is a LISP expression whose value is the ordered pair $<1,2>$. Similarly, (quote (abc . 0)) evaluates to $<abc,0>$.

In LISP, lists of objects are represented with ordered pairs and the atom nil. The empty list is represented by nil. The list obtained by adding the element x to the list y is represented by $<x,y>$. Thus $<3,nil>$ represents the singleton list containing 3. quote notation treats specially pairs whose second component is nil; for example, (quote (3 . nil)) can also be written (quote (3)). The list containing successively 1, 2, 3, and no other element is represented by the nest of three ordered pairs $<1,<2,<3,nil>>>$, and is typically written (quote (1 2 3)).

LISP provides primitive programs for manipulating such objects. For example, the program cons, when applied to two objects x and y, returns the ordered pair $<x,y>$. The programs car and cdr extract the two components of a pair.

Complex manipulations on such data structures are described by defining recursive programs. For example, the program app, which takes two lists and constructs their concatenation, can be defined as follows:

```
(app (lambda (x y)
      (if (equal x (quote nil))
          y
          (cons (car x) (app (cdr x) y))))).
```

This definition is interpreted as follows. Suppose it is desired to apply the program app to two objects x and y. We first ask whether x is the atom nil. If so, the final answer is y. Otherwise, we construct the final answer by applying cons to two objects. The first object is obtained by applying car to x and is thus the first element of the list x. The second object is obtained by recursively applying app to the remaining elements of x and y. Thus, (app (quote (a b c)) (quote (1 2 3))) computes out to the same thing as (quote (a b c 1 2 3)).

It is conceivable that the computation thus described never terminates. For example, in our LISP applying cdr to the non-pair 0 produces 0. Thus the execution of (app 0 (quote nil)) examines the successive cdr's of 0, app never encounters nil, and the computation "runs forever."

We wish to prove a theorem about what can be computed by LISP. We must therefore be precise in our definition of LISP. One way to specify a programming language is to define mathematically an interpreter for the language. That is, we might define a function that determines the value of every expression in the language. However, the value of an arbitrary s-expression, e.g., (app a b), is a function of the values of the variables in it, e.g., a and b, and the definitions of the program names used, e.g., app. Thus the LISP interpreter can be thought of as a function with three inputs: an s-expression, a mapping from variable names to values, and a mapping from program names to definitions. The interpreter either returns the value of the expression or indicates that the computation does not terminate.

2.2. A Quick Introduction to Our Logic

Since we wish to prove our theorem formally (indeed, mechanically), we must define the interpreter in a formal logic. The logic we use is that described in [2].

In what is perhaps a confusing turn of events, the logic happens to be inspired by Pure LISP. For example, to denote the application of the function symbol PLUS to the variable symbols X and Y we write (PLUS X Y) where others might write PLUS(X,Y) or even X+Y.

Axioms provide for a variety of "types." For example, the constants T and F are axiomatized to be distinct. The function EQUAL is axiomatized so that (EQUAL X Y) is T if X=Y and is F if X≠Y. A set of axioms similar to the Peano axioms describes the natural numbers as being inductively constructed from the constant 0 by application of the successor function, named ADD1. Another set of axioms describes the literal atoms as being constructed by the application of the function PACK to objects describing the "print names" of the atoms. Similarly, a set of axioms describes the ordered pairs as being inductively constructed from all other objects by application of the CONS function. One of those axioms is (CAR (CONS X Y)) = X, which specifies the behavior of the function CAR on ordered pairs. Another axiom specifies the function LISTP to return T when given an ordered pair and F otherwise.

We use a notation similar to LISP's quote notation to write down constants. For example, 97 is an abbreviation for a nest of ninety-seven ADD1's around the constant 0. '(97 112 112 . 0) is our abbreviation for the list constant

(CONS 97 (CONS 112 (CONS 112 0))).

We have arbitrarily adopted the ASCII [1] standard as a convention for associating numbers with characters and thus agree that 'app is an abbreviation for the constant term:

(PACK '(97 112 112 . 0)).

'NIL thus abbreviates:

(PACK (CONS 78 (CONS 73 (CONS 76 0)))).

(LIST t_1 t_2 ... t_n) abbreviates

(CONS t_1
    (CONS t_2
        ...
        (CONS t_n 'NIL)...)).

Thus, (LIST A B C) is (CONS A (CONS B (CONS C 'NIL))).

The logic provides for the recursive definition of functions, provided certain conditions are met insuring that there exist one and only one function satisfying the definitional equation. For example, the user of the logic can define the function APP with the following equation:

(APP X Y)
  =
(IF (LISTP X)
    (CONS (CAR X) (APP (CDR X) Y))
    Y).

The definitional principle admits this equation because it can be proved that a certain measure of the arguments decreases in a well-founded sense in the recursion. From these and other syntactic observations it can be shown that one and only one function satisfies this equation. Once admitted, the equation is regarded as a new axiom in which X and Y are implicitly universally quantified.

Various theorems can then be proved about APP. For example, it is a theorem that (APP '(A B C) '(1 2 3)) = '(A B C 1 2 3). It is also a theorem that (APP (APP X Y) Z) = (APP X (APP Y Z)). The latter theorem requires induction to prove.

The reader interested in the details of the logic should see [2, 3].

We wish to use the logic to formalize LISP. We will thus be using formulas in the logic to discuss programs in LISP. We will use logical constants to represent particular LISP expressions. For example, we will use literal atom constants in the logic, such as 'app, 'a, and 'b, to represent LISP program names and variables. We will use list constants, such as '(app a b) and '(quote (a b c)), to represent non-variable LISP s-expressions.

Since the logic and LISP have very similar syntax, use of the one to discuss the other may be confusing. We therefore have followed four conventions. First, instead of displaying LISP s-expressions, e.g., (app a b), we will henceforth display only the logical constants we use to represent them, e.g., '(app a b). Second, despite the practice common to LISP programmers of referring to their programs as "functions," we are careful here to use the word *program* when referring to a LISP program and reserve the word *function* for references to the usually understood mathematical concept. Third, we display all constants denoting LISP s-expressions in lower case. Fourth, variable symbols in the logic that intuitively take s-expressions for values are usually written in lower case. For perfect syntactic consistency with [4], all lower case letters in formulas should be read in upper case.

Thus, '(app a b) is a LISP s-expression and 'app is the name of a LISP program; (APP A B) is a term in the logic and APP is the name of a function.

2.3. *The Function* EVAL

In [4] we formalize LISP in the logic sketched above. We will sketch here that formalization.

The ideal LISP interpreter is a function that takes three inputs: an s-expression X, a mapping VA that associates variable symbols with values, and a mapping PA that associates program symbols with their definitions. The ideal interpreter either returns the value of X under VA and PA or it returns an indication that the computation never terminates. We would like to define such a function under the principle of definition in our logic. However, the principle requires that when $(f$ X VA PA$) = body$ is submitted as a definition, there exist a measure of $<$X,VA,PA$>$ that decreases in a well-founded sense in every call of f in *body*. This prevents us from directly defining the ideal interpreter as a function.

Instead we define the function EVAL which takes four arguments, X, VA, PA, and N. N can be thought of as the amount of "stack space" available to the interpreter; that is, N is the maximum depth of recursion available to interpreted LISP programs defined on PA. (EVAL X VA PA N) returns either the value of X under VA and PA, or else it returns the logical constant (BTM) to denote that it ran out of stack space.

(BTM) is axiomatized as an object of a "new" type; it is not equal to T, F, any number, literal atom, or list. (BTM) is recognized by the function BTMP, which returns T or F according to whether its argument is equal to (BTM).

The mappings associating variable and program names with values and definitions are represented by lists of pairs, called *association lists* or simply *alists*. For example, the logical constant:

    '((a . 1) (b . 2) (c . 3))

is an alist in which the literal atom 'a has the value 1, 'b has the value 2, 'c has the value 3, and all other objects have the value (BTM), by default. The logical function GET is defined to take an object and an alist and return the associated value of the object in the alist.

Here is a brief sketch of the definition of (EVAL x VA PA N). If x is a built-in LISP constant, such as 't, 'f, 'nil, or 97, its value is T, F, NIL, or 97, as appropriate. If x is a variable symbol, its value is looked up in VA. If x is of the form '(quote v), its value is v. If x is of the form '(if *test true-value false-value*), then *test* is evaluated; if the value of *test* is (BTM), then (BTM) is returned, if the value is F, then the value of *false-value* is returned, and otherwise the value of *true-value* is returned. If the function symbol of x is a primitive, e.g., cons or car, the value of x is determined by applying the appropriate function, e.g., CONS or CAR, to the recursively obtained values of the arguments, unless (BTM) is among those values, in which case the answer is (BTM). Otherwise, x is of the form '($f\ a_1\ ...\ a_n$), where f is supposedly defined on PA. If N is 0 or f is not found on PA, the value of x is (BTM). Otherwise, the definition of f specifies a list of parameters names '($x_1\ ...\ x_n$) and an s-expression *body*. EVAL recursively evaluates the arguments, $a_1, ..., a_n$ to obtain n values, $v_1, ..., v_n$. If (BTM) is among the values, the value of x is BTM. Otherwise, EVAL constructs a new variable alist associating each parameter name x_i with the corresponding value v_i, and then recursively evaluates *body* under the new alist, the same PA, and N-1.

We now illustrate EVAL. Let VA be an alist in which 'a has the value '(a b c), 'a0 has the value '(a b c . 0), and 'b has the value '(1 2 3).

    VA = '((a . (a b c))
           (a0 . (a b c . 0))
           (b . (1 2 3))).

Let PA be an alist in which the atom 'app is defined as shown below:

```
PA = '((app (x y)
          (if (equal x nil)
              y
              (cons (car x) (app (cdr x) y)))))
```

Then here are two theorems about EVAL:

```
(EVAL '(app a b) VA PA N) = (IF (LESSP N 4)
                                (BTM)
                                '(a b c 1 2 3)),

(EVAL '(app a0 b) VA PA N) = (BTM).
```

The proof of the second theorem depends on the fact that the CDR of 0 is 0, which is not NIL.

We do not give the formal definition of EVAL here. The interested reader should see [4].

The informal notion that the computation of X under VA and PA "terminates" is formalized by "there exists a stack depth N such that (EVAL x VA PA N) $\neq$ (BTM)." Conversely, the notion that the computation "runs forever" is formalized by "for all N (EVAL x VA PA N) = (BTM)." Thus, we see that the computation of '(app a b) under the VA and PA above terminates, but the computation of '(app a0 b) under VA and PA does not. Since our logic does not contain quantifiers, we will have to derive entirely constructive replacements of these formulas to formulate the Turing completeness of EVAL.

3. Turing Machines. Our formalization of Turing machines follows the description by Rogers [8]. The ideal Turing machine interpreter takes three arguments: a state symbol ST, an infinite tape TAPE (with a distinguished "cell" marking the initial position of the Turing machine), and a Turing machine TM. If the computation terminates, the Turing machine interpreter returns the final tape. Otherwise, the ideal interpreter runs forever.

We cannot define the ideal interpreter directly as a function in our logic. Instead, we define the function TMI of four arguments, ST, TAPE, TM, and K. The fourth argument is the maximum number of state transitions permitted. TMI returns either the final tape (if the machine halts after executing fewer than K instructions) or (BTM). We say the computation of TM (starting in state ST) on TAPE "terminates" if "there is a K such that (TMI ST TAPE TM K) $\neq$ (BTM)" and "runs forever" if "for all K (TMI ST TAPE TM K) = (BTM)."

The primary difference between our formalization of Turing machines and

Rogers' description is that we limit our attention to infinite tapes containing only a finite number of 1's. This is acceptable since Rogers uses only such tapes to compute the partial recursive functions (page 14 of [8]):

> Given a Turing machine, a tape, a cell on that tape, and an initial internal state, the Turing machine carries out a uniquely determined succession of operations, which may or may not terminate in a finite number of steps. We can associate a partial function with each Turing machine in the following way. To represent an input integer x, take a string of x+1 consecutive 1's on the tape. Start the machine in state q0 on the leftmost cell containing a 1. As output integer, take the total number of 1's appearing anywhere on the tape when (and if) the machine stops.

Our conventions for formalizing states, tapes, and Turing machines are made clear in Section .

4. The Statement of the Turing Completeness of LISP. We wish to formulate the idea that LISP can compute anything a Turing machine can compute. Roughly speaking we wish to establish a correspondence between Turing machines, their inputs, and their outputs on the one hand, and LISP programs, their inputs, and their outputs on the other. Then we wish to prove that (a) if a given Turing machine runs forever on given input, then so does its LISP counterpart; and (b) if a given Turing machine terminates on given input, then its LISP counterpart terminates with the same answer.

We will set up a correspondence between Turing machines and program alists (since it is there that LISP programs are defined). The mapping will be done by the function TMI.PA, which takes a Turing machine and returns a corresponding LISP program. We will also set up a correspondence between the state and tape "inputs" of the Turing machine and the s-expression being evaluated by LISP, since the s-expression can be thought of as the "input" to the LISP program defined on the program alist. That mapping is done by the function TMI.X. In our statement of the completeness theorem, we use the NIL variable alist.

Since the value of a terminating Turing machine computation is a tape, and tapes are formalized as list structures that can be manipulated by LISP programs, we will require that the corresponding LISP computation deliver the list structure identical to the final tape.

Our informal characterization of the Turing completeness of LISP is then as follows. Let TM be a Turing machine, ST a state, and TAPE a tape. Let PA_{TM} be the corresponding LISP program (i.e., the result of (TMI.PA TM)), and let $X_{ST,TAPE}$ be the corresponding LISP expression (i.e., the result of (TMI.X ST TAPE)). Then

(a) If the Turing machine computation of TM (starting in ST) on TAPE runs forever, then so does the LISP computation of $X_{ST,TAPE}$ under the NIL variable alist and the program alist PA_{TM}.

(b) If the Turing machine computation of TM (starting in ST) on TAPE terminates with tape Z, then the LISP computation of $X_{ST,TAPE}$ under NIL and PA_{TM} terminates with result Z.

To formalize (a) we replace the informal notions of "runs forever" by the formal ones. The result is:

$$\forall K \ (TMI \ ST \ TAPE \ TM \ K) = (BTM)$$
$$\rightarrow$$
$$\forall N \ (EVAL \ X_{ST,TAPE} \ NIL \ PA_{TM} \ N) = (BTM).$$

This is equivalent to:

$$\exists N \ (EVAL \ X_{ST,TAPE} \ NIL \ PA_{TM} \ N) \neq (BTM)$$
$$\rightarrow$$
$$\exists K \ (TMI \ ST \ TAPE \ TM \ K) \neq (BTM).$$

The existential quantification on N in the hypothesis can be transformed into implicit universal quantification on the outside. The existential quantification on K in the conclusion can be removed by replacing K with a term that delivers a suitable K as a function of all the variables whose (implicit) quantifiers govern the occurrence of K. The result is:

(a') $(EVAL \ X_{ST,TAPE} \ NIL \ PA_{TM} \ N) \neq (BTM)$
$\rightarrow$
$(TMI \ ST \ TAPE \ TM \ (TMI.K \ ST \ TAPE \ TM \ N)) \neq (BTM).$

That is, we prove that a suitable K exists by constructively exhibiting one.

To formalize part (b) observe that it is equivalent to
    If the computation of TM (starting in ST) on TAPE terminates within K steps with tape Z, then there exists an N such that Z = $(EVAL \ X_{ST,TAPE} \ NIL \ PA_{TM} \ N)$.
Thus, (b) becomes:
    $(TMI \ ST \ TAPE \ TM \ K) \neq (BTM)$
    $\rightarrow$
    $\exists N \ (TMI \ ST \ TAPE \ TM \ K) = (EVAL \ X_{ST,TAPE} \ NIL \ PA_{TM} \ N).$

We remove the existential quantification on N by replacing N with a particular term, $(TMI.N \ ST \ TAPE \ TM \ K)$, that delivers an appropriate N as a function of the other variables.

Our formal statement of the Turing completeness of LISP is:

```
Theorem.   TURING.COMPLETENESS.OF.LISP:
(IMPLIES (AND (STATE ST)
              (TAPE TAPE)
              (TURING.MACHINE TM))
         (AND (IMPLIES (NOT (BTMP (EVAL (TMI.X ST TAPE)
                                        NIL
                                        (TMI.PA TM)
                                        N)))
                       (NOT (BTMP (TMI ST TAPE TM
                                       (TMI.K ST TAPE
                                              TM N)))))
              (IMPLIES (NOT (BTMP (TMI ST TAPE TM K)))
                       (EQUAL (TMI ST TAPE TM K)
                              (EVAL (TMI.X ST TAPE)
                                    NIL
                                    (TMI.PA TM)
                                    (TMI.N ST TAPE
                                           TM K)))))).
```

Before proving this theorem we must define the functions TMI.X, TMI.PA, TMI.K and TMI.N.

We find the above formal statement of the completeness result to be intuitively appealing. However, the reader should study it to confirm that it indeed captures the intuitive notion. Some earlier formalizations by us were in fact inadequate. For example, one might wonder why we talk of the correspondences set up by TMI.X and TMI.PA instead of stating part (b) simply as:

> For every Turing machine TM, there exists a program alist PA, such that for every state ST and tape TAPE, there exists an X such that: if the computation of TM (starting in ST) on TAPE terminates with tape Z then the computation of X under NIL and PA terminates with Z.

However, observe that the existentially quantified X -- the expression EVAL is to evaluate -- can be chosen after TM, ST, and TAPE are fixed. Thus, a suitable X could be constructed by running the ideal Turing machine interpreter on the given TM, ST, and TAPE until it halts (as it must do by hypothesis), getting the final tape, and embedding it in a 'quote form. Thus, when EVAL evaluates X it will return the final tape. Indeed, an EVAL that only knew how to interpret 'quote would satisfy part (b) of the above statement of Turing completeness! Rearranging the quantifiers provides no relief. If the quantification on PA is made the innermost, one can appeal to a similar trick and embed the correct final answer in a 'quote form inside the definition of a single dummy program.

Our statement does not admit such short cuts. Since TMI.PA is a function only of TM, one must invent a suitable LISP program from TM without knowing what its input will be. Since TMI.X is a function only of ST and TAPE, one must invent a suitable input for the LISP program without knowing which Turing machine will run it.

Proving the Turing completeness of EVAL is not entirely academic. The definition of EVAL requires about two pages. Its correctness as a formalization of Pure LISP is not nearly as obvious as the correctness of our formalization of Turing machines, which requires only half a page of definitions. The completeness result will assure us that EVAL -- whether a formalization of LISP or not -- is as powerful as the Turing machine paradigm.

For our proof of the Turing completeness of LISP we define (TMI.PA TM) to be a program alist that contains a LISP encoding of the ideal Turing machine interpreter together with a single dummy program that returns the Turing machine to be interpreted. Our definition of (TMI.X ST TAPE) returns the LISP s-expression that calls (the LISP version of) the ideal Turing machine interpreter on ST, TAPE, and the Turing machine TM. The heart of our proof is a "correspondence lemma" establishing that the LISP version of the ideal Turing machine interpreter behaves just like the ideal interpreter.

The remainder of this document presents formally the ideas sketched informally above. In the following sections we define Turing machines in our logic and develop the completeness proof. We simultaneously present the input necessary to lead our mechanical theorem-prover to the proof. Indented formulas preceded by the words "Definition," "Theorem," or "Disable" are actually commands to the theorem-prover to define functions, prove theorems, or cease to use certain facts. During our discussion we so present every command used to lead the system to the completeness proof, starting from the library of lemmas and definitions produced by our proof of the unsolvability of the halting problem [4].

5. The Formalization of Turing Machines.

We now describe our formalization of Turing machines, following [8]. A tape is a sequence of 0's and 1's, containing a finite number of 1's, together with some pointer that indicates which cell of the tape the machine is "on." The symbol in that cell (a 0 or a 1) is the "current symbol."

We represent a tape by a pair. The CAR of the pair is a list of 0's and 1's representing the left half of the tape, in reverse order. The CDR of the pair a list of 0's and 1's representing the right half of the tape. The current symbol is the first one on the right half of the tape. To represent the infinite half-tape of 0's we use 0, since both the CAR and the CDR of 0 are 0. The function TAPE, defined below, returns T or F according to whether its argument is a Turing machine tape.

```
Definition.
(SYMBOL X) = (MEMBER X '(0 1)).

Definition.
(HALF.TAPE X)
  =
(IF (NLISTP X)
    (EQUAL X 0)
    (AND (SYMBOL (CAR X))
         (HALF.TAPE (CDR X)))).

Definition.
(TAPE X)
  =
(AND (LISTP X)
     (HALF.TAPE (CAR X))
     (HALF.TAPE (CDR X))).
```

For example, to represent the tape below (containing all 0's to the extreme left and right) with the machine on the cell indicated by the "M"

$$\overset{M}{\ldots 0\ 1\ 0\ 1\ 1\ 0\ 1\ 1\ 1\ 0\ 1\ 1\ 1\ 1\ 0\ 1\ 1\ 1\ 1\ 0\ \ldots}$$

we can use the following:

'((1 1 1 0 1 1 0 1 . 0) . (0 1 1 1 1 0 1 1 1 1 . 0)).

The operations on a tape are to move to the left, move to the right, or replace the current symbol with a 0 or a 1. The reader should briefly contemplate the formal transformations induced by these operations. For example, if the current tape is given by (CONS *left* (CONS *s right*)) then the result of moving to the right is given by (CONS (CONS *s left*) *right*).

A Turing machine is a list of 4-tuples which, given an initial state and tape, uniquely determines a succession of operations and state changes. If the current state is q and the current symbol is s and the machine contains a four-tuple *(q s op q')*, then q' becomes the new current state after the tape is modified as described by *op*; *op* can be 'L, which means move left, 'R, which means move right, or a symbol to write on the tape at the current position. If there is no matching four-tuple in the machine, the machine halts, returning the tape. The function TURING.MACHINE, defined below, returns T or F according to whether its argument is a well-formed Turing machine:

```
Definition.
(OPERATION X) = (MEMBER X '(L R 0 1)).

Definition.
(STATE X) = (LITATOM X).

Definition.
(TURING.4TUPLE X)
   =
(AND (LISTP X)
     (STATE (CAR X))
     (SYMBOL (CADR X))
     (OPERATION (CADDR X))
     (STATE (CADDDR X))
     (EQUAL (CDDDDR X) NIL)).
```

(CAAR x) is an abbreviation of (CAR (CAR x)), (CADR x) of (CAR (CDR x)), (CADDR x) of (CAR (CDR (CDR x))), etc.

```
Definition.
(TURING.MACHINE X)
   =
(IF (NLISTP X)
    (EQUAL X NIL)
    (AND (TURING.4TUPLE (CAR X))
         (TURING.MACHINE (CDR X)))).
```

An example Turing machine in this representation is:

```
TM:   '((q0 1 0 q1)
        (q1 0 r q2)
        (q2 1 0 q3)
        (q3 0 r q4)
        (q4 1 r q4)
        (q4 0 r q5)
        (q5 1 r q5)
        (q5 0 1 q6)
        (q6 1 r q6)
        (q6 0 1 q7)
        (q7 1 1 q7)
        (q7 0 1 q8)
        (q8 1 1 q1)
        (q1 1 1 q1)).
```

This machine is the one shown on page 14 of [8].

Our Turing machine interpreter, TMI, is defined in terms of two other defined functions. The first, INSTR, searches up the Turing machine description for the current state and symbol. If it finds them, it returns the pair whose CAR is the operation and whose CADR is the new state. Otherwise, it

TURING COMPLETENESS OF LISP

returns F. The second function, NEW.TAPE, takes the operation and the current tape and returns the new tape configuration.

```
Definition.
(INSTR ST SYM TM)
    =
(IF (LISTP TM)
    (IF (EQUAL ST (CAAR TM))
        (IF (EQUAL SYM (CADAR TM))
            (CDDAR TM)
            (INSTR ST SYM (CDR TM)))
        (INSTR ST SYM (CDR TM)))
    F).

Definition.
(NEW.TAPE OP TAPE)
    =
(IF (EQUAL OP 'L)
    (CONS (CDAR TAPE)
          (CONS (CAAR TAPE) (CDR TAPE)))
    (IF (EQUAL OP 'R)
        (CONS (CONS (CADR TAPE) (CAR TAPE))
              (CDDR TAPE))
        (CONS (CAR TAPE)
              (CONS OP (CDDR TAPE))))).
```

Our Turing machine interpreter is defined as follows:

```
Definition.
(TMI ST TAPE TM N)
    =
(IF (ZEROP N)
    (BTM)
    (IF (INSTR ST (CADR TAPE) TM)
        (TMI (CADR (INSTR ST (CADR TAPE) TM))
             (NEW.TAPE (CAR (INSTR ST (CADR TAPE) TM))
                       TAPE)
             TM
             (SUB1 N))
        TAPE)).
```

We now illustrate TMI. Let TM be the example Turing machine shown above. The partial recursive function computed by the machine is that defined by $f(x)=2x$.

Let TAPE be the tape '(0 . (1 1 1 1 . 0)), in which the current cell is the leftmost 1 in a block of 4+1 consecutive 1's. Then the following is a theorem:

```
(TMI 'q0 TAPE TM N)
    =
```

```
    (IF (LESSP N 78)
        (BTM)
        '((0 0 0 0 . 0) . (0 0 1 1 1 1 1 1 1 1 . 0))).
```
Observe that there are 8 1's on the final tape.

Our TMI differs from Rogers' interpreter in the following minor respects. Rogers uses B (blank) instead of our 0. In his description of the interpreter, Rogers permits tapes containing infinitely many 1's. But his characterization of partial recursive functions does not use this fact and we have limited our machines to tapes containing only a finite number of 1's. Finally, Rogers' interpreter does certain kinds of syntax checking (e.g., insuring that *op* is either L, R, 0, or 1) that ours does not, with the result that our interpreter will execute more machines on more tapes than Rogers'.

6. The Definitions of TMI.PA, TMI.X, TMI.N, **and** TMI.K. In this section we define the LISP analogues of INSTR, NEW.TAPE, and TMI, and we then define TMI.PA, TMI.X, TMI.N, and TMI.K.

To define a LISP program on the program alist argument of EVAL one adds to the alist a pair of the form '(*name* . (*args body*)), where *name* is the name of the program, *args* is a list of the formal parameters, and *body* is the s-expression describing the computation to be performed. We wish to define the LISP counterpart of the ideal Turing machine interpreter. We will call the program 'tmi. It is defined in terms of two other programs, 'instr and 'new.tape. The three functions defined below return as their values the (*args body*) part of the LISP definitions.

```
        Definition.
        (INSTR.DEFN)
        =
        '((st sym tm)
          (if (listp tm)
              (if (equal st (car (car tm)))
                  (if (equal sym (car (cdr (car tm))))
                      (cdr (cdr (car tm)))
                      (instr st sym (cdr tm)))
                  (instr st sym (cdr tm)))
              f)).
```

```
Definition.
(NEW.TAPE.DEFN)
  =
'((op tape)
  (if (equal op  (quote l))
      (cons (cdr (car tape))
            (cons (car (car tape)) (cdr tape)))
      (if (equal op (quote r))
          (cons (cons (car (cdr tape)) (car tape))
                (cdr (cdr tape)))
          (cons (car tape)
                (cons op (cdr (cdr tape))))))).

Definition.
(TMI.DEFN)
  =
'((st tape tm)
  (if (instr st (car (cdr tape)) tm)
      (tmi (car (cdr (instr st (car (cdr tape)) tm)))
           (new.tape (car (instr st (car (cdr tape))
                                     tm))
                     tape)
           tm)
      tape)).
```

Observe that the LISP program 'tmi takes only the three arguments st, tape, and tm; it does not take the number of instructions to execute. Thus, the program 'tmi is the LISP analogue of the ideal Turing machine interpreter, rather than our function TMI, and on some inputs "runs forever."

```
Definition.
(KWOTE X)
  =
(LIST 'quote X).
```

(KWOTE 'a) is '(quote a). In general (KWOTE X) returns a LISP s-expression that when evaluated returns X.

Here now is the function, TMI.PA, that when given a Turing machine returns the corresponding LISP program environment:

```
Definition.
(TMI.PA TM)
  =
(LIST (LIST 'tm NIL (KWOTE TM))
      (CONS 'instr (INSTR.DEFN))
      (CONS 'new.tape (NEW.TAPE.DEFN))
      (CONS 'tmi (TMI.DEFN))).
```

Note that (TMI.PA TM) is a program alist containing definitions for 'instr, 'new.tape, and 'tmi, as well as the dummy program 'tm which, when called, returns the Turing machine TM to be interpreted. Thus, the Turing machine is available during the LISP evaluation of the s-expression returned by TMI.X. But it is not available to TMI.X itself.

Here is the function, TMI.X, that, when given a Turing machine state and tape, returns the corresponding input for the LISP program 'tmi.

```
Definition.
(TMI.X ST TAPE)
   =
(LIST 'tmi
      (KWOTE ST)
      (KWOTE TAPE)
      '(tm)).
```

The form returned by TMI.X is a call of (the LISP version of) the ideal interpreter, 'tmi, on arguments that evaluate, under (TMI.PA TM), to ST, TAPE, and TM.

We now define TMI.K, which is supposed to return a number of Turing machine steps sufficient to insure that TMI terminates gracefully if EVAL terminates in stack space N.

```
Definition.
(TMI.K ST TAPE TM N)
   =
(DIFFERENCE N (ADD1 (LENGTH TM))).
```

Finally, we define TMI.N, which returns a stack depth sufficient to insure that EVAL terminates gracefully if TMI terminates in K steps.

```
Definition.
(TMI.N ST TAPE TM K)
   =
(PLUS K (ADD1 (LENGTH TM))).
```

We have now completed all the definitions necessary to formally state TURING.COMPLETENESS.OF.LISP. The reader may now wish to prove the theorem himself.

7. The Formal Proof.

7.1. *Some Preliminary Remarks about* BTM. We wish to establish a correspondence between calls of the LISP program 'tmi and the function TMI. We will do so by first noting the correspondence between primitive LISP expressions, such as '(car (cdr x)), and their logical counterparts, e.g., (CAR (CDR X)). These primitive correspondences are explicit in the definition of EVAL. We then prove correspondence lemmas for the defined LISP programs 'instr and 'new.tape, before finally turning to 'tmi.

Unfortunately, the correspondence, even at the primitive level, is not as neat as one might wish because of the role of (BTM) in the LISP computations. For example, the value of the LISP program 'cdr applied to some object is the CDR of the object *provided* the object is not (BTM). Thus, if the LISP variable symbol 'x is bound to some object, X, then the value of the LISP expression '(car (cdr x)) is (CAR (CDR X)) only if X is not (BTM) and its CDR is not (BTM). If 'x were bound to the pair (CONS 3 (BTM)), then '(car (cdr x)) would evaluate to (BTM) while (CAR (CDR X)) would be 0, since the CAR of the nonlist (BTM) is 0. LISP programs cannot construct an object like (CONS 3 (BTM)), for if a LISP program tried to 'cons something onto (BTM) the result would be (BTM). But such "uncomputable" objects can be constructed in the logic and might conceivably be found in the input to EVAL.

Therefore, the correspondence lemmas for our programs 'instr, 'new.tape, and 'tmi contain conditions that prohibit the occurrence of (BTM) within the objects to which the programs are applied. These conditions are stated with the function CNB (read "contains no (BTM)'s"). To use such correspondence lemmas when we encounter calls of these programs we must be able to establish that their arguments contain no (BTM)'s. This happens to be the case because in the main theorem 'tmi is called on a STATE, TAPE, and TURING.MACHINE and all subsequent calls are on components of those objects, none of which contains a (BTM).

The correspondence lemmas for recursive programs are complicated in a much more insidious way by the possibility of nontermination or mere "stack overflow." The term

 (LIST 'instr st sym tmc),

describes an arbitrary call of 'instr, when the logical variables st, sym and tmc are instantiated with LISP s-expressions. Let st, sym, and tm be the values of st, sym, and tmc, respectively, and suppose further that those values contain no (BTM)'s. Is the value of (LIST 'instr st sym tmc) equal to (INSTR st sym tm)? It depends on the amount of stack depth available. If the depth exceeds the length of tm, $|tm|$, then the answer computed is (INSTR st sym tm). If the depth is less than or equal to $|tm|$ is the answer computed (BTM)? Not necessarily. It depends on where (and whether) the 4-tuple for st and sym occurs in tm. If it occurs early enough, 'instr will compute the same answer as INSTR even with insufficient stack space to explore the entire tm.

In stating the correspondence lemma for 'instr we could add a hypothesis that the amount of stack depth available exceeds $|tm|$. However, not all calls of 'instr arising from the computation described in the main theorem provide adequate stack space. Part (a) of the main theorem forces us to deal explicitly with the conditions under which 'tmi computes (BTM). Thus it is necessary to know not merely when 'instr computes INSTR but also when 'instr computes (BTM).

We therefore will define a "derived function," INSTRN, in the logic that is like INSTR except that it takes a stack depth argument in addition to its other arguments and returns (BTM) when it exhausts the stack, just as does the evaluation of 'instr. We will then prove a correspondence lemma between 'instr and INSTRN. We will take a similar approach to 'tmi, defining the derived function TMIN that takes a stack depth and uses INSTRN where TMI uses INSTR. After proving the correspondence lemma between 'tmi and TMIN we will have eliminated all consideration of EVAL from the problem at hand and then focus our attention on the relation between TMIN and TMI.

Before proving any correspondence lemmas we defined "contains no (BTM)'s" and had the machine prove a variety of general purpose lemmas about arithmetic, EVAL, and CNB. These details are merely distracting to the reader, as they have nothing to do with Turing machines. We have therefore put those events in the appendix of this document.

7.2. *The Correspondence Lemma for* 'instr.

```
Definition.
(INSTRN ST SYM TM N)
    =
(IF (ZEROP N)
    (BTM)
    (IF (LISTP TM)
        (IF (EQUAL ST (CAAR TM))
            (IF (EQUAL SYM (CADAR TM))
                (CDDAR TM)
                (INSTRN ST SYM (CDR TM) (SUB1 N)))
            (INSTRN ST SYM (CDR TM) (SUB1 N)))
        F)).
```

The proof of the correspondence between 'instr and INSTRN requires a rather odd induction, in which certain variables are replaced by constants. We tell the theorem-prover about this induction by defining a recursive function that mirrors the induction we desire.

Definition.
```
(EVAL.INSTR.INDUCTION.SCHEME st sym tmc VA TM N)
   =
(IF (ZEROP N)
    T
    (EVAL.INSTR.INDUCTION.SCHEME
        'st
        'sym
        '(cdr tm)
        (LIST (CONS 'st
                    (EVAL st VA (TMI.PA TM) N))
              (CONS 'sym
                    (EVAL sym VA (TMI.PA TM) N))
              (CONS 'tm
                    (EVAL tmc VA (TMI.PA TM) N)))
        TM
        (SUB1 N))).
```

This definition is accepted under the principle of definition because N decreases in the recursion. We discuss below the sense in which this definition describes an induction.

Here is the correspondence lemma for 'instr. Note that in the "Hint" we explicitly tell the machine to induct as suggested by EVAL.INSTR.INDUCTION.SCHEME.

```
Theorem.  EVAL.INSTR (rewrite):
(IMPLIES
    (AND (CNB (EV 'AL st VA (TMI.PA TM) N))
         (CNB (EV 'AL sym VA (TMI.PA TM) N))
         (CNB (EV 'AL tmc VA (TMI.PA TM) N)))
    (EQUAL (EV 'AL (LIST 'instr st sym tmc)
               VA
               (TMI.PA TM)
               N)
           (INSTRN (EV 'AL st VA (TMI.PA TM) N)
                   (EV 'AL sym VA (TMI.PA TM) N)
                   (EV 'AL tmc VA (TMI.PA TM) N)
                   N)))
Hint:  Induct as for
(EVAL.INSTR.INDUCTION.SCHEME st sym tmc VA TM N).
```

Since the correspondence lemmas are the heart of the proof, we will dwell on this one briefly.

As noted, an arbitrary call of 'instr has the form (LIST 'instr st sym tmc), where st, sym and tmc are arbitrary LISP *expressions*. Suppose the call is evaluated in an arbitrary variable alist, VA, the program alist (TMI.PA

TM), and with an arbitrary stack depth N available. Let st, sym, and tm be the results of evaluating the expressions st, sym, and tmc respectively. Then the arbitrary call of 'instr returns (INSTRN st sym tm N) provided only that st, sym, and tm contain no (BTM)'s.

The reader of [4] will recall that (EVAL X VA PA N) is defined as (EV 'AL X VA PA N), since EVAL is mutually recursive with a function called EVLIST which evaluates a list of s-expressions. Thus, formally, both EVAL and EVLIST are defined in terms of a more primitive function EV which computes either, according to a flag supplied as its first argument. We state our lemmas in terms of EV so they will be more useful as rewrite rules. (Terms beginning with EVAL will be expanded to EV terms by the definition of EVAL, and then our lemmas will apply.)

The program alist for the 'instr correspondence lemma need not be (TMI.PA TM) for the lemma to hold; it is sufficient if the program alist merely contains the expected definition for 'instr. However, as (TMI.PA TM) is the only program alist we will need in this problem it was simplest to state the lemma this way.

The base case of the suggested induction is (ZEROP N), that is, N is 0 or not a natural number. In the induction step, we assume N>0. The induction hypothesis is obtained by instantiating the conjecture with the following substitution -- derived from the recursive call in the definition of EVAL.INSTR.INDUCTION.SCHEME:

```
   variable    replacement term

   st          'st
   sym         'sym
   tmc         '(cdr tm)
   VA          (LIST (CONS 'st
                           (EVAL st VA (TMI.PA TM) N))
                     (CONS 'sym
                           (EVAL sym VA (TMI.PA TM) N))
                     (CONS 'tm
                           (EVAL tmc VA (TMI.PA TM) N)))
   TM          TM
   N           (SUB1 N) .
```

We invite the reader to do the theorem-prover's job and construct the proof of the lemma from the foregoing lemmas, definitions and hint.

Once proved, this lemma essentially adds 'instr to the set of "primitive" LISP programs with logical counterparts. That is, when the theorem-prover sees:

```
(EVAL '(instr st sym tm)
       VA (TMI.PA TM) N),
```
it will replace it by:

```
(INSTRN ST SYM TM N)
```

provided 'st, 'sym, and 'tm are bound in VA to ST, SYM, and TM (respectively) and ST, SYM, and TM contain no (BTM)'s. This eliminates a call of EVAL.

7.3. *The Correspondence Lemma for* `'new.tape`.

```
Theorem.   EVAL.NEW.TAPE (rewrite):
(IMPLIES
  (AND (CNB (EV 'AL op VA (TMI.PA TM) N))
       (CNB (EV 'AL tape VA (TMI.PA TM) N)))
  (EQUAL (EV 'AL (LIST 'new.tape op tape)
             VA
             (TMI.PA TM)
             N)
         (IF (ZEROP N)
             (BTM)
             (NEW.TAPE (EV 'AL op VA (TMI.PA TM) N)
                       (EV 'AL tape VA
                           (TMI.PA TM) N))))).
```

7.4. *The Correspondence Lemma for* `'tmi`. We now consider the correspondence lemma for `'tmi`. As we did for `'instr`, consider an arbitrary call of `'tmi`, (LIST 'tmi st tape tmc). Let *st, tape,* and *tm* be the values of the st, tape, and tmc expressions. Suppose the stack depth available is N. At first glance, the evaluation of (LIST 'tmi st tape tmc) is just (TMI *st tape tm* N), since `'tmi` causes EVAL to recurse down the stack exactly once for every time TMI decrements its step count N. But again we must carefully note that if N is sufficiently small, then `'instr` will return (BTM) while trying to fetch the next op and state. We therefore define the derived function TMIN, which is just like TMI except that instead of using INSTR it uses INSTRN. We then prove the correspondence lemma between `'tmi` and TMIN; but first we prove two simple lemmas about INSTRN and NEW.TAPE.

```
Theorem.   CNB.INSTRN (rewrite):
(IMPLIES (AND (NOT (BTMP (INSTRN ST SYM TM N)))
              (CNB TM))
         (CNB (INSTRN ST SYM TM N))).
```

```
Theorem.   CNB.NEW.TAPE (rewrite):
(IMPLIES (AND (CNB OP) (CNB TAPE))
         (CNB (NEW.TAPE OP TAPE))).
```

Disable NEW.TAPE.

When we disable a function name we tell the theorem-prover not to consider using the definition of the disabled function in future proofs unless the function symbol is applied to explicit constants.

Here is the derived function for the program 'tm1.

```
Definition.
(TMIN ST TAPE TM N)
  =
(IF (ZEROP N)
    (BTM)
    (IF (BTMP (INSTRN ST (CADR TAPE) TM (SUB1 N)))
        (BTM)
        (IF (INSTRN ST (CADR TAPE) TM (SUB1 N))
            (TMIN (CADR (INSTRN ST (CADR TAPE)
                                TM (SUB1 N)))
                  (NEW.TAPE
                     (CAR (INSTRN ST
                                (CADR TAPE)
                                TM (SUB1 N)))
                     TAPE)
                  TM
                  (SUB1 N))
            TAPE))).
```

The induction scheme we use for the 'tm1 correspondence lemma is suggested by the following definition:

```
Definition.
(EVAL.TMI.INDUCTION.SCHEME st tape tmc VA TM N)
  =
(IF (ZEROP N)
    T
    (EVAL.TMI.INDUCTION.SCHEME
      '(car (cdr (instr st (car (cdr tape)) tm)))
      '(new.tape (car (instr st (car (cdr tape)) tm))
                 tape)
      'tm
      (LIST (CONS 'st
                  (EV 'AL st VA (TMI.PA TM) N))
            (CONS 'tape
                  (EV 'AL tape VA (TMI.PA TM) N))
            (CONS 'tm
                  (EV 'AL tmc VA (TMI.PA TM) N)))
      TM
      (SUB1 N))).
```

Here is the correspondence lemma for 'tmi, together with the hint for how to prove it.

```
    Theorem.   EVAL.TMI (rewrite):
    (IMPLIES
        (AND (CNB (EV 'AL st VA (TMI.PA TM) N))
             (CNB (EV 'AL tape VA (TMI.PA TM) N))
             (CNB (EV 'AL tmc VA (TMI.PA TM) N)))
        (EQUAL (EV 'AL
                   (LIST 'tmi st tape tmc)
                   VA
                   (TMI.PA TM)
                   N)
               (TMIN (EV 'AL st VA (TMI.PA TM) N)
                     (EV 'AL tape VA (TMI.PA TM) N)
                     (EV 'AL tmc VA (TMI.PA TM) N)
                     N)))
    Hint:   Induct as for
            (EVAL.TMI.INDUCTION.SCHEME st tape tmc VA TM N).
```

7.5. EVAL of TMI.X.

EVAL occurs twice in the completeness theorem. The two occurrences are:

```
    (EVAL (TMI.X ST TAPE) NIL (TMI.PA TM) N)
```

and

```
    (EVAL (TMI.X ST TAPE) NIL
          (TMI.PA TM) (TMI.N ST TAPE TM K)).
```

But since (TMI.X ST TAPE) is just

```
    (LIST 'tmi (LIST 'quote ST) (LIST 'quote TAPE) '(tm)).
```

we can eliminate both uses of EVAL from the main theorem by appealing to the correspondence lemma for 'tmi.

```
    Theorem.   EVAL.TMI.X (rewrite):
    (IMPLIES (AND (CNB ST) (CNB TAPE) (CNB TM))
             (EQUAL (EV 'AL
                        (TMI.X ST TAPE)
                        NIL
                        (TMI.PA TM)
                        N)
                    (IF (ZEROP N)
                        (BTM)
                        (TMIN ST TAPE TM N)))).
```

The test on whether N is 0 is necessary to permit us to use the correspondence lemma for 'tmi. If N=0 then the evaluation of the third argument in the call of 'tmi returns (BTM) and we cannot relieve the hypothesis that the value of the actual expression contains no (BTM).

Disable TMI.X.

7.6. *The Relation Between* TMI *and* TMIN. By using the lemma just proved the main theorem becomes:

```
(IMPLIES
  (AND (STATE ST)
       (TAPE TAPE)
       (TURING.MACHINE TM))
  (AND
    (IMPLIES (AND (NOT (ZEROP N))
                  (NOT (BTMP (TMIN ST TAPE TM N))))
             (NOT (BTMP (TMI ST TAPE TM
                             (TMI.K ST TAPE TM N)))))
    (IMPLIES (NOT (BTMP (TMI ST TAPE TM K)))
             (EQUAL (TMI ST TAPE TM K)
                    (TMIN ST TAPE TM
                          (TMI.N ST TAPE TM K)))))).
```

That is, EVAL has been eliminated and we must prove two facts relating TMIN and TMI. We seek a general statement of the relation between TMI and TMIN. It turns out there is a very simple one. Given certain reasonable conditions on the structure of TM,

```
(TMI ST TAPE TM K) = (TMIN ST TAPE TM
                           (PLUS K
                                 (ADD1 (LENGTH TM)))).
```

Note that once this result has been proved, the main theorem follows from the definitions of TMI.K and TMI.N. But let us briefly consider why this result is valid.

Clearly, if (TMI ST TAPE TM K) terminates normally within K steps, then TMIN will also terminate with the same answer with K+|TM|+1 stack depth. The reason TMIN needs more stack depth than TMI needs steps is that on the last step, TMIN will need enough space to run INSTRN all the way down TM to confirm that the current state is a final state. Of course, we use the facts that, given enough stack space, INSTRN is just INSTR and does not return (BTM):

```
Theorem.  INSTRN.INSTR (rewrite):
(IMPLIES (LESSP (LENGTH TM) N)
         (EQUAL (INSTRN ST SYM TM N)
                (INSTR ST SYM TM))).

Theorem.  NBTMP.INSTR (rewrite):
(IMPLIES (TURING.MACHINE TM)
         (NOT (BTMP (INSTR ST SYM TM)))).
```

It is less clear that if TMI returns (BTM) after executing K steps, then the TMIN expression also returns (BTM), even though its stack depth is K+|TM|+1. What prevents TMIN, after recursing K times to take the first K steps, from using some of the extra stack space to take a few more steps, find a terminal state, and halt gracefully? In fact, TMIN may well take a few more steps (if the state transitions it runs through occur early in TM so that it does not exhaust its stack looking for them). However, to confirm that it has reached a terminal state it must make an exhaustive sweep through TM, which will of necessity use up all the stack space. Another way of putting it is this: if TMIN is to terminate gracefully, INSTRN must return F, but INSTRN will not return F unless given a stack depth greater than |TM|.

```
Theorem.  INSTRN.NON.F (rewrite):
(IMPLIES (AND (TURING.MACHINE TM)
              (LEQ N (LENGTH TM)))
         (INSTRN ST SYM TM N)).
```

We therefore conclude:

```
Theorem.  TMIN.BTM (rewrite):
(IMPLIES (AND (TURING.MACHINE TM)
              (LEQ N (LENGTH TM)))
         (EQUAL (TMIN ST TAPE TM N) (BTM))).
```

So it is easy now to prove:

```
Theorem.  TMIN.TMI (rewrite):
(IMPLIES (TURING.MACHINE TM)
         (EQUAL (TMI ST TAPE TM K)
                (TMIN ST TAPE TM
                      (PLUS K (ADD1 (LENGTH TM)))))).
```

The reader is reminded that the lemmas being displayed here are not merely read by the theorem-prover. They are proved.

7.7. *Relating* CNB *to Turing Machines.*

All that remains is to establish the obvious connections between the predicates STATE, TAPE, and TURING.MACHINE and the concept of "contains no (BTM)'s."

```
Theorem.  SYMBOL.CNB (rewrite):
(IMPLIES (SYMBOL SYM) (CNB SYM)).

Disable SYMBOL.
```

Theorem. HALF.TAPE.CNB (rewrite):
(IMPLIES (HALF.TAPE X) (CNB X)).

Theorem. TAPE.CNB (rewrite):
(IMPLIES (TAPE X) (CNB X)).

Disable TAPE.

Theorem. OPERATION.CNB (rewrite):
(IMPLIES (OPERATION OP) (CNB OP)).

Disable OPERATION.

Theorem. TURING.MACHINE.CNB (rewrite):
(IMPLIES (TURING.MACHINE TM) (CNB TM)).

Disable TURING.MACHINE.

7.8. *The Main Theorem.* We are now able to prove that LISP is Turing complete.

```
Theorem.   TURING.COMPLETENESS.OF.LISP:
(IMPLIES
  (AND (STATE ST)
       (TAPE TAPE)
       (TURING.MACHINE TM))
  (AND
    (IMPLIES (NOT (BTMP (EVAL (TMI.X ST TAPE)
                              NIL
                              (TMI.PA TM)
                              N)))
             (NOT (BTMP (TMI ST TAPE TM
                             (TMI.K ST TAPE TM N)))))
    (IMPLIES (NOT (BTMP (TMI ST TAPE TM K)))
             (EQUAL (TMI ST TAPE TM K)
                    (EVAL (TMI.X ST TAPE)
                          NIL
                          (TMI.PA TM)
                          (TMI.N ST TAPE TM K))))))
```

Q.E.D.

8. Conclusion.

We have described a mechanically checked proof of the Turing completeness of Pure LISP. While the result is obvious to all LISP programmers, it was not obvious, even to us, that a mechanically checked proof could be constructed so easily.

It is unusual to see a completely formal characterization of any computational paradigm. We were particularly heartened to see how simply the Turing machine paradigm could be expressed formally in our constructive logic. Because we are constructivists at heart, we found the step counter intuitively appealing. Furthermore, it did not obstruct the proofs. In any case, one's formalism must be able to express the notion that the interpreter "runs forever." We invite adherents of other mechanized logical systems to construct and mechanically check the entire proof -- from the definition of EVAL and Turing machines through the completeness theorem -- for comparison with this one.

Another aspect of the proof that delighted us was the fact that our Pure LISP interpreter -- defined with its stack depth parameter to fit it into our constructive logic -- was used to simulate the ideal Turing machine interpreter, not our TMI with its step counter.

It should be noted that our use of the "derived functions" INSTRN and TMIN to factor EVAL out of the proof is very similar to McCarthy's notion of functional semantics. Our derived functions capture the input/output semantics of those programs in addition to such intensional properties as their termination and stack overflow properties. It would have been much harder to construct the final sequence of proofs (in which we argued about termination conditions for 'instr and 'tmi) had EVAL been involved.

Finally we would like to remind the reader of our conjecture that it may be practical, with the help of automatic theorem-provers, to do much mathematics formally.

Some readers may feel that the proof was presented at such a low level of detail that it could have been checked by a far less sophisticated theorem-prover. The reader is urged to review the 17 named theorems displayed in Section and produce a *proof* of each. We stress the word "proof" because we have found that many people imagine that they have a proof of a formula when in fact all they have is an intuitive grasp of what the formula says and find themselves in agreement. We would be most interested if any reader successfully checks the entire proof here with another mechanical theorem-prover or proof checker.

Other readers may feel that the proof presented here was far from formal; that is true and exactly our point. The task required of the user in this proof was to describe the proof at roughly the level we are accustomed to presenting proofs. It was the machine that constructed the formal proof.

Appendix A. Some Elementary Lemmas used in the Proof. After defining the functions TMI.PA, TMI.X, TMI.N, and TMI.K used in the statement of the completeness theorem, but before proving the first correspondence lemma, we had the theorem-prover do some preliminary work with arithmetic, EVAL, CNB, and TMI.PA. In all, 22 commands are listed in this appendix: 16 lemmas, 1 definition, and 5 disable commands. All of the results are elementary and are unconcerned with Turing machines themselves. Indeed, as the reader will see, the arithmetic, EVAL, and CNB results are of a completely general and basic nature and will be of use in other proofs about EVAL. The one lemma about TMI.PA is strictly unnecessary for the completeness proof, but makes the output less voluminous. Nevertheless, in the interests of providing a complete record of the completeness proof, we list the commands here.

A.1. *Elementary Lemmas about Arithmetic.*

```
Theorem.   LENGTH.0 (rewrite):
(EQUAL (EQUAL (LENGTH X) 0)
       (NLISTP X)).

Theorem.   PLUS.EQUAL.0 (rewrite):
(EQUAL (EQUAL (PLUS I J) 0)
       (AND (ZEROP I) (ZEROP J))).

Theorem.   PLUS.DIFFERENCE (rewrite):
(EQUAL (PLUS (DIFFERENCE I J) J)
       (IF (LEQ I J) (FIX J) I)).

Disable DIFFERENCE.
```

A.2. *Elementary Lemmas about* EVAL. These lemmas all follow directly from the definition of EV. Indeed, taken together they are virtually equivalent to the definition. However, by making them available as rewrite rules we speed up the simplification of EV expressions in which the s-expression being evaluated is a constant. For example, after proving the lemmas, the EVAL expression

```
(EVAL '(if (listp x)
           (cons (car x) (cons op (cdr (cdr x))))
           f)
      (LIST (CONS 'x X) (CONS 'op OP))
      PA
      N)
```

is immediately simplified to:

```
(IF (LISTP X)
    (CONS (CAR X) (CONS OP (CDR (CDR X))))
    F)
```

provided only that OP, X, (CAR X), (CDR X), and (CDDR X) are not (BTM). Were these lemmas not available as rewrite rules the EVAL expression would still simplify as above, but it would take longer as it would involve the heuristics controlling the opening up (many times) of the recursive function EV.

```
Theorem.    EVAL.FN.0 (rewrite):
(IMPLIES (AND (ZEROP N)
              (NOT (EQUAL fn 'QUOTE))
              (NOT (EQUAL fn 'IF))
              (NOT (SUBRP fn))
              (EQUAL VARGS
                     (EV 'LIST args VA PA N)))
         (EQUAL (EV 'AL (CONS fn args) VA PA N)
                (BTM))).
```

This lemma says that the value of an arbitrary LISP call of the program fn on a list of expressions, args, is (BTM) if the stack depth is 0 and fn is not one of the LISP primitives.

```
Theorem.    EVAL.FN.1 (rewrite):
(IMPLIES
  (AND (NOT (ZEROP N))
       (NOT (EQUAL fn 'QUOTE))
       (NOT (EQUAL fn 'IF))
       (NOT (SUBRP fn))
       (EQUAL VARGS
              (EV 'LIST args VA PA N)))
  (EQUAL
     (EV 'AL (CONS fn args) VA PA N)
     (IF (BTMP VARGS)
         (BTM)
         (IF (BTMP (GET fn PA))
             (BTM)
             (EV 'AL
                 (CADR (GET fn PA))
                 (PAIRLIST (CAR (GET fn PA)) VARGS)
                 PA
                 (SUB1 N)))))).
```

This lemma characterizes the result of an arbitrary LISP call of a nonprimitive program fn on args when the stack depth is non-0. If either the evaluation of args produces (BTM) or no definition for fn is found in the program alist, the result is (BTM). Otherwise, the result is obtained by evaluating the body of fn in an environment in which its formal parameters are bound to the values of the actuals and the stack depth is decremented by one.

The remaining EV lemmas characterize the behavior of EVAL on the primitive programs (e.g., the program names recognized by the function SUBRP, which include cons, car, and cdr, the programs if and quote, and atomic symbols) and characterize EVLIST on NIL and the general nonempty list.

```
Theorem.   EVAL.SUBRP (rewrite):
(IMPLIES (AND (SUBRP fn)
              (EQUAL VARGS
                     (EV 'LIST args VA PA N)))
         (EQUAL (EV 'AL (CONS fn args) VA PA N)
                (IF (BTMP VARGS)
                    (BTM)
                    (APPLY.SUBR fn VARGS)))).

Theorem.   EVAL.IF (rewrite):
(IMPLIES
      (EQUAL VX1 (EV 'AL x1 VA PA N))
      (EQUAL (EV 'AL (LIST 'if x1 x2 x3) VA PA N)
             (IF (BTMP vx1)
                 (BTM)
                 (IF vx1
                     (EV 'AL x2 VA PA N)
                     (EV 'AL x3 VA PA N))))).

Theorem.   EVAL.QUOTE (rewrite):
(EQUAL (EV 'AL (LIST 'quote x) VA PA N)
       x).

Theorem.   EVAL.NLISTP (rewrite):
(AND (EQUAL (EV 'AL 0 VA PA N) 0)
     (EQUAL (EV 'AL (ADD1 N) VA PA N)
            (ADD1 N))
     (EQUAL (EV 'AL (PACK X) VA PA N)
            (IF (EQUAL (PACK X) 't)
                T
                (IF (EQUAL (PACK X) 'f)
                    F
                    (IF (EQUAL (PACK X) nil)
                        NIL
                        (GET (PACK X) VA)))))).

Theorem.   EVLIST.NIL (rewrite):
(EQUAL (EV 'LIST NIL VA PA N) NIL).
```

Theorem.   EVLIST.CONS (rewrite):
(IMPLIES
   (AND (EQUAL VX (EV 'AL X VA PA N))
        (EQUAL VL (EV 'LIST L VA PA N)))
   (EQUAL (EV 'LIST (CONS X L) VA PA N)
          (IF (BTMP VX)
              (BTM)
              (IF (BTMP VL) (BTM) (CONS VX VL))))).

We now disable SUBRP and EV so that their definitions are not even considered. Such expressions as (SUBRP 'cons) and (SUBRP 'tmi) will still rewrite to T and F respectively, because even disabled definitions are evaluated on explicit constants. EV expressions will henceforth be rewritten only by the lemmas above.

Disable SUBRP.

Disable EV.

A.3. CNB.

Definition.
(CNB X)
   =
(IF (LISTP X)
    (AND (CNB (CAR X)) (CNB (CDR X)))
    (NOT (BTMP X))).

In the same spirit as our EVAL lemmas, we now prove a set of lemmas that let CNB expressions rewrite efficiently.

Theorem.   CNB.NBTM (rewrite):
(IMPLIES (CNB X) (NOT (BTMP X))).

Theorem.   CNB.CONS (rewrite):
(AND (EQUAL (CNB (CONS A B))
            (AND (CNB A) (CNB B)))
     (IMPLIES (CNB X) (CNB (CAR X)))
     (IMPLIES (CNB X) (CNB (CDR X)))).

Theorem.   CNB.LITATOM (rewrite):
(IMPLIES (LITATOM X) (CNB X)).

Theorem.   CNB.NUMBERP (rewrite):
(IMPLIES (NUMBERP X) (CNB X)).

Disable CNB.

Note how these lemmas are used by the simplifier. Suppose the simplifier encounters the test (BTMP (CDR (CDR (CAR TM)))), as it does in the proof of the correspondence lemma for 'instr. Suppose the hypothesis of the conjecture being proved contains (CNB TM). Then the BTMP expression above is simplified to F by backwards chaining through the lemmas just proved:

```
(CNB TM)
   → (CNB (CAR TM))
       → (CNB (CDR (CAR TM)))
           → (CNB (CDR (CDR (CAR TM))))
               → (BTMP (CDR (CDR (CAR TM)))) = F.
```

A.4. *Shutting Off Some Verbose Output.*

Consider (TMI.PA TM). It is equal to a LIST expression containing three large list constants, namely, the constants defined to be (INSTR.DEFN), (NEW.TAPE.DEFN), and (TMI.DEFN). Left to its own devices, the theorem-prover chooses to eliminate all mention of TMI.PA, INSTR.DEFN, NEW.TAPE.DEFN, and TMI.DEFN by replacing them by their definitions. Thus, (TMI.PA TM) expands into a very large expression and this expression is printed many times during the display of the machine's proofs.

However, the only use that is ever made of the expression is to use GET to fetch the definitions of 'tm, 'new.tape, 'instr, and 'tmi. Rather than suffer through pages of output devoted to (TMI.PA TM), we here prove the four relevant properties of it, namely that GET returns the appropriate constant when given one of the known program names, and then disable TMI.PA so that it is no longer expanded into its verbose form. Again, these lemmas are not necessary to the proof, but they make the production of the proof much less taxing on the user.

```
      Theorem.   GET.TMI.PA (rewrite):
      (AND (EQUAL (GET 'tm (TMI.PA TM))
                  (LIST NIL (KWOTE TM)))
           (EQUAL (GET 'instr (TMI.PA TM))
                  (INSTR.DEFN))
           (EQUAL (GET 'new.tape (TMI.PA TM))
                  (NEW.TAPE.DEFN))
           (EQUAL (GET 'tmi (TMI.PA TM))
                  (TMI.DEFN))).

      Disable TMI.PA.
```

Bibliography

1. American National Standards Institute, Inc. Code for Information Interchange. Tech. Rept. ANSI X3.4-1977, American National Standards Institute, Inc., 1430 Broadway, N.Y. 10018, 1977.

2. R. S. Boyer and J S. Moore. *A Computational Logic.* Academic Press, New York, 1979.

3. R. S. Boyer and J S. Moore. Metafunctions: Proving Them Correct and Using Them Efficiently as New Proof Procedures. In *The Correctness Problem in Computer Science*, R. S. Boyer and J S. Moore, Eds., Academic Press, London, 1981.

4. R. S. Boyer and J S. Moore. A Mechanical Proof of the Unsolvability of the Halting Problem. Technical Report ICSCA-CMP-28, University of Texas at Austin, 1982. To appear in the *Journal of the Association for Computing Machinery.*

5. R. S. Boyer and J S. Moore. Proof-Checking, Theorem-Proving, and Program Verification. Technical Report ICSCA-CMP-35, Institute for Computing Science and Computer Appplications, University of Texas at Austin, 1983. To appear in the *Automated Theorem Proving* volume of the Contemporary Mathematics Series of the American Mathematical Society.

6. A. Church. The Calculi of Lambda-Conversion. In *Annals of Mathematical Studies*, Princeton University Press, Princeton, New Jersey, 1941.

7. J. McCarthy, et al. *LISP 1.5 Programmer's Manual.* The MIT Press, Cambridge, Massachusetts, 1965.

8. H. Rogers, Jr.. *Theory of Recursive Functions and Effective Computability.* McGraw-Hill Book Company, New York, 1967.

AUTOMATING HIGHER-ORDER LOGIC

Peter B. Andrews[1], Dale A. Miller[2],
Eve Longini Cohen[3], Frank Pfenning[1]

Abstract

An automated theorem-proving system called TPS for proving theorems of first or higher-order logic in automatic, semi-automatic, or interactive mode has been developed. As its logical language TPS uses Church's formulation of type theory with λ-notation, in which most theorems of mathematics can be expressed very directly. As an interactive tool TPS has many features which facilitate writing formal proofs and manipulating and displaying formulas of logic in traditional notations. In automatic mode TPS combines theorem-proving methods for first-order logic with Huet's unification algorithm for typed λ-calculus, finds acceptable general matings (which represent the essential syntactic combinatorial information implicit in proofs), and constructs proofs in natural deduction style. Among the theorems which can be proved completely automatically is $\sim \exists G_{o\iota} \forall F_{o\iota} \exists J_\iota [[G J] = F]$, which expresses Cantor's Theorem that a set has more subsets than members by asserting that there is no function G from individuals to sets which has every set F of individuals in its range. The computer substitutes for $F_{o\iota}$ the formula $[\lambda W_\iota \sim G_{o\iota} W W]$, which denotes the set $\{W \mid W \notin G W\}$ and expresses the key idea in the classical diagonal argument.

The methods presently used by TPS in automatic mode are in principle complete for first-order logic, but not for higher-order logic. A recently proved extension of Herbrand's Theorem to type theory is presented. It is anticipated that this metatheorem will provide a basis for extending the capabilities of TPS.

---

[1] Mathematics Department, Carnegie-Mellon University, Pittsburgh, Pa. 15213

[2] Department of Computer and Information Sciences, Moore School/D2, University of Pennsylvania, Philadelphia, Pa. 19104

[3] The Aerospace Corporation, Information Sciences Research Office, M-105, PO Box 92957, Los Angeles, Ca. 90009

This work is supported by NSF grant MCS81-02870.

© 1984 American Mathematical Society
0271-4132/84 $1.00 + $.25 per page

§1 Introduction

We have barely begun to automate higher-order logic, but we have made a start, and this paper will give some indication of what we have done so far, and what we may do in the future. After setting the stage in this section, we shall give an example of an automatic proof in §2. We shall discuss a metatheorem which may provide a foundation for further progress in §3, and prove it in §4. The ideas in §§1-2 have been discussed previously in conferences on automated deduction, and more details can be found in [3], [4], and [10].

Theorems and proofs are expressed in some language. When one is automating the theorem-proving process, it's convenient to use a language of symbolic logic, because such a language represents a nice compromise between human languages and computer languages. It's comprehensible to people trained in logic, but has a precise syntax and clear representation of logical ideas which facilitates computer manipulations.

We wanted to automate the logic of a language which is generally adequate for expressing mathematical ideas in a direct and natural way, and chose to use an elegant formulation of type theory (otherwise known as higher-order logic) due to Alonzo Church [6]. Of course, some people would prefer to use axiomatic set theory as a language for formalizing mathematics, and thus stay within the framework of first-order logic. We won't take time to debate this issue now, but let's note that mathematicians do make intuitive distinctions between different types of mathematical objects, and it's useful to have these distinctions represented explicitly in the computer. Also, in Church's type theory one doesn't need axioms for set existence, since sets are represented explicitly by λ-expressions. Another advantage of using typed λ-calculus is the existence of powerful unification algorithms for expressions of this language which were developed by Gérard Huet [8] and by Pietrzykowski and Jensen [9]. We shall see how Huet's algorithm can be used in §2.

For the convenience of the reader we provide a brief introduction to the main features of the language in [6], which we shall call T. T uses λ-notation for functions, which can easily be explained by an example. If F is a function and $F(x) = x^2 + x + 3$ for all x in the domain of F, then we write $[\lambda x \cdot x^2 + x + 3]$ as a notation for the function F itself. We denote the result of applying this function to the argument 5 by $[[\lambda x \cdot x^2 + x + 3] \, 5]$, which we convert to $[5^2 + 5 + 3]$ by a syntactic operation called λ-*contraction*.

Expressions of T have *types*, and we often indicate the types of variables and constants by attaching subscripts called *type symbols* to them. If F is a function which maps elements of type β to elements of type α, then F is said to have type $(\alpha\beta)$. Truth values and statements have type o. If $S_{o\gamma}$ (a function which maps elements of type γ to truth values) maps an element X_γ to truth, we write $[S_{o\gamma} X_\gamma]$ to indicate that $[S_{o\gamma} X_\gamma]$ is true, and say that X_γ is *in the set* $S_{o\gamma}$ or that X_γ *has the property* $S_{o\gamma}$. Thus sets and properties are identified with functions which map elements to truth values, and have types of the form $(o\gamma)$. It can be seen that if A_o is a statement, $[\lambda X_\gamma A_o]$ is another notation for the set $\{X_\gamma \mid A_o\}$ of all elements X_γ for which the statement A_o is true.

We use the convention of association to the left for omitting parentheses and brackets, so that $\alpha\beta\gamma$ is regarded as an abbreviation for $((\alpha\beta)\gamma)$. Also, a dot denotes a left bracket whose mate is as far to the right as possible without changing the pairing of brackets already present.

The proofs we construct are in *natural deduction* format, and here is an example of a very simple proof in this format:

(1) 1 $\vdash \forall x_\iota . [P_{o\iota} x] \wedge . Q_{o\iota} x$ Hyp
(2) 1 $\vdash [P_{o\iota} H_\iota] \wedge . Q_{o\iota} H$ $\forall$I: H_ι 1
(3) 1 $\vdash P_{o\iota} H_\iota$ RuleP: 2
(4) 1 $\vdash \exists y_\iota . P_{o\iota} y$ $\exists$G: H_ι 3
(5) $\vdash [\forall x_\iota . [P_{o\iota} x] \wedge . Q_{o\iota} x] \supset . \exists y_\iota . P y$
                                                                    Deduct: 4

The proof is a sequence of *lines*. Each line has the form

(n) $\mathcal{H}$ $\vdash$ A   J

where n is a number serving as a *label* for the line, $\mathcal{H}$ is a (possibly empty) sequence containing the numbers of lines which are assumed as *hypotheses* for that line, A is the statement being *asserted* in that line, and J is the *justification* for that line. J indicates what rule of inference was applied to obtain the given line, and how it was used. The rules of inference for this system of natural deduction are listed in the Appendix.

At Carnegie-Mellon we have developed an automated theorem-proving system called TPS which can be used in both interactive and automatic modes, and various combinations of these, to construct such proofs. Considerable effort has been devoted to making TPS easy to interact with.

Certain computer terminals have been equipped with extra character generators so that a wide variety of symbols can be displayed on their screens. Thus, traditional notations of logic and mathematics appear on the terminal screens as well as on printed output. TPS can handle expressions of higher-order logic as well as first-order logic, but it is logically complete in automatic mode only for first-order logic.

TPS is both a system under development and a research tool. It enables us to test our ideas, and to look efficiently at examples which reveal new problems and suggest new ideas. While we would ultimately like to build into TPS a variety of facilities useful in automating logic, our present focus is on an approach to theorem-proving which involves analyzing the essential logical structure of a theorem [4], and using the information thus obtained to construct a proof without further search [3].

§2 An Example of an Automatic Proof

In order to convey a general idea of what TPS is like, we shall show how it works on an example. We shall present the actual output produced by the computer as it proves the theorem, with certain material edited out, and with explanatory comments added in italics. (Since the program is still under development, the reader may notice that it could be made more elegant in certain respects.) Lines preceded by a * below are typed in by the human operator, and all other lines (except explanations) are produced by TPS.

TPS has a list of theorems with attached comments stored in its memory, and we start by asking it to state THM5:

*STATE THM5

$\sim .\exists G_{o\iota\iota} \forall F_{o\iota} \exists J_{\iota} .[G J] = F$

(THIS IS CANTOR'S THEOREM FOR SETS)
(THE POWER SET OF A SET HAS MORE MEMBERS THAN THE SET)

THM5 is a rather special statement of Cantor's Theorem in which the types attached to the variables play a very significant role. Let S be the set which we wish to show is smaller than its power set, and let an object have type ι iff it is a member of S. Thus the objects of type $(o\iota)$ are simply the subsets of S, and the objects of type $((o\iota)\iota)$ are the functions which map members of S to subsets of S. Thus THM5 simply says that there is no function G which maps members of S to subsets of S and which has every subset F of S in its range.

Statements of the form [X = Y] are actually abbreviations for $[=_{o\alpha\alpha} X_\alpha Y_\alpha]$, where $=_{o\alpha\alpha}$ is defined as $\lambda X_\alpha \lambda Y_\alpha \forall Q_{o\alpha} .[Q X] \supset .Q Y$. (Thus, X and Y are equal iff every property Q of X is also a property of Y.) If we instantiate the definition of = in THM5, here is what we get:

$\sim .\exists G_{o\iota\iota} \forall F_{o\iota} \exists J_\iota \forall Q_{o(o\iota)} .[Q .G J] \supset .Q F$

We initiate the process of having TPS prove THM5 with the PLAN command:

*PLAN THM5

(RunTime 0.10091015 seconds ConsCount 9)

++
(100)

(100)   ⊢ 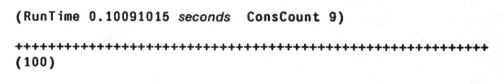   PLAN1

TPS displays the proof it is constructing in a portion of the terminal screen called the <u>proof window</u>, and the inital contents of the proof window are shown above. The proof will be constructed from the bottom up as well as from the top down, and TPS starts off by making the theorem to be proved the last line of the proof, and giving this line the number 100. Of course, this line of the proof is really unjustified at this stage, and PLAN1 is just an indication that TPS plans to prove line 100. At the top of the proof window is the <u>status line</u>, which is a list of the numbers of lines which are still <u>active</u> (in need of further processing).

We could now say GO and TPS would find a proof completely automatically. However, we wish to force TPS to construct an indirect proof, since this will be a little easier for us to grasp intuitively. (A heuristic which would enable TPS to discover that an indirect proof is most appropriate for this theorem is discussed on page 284 of [3], but it is not yet implemented in TPS.) Therefore we instruct TPS to set up an indirect proof:

*P-INDIRECT

(99 1)

(1)	1	$\vdash \sim .\sim .\exists G_{o\iota\iota} \forall F_{o\iota} \exists J_{\iota} .[G\ J\ ] = F$	Hyp
(99)	1	$\vdash \perp$	PLAN2
(100)		$\vdash \sim .\exists G_{o\iota\iota} \forall F_{o\iota} \exists J_{\iota} .[G\ J\ ] = F$	Indirect: 99

TPS has expanded the proof. Line 100 now has a justification and is no longer active, and TPS plans to prove line 99, which asserts that a contradiction (symbolized by $\perp$) follows from line 1, which is the negation of THM5.

From here on the proof will proceed automatically. The first few steps involve applications of rules of inference whose appropriateness is rather obvious.

*GO

Evaluating D-NEG 1

++
(99 2)

(1)	1	$\vdash \sim .\sim .\exists G_{o\iota\iota} \forall F_{o\iota} \exists J_{\iota} .[G\ J\ ] = F$	Hyp
(2)	1	$\vdash \exists G_{o\iota\iota} \forall F_{o\iota} \exists J_{\iota} .[G\ J\ ] = F$	RuleP: 1
(99)	1	$\vdash \perp$	PLAN2
(100)		$\vdash \sim .\exists G_{o\iota\iota} \forall F_{o\iota} \exists J_{\iota} .[G\ J\ ] = F$	Indirect: 99

Evaluating P-CHOOSE 99 2
++
(98 3)

(1)	1	$\vdash \sim .\sim .\exists G_{o\iota\iota} \forall F_{o\iota} \exists J_{\iota} .[G\ J\ ] = F$	Hyp
(2)	1	$\vdash \exists G_{o\iota\iota} \forall F_{o\iota} \exists J_{\iota} .[G\ J\ ] = F$	RuleP: 1
(3)	3	$\vdash \forall F_{o\iota} \exists J_{\iota} .[G_{o\iota\iota}\ J\ ] = F$	Choose: $G_{o\iota\iota}$ 2
(98)	1,3	$\vdash \perp$	PLAN2
(99)	1	$\vdash \perp$	RuleC: 2 98
(100)		$\vdash \sim .\exists G_{o\iota\iota} \forall F_{o\iota} \exists J_{\iota} .[G\ J\ ] = F$	Indirect: 99

The problem, (98 3), can not be reduced.

TPS has discovered that it can progress no further by trivial applications of rules of logic, and prepares to proceed with the MatingSearch program. It notes that the essential problem is to derive a contradiction from line 3, and (rather inelegantly) adds line 97 to the proof (as shown below) to facilitate its internal processing.

```
Evaluating  FINDPLAN 98 (3)

(RunTime 3.7734394  ConsCount 22311)
++++++++++++++++++++++++++++++++++++++++++++++++++++++++++++
(97 3)
```

(1)	1	⊢	$\sim .\sim .\exists G_{o\iota} \forall F_{o\iota} \exists J_\iota .[G\ J\ ] = F$	Hyp
(2)	1	⊢	$\exists G_{o\iota} \forall F_{o\iota} \exists J_\iota .[G\ J\ ] = F$	RuleP: 1
(3)	3	⊢	$\forall F_{o\iota} \exists J_\iota .[G_{o\iota}\ J\ ] = F$	Choose: $G_{o\iota}$ 2
(97)	3	⊢	⊥	PLAN3
(98)	1,3	⊢	⊥	Deduct,RuleP: 97
(99)	1	⊢	⊥	RuleC: 2 98
(100)		⊢	$\sim .\exists G_{o\iota} \forall F_{o\iota} \exists J_\iota .[G\ J\ ] = F$	Indirect: 99

Next TPS processes line 3 by instantiating the definition of =, writing the ⊃ thus introduced in terms of $\sim$ and $\vee$, and skolemizing to eliminate the existential quantifier. The variable J_ι is replaced by the term $[J|A_{\iota(o\iota)}\ F_{o\iota}]$, where $J|A_{\iota(o\iota)}$ is a skolem function. TPS displays the resulting formula, and the names it has attached to its literals, as follows:

[1] $\forall F_{o\iota} \forall Q_{o(o\iota)} .[\sim .Q\ .G_{o\iota} .J|A_{\iota(o\iota)}\ F\ ] \vee .Q\ F$

[1] $\forall F_{o\iota} \forall Q_{o(o\iota)}$ .LIT2 $\vee$ LIT3

From [1] TPS must derive a contradiction by instantiating the quantifiers appropriately.

```
Path with no potential mates: (LIT2)
No proof on this level
(RunTime 5.5491445  ConsCount 31522)
```

TPS applied its search process, which will be described below, and quickly discovered that no single instantiation of the two quantifiers can produce a contradiction.

Replicate outermost quantifiers:

[2] $[\forall F^1_{o\iota} \; \forall Q^1_{o(o\iota)} \; . [\sim \; . Q^1 \; . G_{o\iota\iota} \; . J | A_{\iota(o\iota)} \; F^1 \;] \; \vee \; . Q^1 \; F^1 \;]$
$\wedge \; . \forall F^2_{o\iota} \; \forall Q^2_{o(o\iota)} \; . [\sim \; . Q^2 \; . G \; . J | A \; F^2 \;] \; \vee \; . Q^2 \; F^2$

[2] $[\forall F^1_{o\iota} \; \forall Q^1_{o(o\iota)} \; . LIT2^1 \; \vee \; LIT3^1 \;]$
$\wedge \; . \forall F^2_{o\iota} \; \forall Q^2_{o(o\iota)} \; . LIT2^2 \; \vee \; LIT3^2$

The matingsearch process is described in [4], but we shall briefly summarize the basic concepts which underlie it. A _mating_ $\mathcal{M}$ for a wff W (such as the matrix of [2]) is a relation between occurrences of literals of W such that there is a substitution θ which makes literals with mated occurrences complementary. Thus, if A $\mathcal{M}$ B, then $\theta B = \sim \theta A$. θ is called the _unifying substitution associated with_ $\mathcal{M}$. A _vertical path_ through W is a sequence of occurrences of literals whose conjunction is one of the conjuncts in the disjunctive normal form of W. The vertical paths through [2] are ($LIT2^1 \; LIT2^2$), ($LIT2^1 \; LIT3^2$), ($LIT3^1 \; LIT2^2$), and ($LIT3^1 \; LIT3^2$). A mating for W is _acceptable_ iff every vertical path contains a mated pair of literals. Thus, W becomes a contradiction when a substitution associated with an acceptable mating is applied to it.

TPS searches for an acceptable mating and its associated substitution in a systematic way. It tries various possibilities, and backtracks when it finds that the subtitutions required to mate various literal-pairs are incompatible. The processes of finding appropriate pairs of literals to mate, and searching for unifying substitutions, are carried on more or less simultaneously, and interact with each other. The search processes are controlled by various heuristics.

```
Path with no mates: (LIT2¹ LIT2²)
    try this arc: (LIT2¹ LIT2²)

Partial Mating 0:
```

    (LIT2^1)-(LIT2^2)

Path with no mates: (LIT3^1 LIT2^2)
    try this arc: (LIT3^1 LIT2^2)

Partial Mating 0:
    (LIT2^1 LIT3^1)-(LIT2^2)

Path with no mates: (LIT3^1 LIT3^2)
    try this arc: (LIT3^1 LIT3^2)

Partial Mating 0:
    (LIT2^1 LIT3^1)-(LIT2^2 LIT3^2)

Mating 0 is complete.
(RunTime 9.7913925   ConsCount 44797)

Mating 0 is the mating which contains the three literal-pairs referred to as arcs above, plus the induced arc (LIT2^1 LIT3^2). TPS actually found this mating without any need for backtracking, though backtracking usually does occur when dealing with formulas with a more complicated propositional structure. In essence, TPS has found that a contradiction will be obtained if the quantifiers in [2] are instantiated in such a way that [2] reduces to a formula having the form

$$[\ D \ \vee \ D\ ]$$
$$\wedge \ [\sim D \ \vee \ \sim D].$$

Thus far TPS has actually done only a partial and superficial check that a substituion associated with Mating 0 exists. It next searches in earnest for such a substitution, using Gérard Huet's elegant unification algorithm [8]. TPS owes much of its power to this algorithm, but we omit the lengthy details of the unification process. Actually, TPS could be working on an incorrect mating, for which the unification algorithm would never terminate, so TPS alternates between generating new matings and continuing work on all of their associated unification problems. Eventually it reaches a Terminal Success Node in its search for a unifier associated with Mating 0:

N0-2-0-0-0-0-0-0:=   TSN
$F^1_{o\iota}$ $\Leftarrow \lambda W1_\iota \ . \sim \ . G_{o\iota}$ W1 W1
$F^2_{o\iota}$ $\Leftarrow \lambda W1_\iota \ . \sim \ . G_{o\iota}$ W1 W1

$Q^1_{o(o\iota)} \Leftarrow \lambda W1_{o\iota} \;\; .W1 \;\; .J|A_{\iota(o\iota)} \;\; .\lambda W2_\iota \;\; .\sim \;\; .G_{o\iota\iota} \;\; W2 \;\; W2$

$Q^2_{o(o\iota)} \Leftarrow \lambda W1_{o\iota} \;\; .\sim \;\; .W1 \;\; .J|A_{\iota(o\iota)} \;\; .\lambda W2_\iota \;\; .\sim \;\; .G_{o\iota\iota} \;\; W2 \;\; W2$

```
Unification complete
(RunTime 25.365179   ConsCount 74157)
```

Having finished its search, TPS prepares to use the information it has found to complete the natural deduction proof. First it notices that the substitutions for F^1 and F^2 are the same, which means that it would have been more appropriate to duplicate the quantifier $\forall Q$ of formula [1] instead of $\forall F$. TPS makes the appropriate changes, replacing formula [2] by formula [3] below:

```
The replication has been simplified.
Old:    ( F_oι    2 )     New:    ( Q_o(oι)    2 )
```

[3]
$\forall F_{o\iota} \;\; . \quad [\forall Q^1_{o(o\iota)} \;\; .[\sim \;\; .Q^1 \;\; .G_{o\iota\iota} \;\; .J|A_{\iota(o\iota)} \;\; F \;\;] \;\lor\; .Q^1 \;\; F \;]$

$\qquad\qquad \land \;\; .\forall Q^2_{o(o\iota)} \;\; .[\sim \;\; .Q^2 \;\; .G \;\; .J|A \;\; F \;\;] \;\lor\; .Q^2 \;\; F$

The substitution has also been changed.

```
New:
```
$F_{o\iota} \;\Leftarrow\; \lambda W1_\iota \;\; .\sim \;\; .G_{o\iota\iota} \;\; W1 \;\; W1$

$Q^1_{o(o\iota)} \;\Leftarrow\; \lambda W1_{o\iota} \;\; .W1 \;\; .J|A_{\iota(o\iota)} \;\; .\lambda W2_\iota \;\; .\sim \;\; .G_{o\iota\iota} \;\; W2 \;\; W2$

$Q^2_{o(o\iota)} \;\Leftarrow\; \lambda W1_{o\iota} \;\; .\sim \;\; .W1 \;\; .J|A_{\iota(o\iota)} \;\; .\lambda W2_\iota \;\; .\sim \;\; .G_{o\iota\iota} \;\; W2 \;\; W2$

TPS displays the result of instantiating the quantifiers in [3] with the new substitution and λ-reducing as formula [4]:

[4]
$\quad [\quad [\sim \;\; .G_{o\iota\iota} \;\; [J|A_{\iota(o\iota)} \;\; .\lambda W1_\iota \;\; .\sim \;\; .G \;\; W1 \;\; W1 \;]$
$\qquad\qquad\qquad\qquad\qquad .J|A \;\; .\lambda W2_\iota \;\; .\sim \;\; .G \;\; W2 \;\; W2 \;]$
$\qquad \lor \;\; .\sim \;\; .G \;\; [J|A \;\; .\lambda W2 \;\; .\sim \;\; .G \;\; W2 \;\; W2 \;]$
$\qquad\qquad\qquad\qquad .J|A \;\; .\lambda W2 \;\; .\sim \;\; .G \;\; W2 \;\; W2 \;]$
$\quad \land \;\; . \quad [\sim \;\; .\sim \;\; .G \;\; [J|A \;\; .\lambda W1 \;\; .\sim \;\; .G \;\; W1 \;\; W1 \;]$
$\qquad\qquad\qquad\qquad .J|A \;\; .\lambda W2 \;\; .\sim \;\; .G \;\; W2 \;\; W2 \;]$
$\qquad \lor \;\; .\sim \;\; .\sim \;\; .G \;\; [J|A \;\; .\lambda W2 \;\; .\sim \;\; .G \;\; W2 \;\; W2 \;]$
$\qquad\qquad\qquad\qquad .J|A \;\; .\lambda W2 \;\; .\sim \;\; .G \;\; W2 \;\; W2$

Note that [4] is indeed a contradiction having the form anticipated above.

TPS now eliminates the skolem terms from the substitution terms and transforms the information it has obtained into a plan for proving line 97, which it calls PLAN3. Thus, PLAN3 is no longer an empty label indicating that TPS plans to prove line 97; it is now the name of a data structure which contains all the information necessary to carry out the proof. PLAN3 is exhibited by displaying the expanded form of the formula to be proved, and the substitution which (as will be seen in §3) reduces it to a tautology.

PLAN3 is:

$$\sim .\forall F_{o\iota} \exists J_{\iota} . \quad [\forall Q^1_{o(o\iota)} .[Q^1 .G_{o\iota\iota} J] \supset .Q^1 F]$$
$$\text{ATM2}^1 \qquad \text{ATM3}^1$$
$$\wedge .\forall Q^2_{o(o\iota)} .[Q^2 .G J] \supset .Q^2 F$$
$$\text{ATM2}^2 \qquad \text{ATM3}^2$$

The substitution is:

$$F_{o\iota} \Leftarrow \lambda W1_{\iota} .\sim .G_{o\iota\iota} W1\ W1 \qquad Q^1_{o(o\iota)} \Leftarrow \lambda W1_{o\iota} .W1\ J_{\iota}$$

$$Q^2_{o(o\iota)} \Leftarrow \lambda W1_{o\iota} .\sim .W1\ J_{\iota}$$

Now TPS continues the construction of the natural deduction proof. The substitution terms in PLAN3 are used to instantiate the quantifiers in the proof below.

```
Evaluating   D-ALL 3 λW1ι .~ .Goιι W1 W1   97
++++++++++++++++++++++++++++++++++++++++++++++++++++++++
(97 4)
```

(1)	1	⊢	$\sim .\sim .\exists G_{o\iota\iota} \forall F_{o\iota} \exists J_{\iota} .[G\ J] = F$	Hyp
(2)	1	⊢	$\exists G_{o\iota\iota} \forall F_{o\iota} \exists J_{\iota} .[G\ J] = F$	RuleP: 1
(3)	3	⊢	$\forall F_{o\iota} \exists J_{\iota} .[G_{o\iota\iota} J] = F$	Choose: $G_{o\iota\iota}$ 2
(4)	3	⊢	$\exists J_{\iota} .[G_{o\iota\iota} J] = .\lambda W1_{\iota} .\sim .G\ W1\ W1$	
			∀I: $[\lambda W1_{\iota} .\sim .G_{o\iota\iota} W1\ W1]$ 3	
(97)	3	⊢ ⊥		PLAN3
(98)	1,3	⊢ ⊥		Deduct,RuleP: 97
(99)	1	⊢ ⊥		RuleC: 2 98
(100)		⊢	$\sim .\exists G_{o\iota\iota} \forall F_{o\iota} \exists J_{\iota} .[G\ J] = F$	Indirect: 99

Evaluating   P-CHOOSE 97 4

Evaluating   D-DEF1 5

Evaluating   D-ALL 6 $\lambda W1_{o\iota}$ .W1 J_ι 96

Evaluating   D-ALL 6 $\lambda W1_{o\iota}$.$\sim$ .W1 J_ι 96
++
(96 8 7)

(1)	1	$\vdash$	$\sim$.$\sim$.$\exists G_{o\iota\iota} \forall F_{o\iota} \exists J_\iota$.[G J] = F	Hyp
(2)	1	$\vdash$	$\exists G_{o\iota\iota} \forall F_{o\iota} \exists J_\iota$.[G J] = F	RuleP: 1
(3)	3	$\vdash$	$\forall F_{o\iota} \exists J_\iota$.[$G_{o\iota\iota}$ J] = F	Choose: $G_{o\iota\iota}$ 2
(4)	3	$\vdash$	$\exists J_\iota$.[$G_{o\iota\iota}$ J] = .$\lambda W1_\iota$.$\sim$ .G W1 W1	
				$\forall$I: [$\lambda W1_\iota$.$\sim$.$G_{o\iota\iota}$ W1 W1] 3
(5)	5	$\vdash$	[$G_{o\iota\iota}$ J_ι] = .$\lambda W1_\iota$.$\sim$ .G W1 W1	
				Choose: J_ι 4
(6)	5	$\vdash$	$\forall Q_{o(o\iota)}$.[Q .$G_{o\iota\iota}$ J_ι] $\supset$ .Q .$\lambda W1_\iota$.$\sim$ .G W1 W1	Def: 5
(7)	5	$\vdash$	[$G_{o\iota\iota}$ J_ι J] $\supset$.$\sim$ .G J J	
				$\forall$I: [$\lambda W1_{o\iota}$ .W1 J_ι] 6
(8)	5	$\vdash$	[$\sim$.$G_{o\iota\iota}$ J_ι J] $\supset$.$\sim$.$\sim$ .G J J	
				$\forall$I: [$\lambda W1_{o\iota}$.$\sim$ .W1 J_ι] 6
(96)	3,5	$\vdash$	$\bot$	PLAN3
(97)	3	$\vdash$	$\bot$	RuleC: 4 96
(98)	1,3	$\vdash$	$\bot$	Deduct,RuleP: 97
(99)	1	$\vdash$	$\bot$	RuleC: 2 98
(100)		$\vdash$	$\sim$.$\exists G_{o\iota\iota} \forall F_{o\iota} \exists J_\iota$.[G J] = F	Indirect: 99

After each application of a rule of inference, TPS checks to see whether the line it is planning to prove follows by RuleP from the other active lines. TPS now notices that line 96 follows from lines 7 and 8 by this rule.

```
Evaluating   RULEP 96 (8 7)
++++++++++++++++++++++++++++++++++++++++++++++++++++++++++
(1)    1       ⊢  ~ .~ .∃G_{oιι} ∀F_{oι} ∃J_ι .[G J] = F        Hyp
(2)    1       ⊢  ∃G_{oιι} ∀F_{oι} ∃J_ι .[G J] = F              RuleP: 1
(3)    3       ⊢  ∀F_{oι} ∃J_ι .[G_{oιι} J] = F                 Choose: G_{oιι}  2
(4)    3       ⊢  ∃J_ι .[G_{oιι} J] = .λW1_ι .~ .G W1 W1
                                       ∀I: [λW1_ι .~ .G_{oιι} W1 W1 ] 3
(5)    5       ⊢  [G_{oιι} J_ι] = .λW1_ι .~ .G W1 W1
                                                          Choose: J_ι  4
(6)    5       ⊢     ∀Q_{o(oι)} .[Q .G_{oιι} J_ι]
                    ⊃ .Q .λW1_ι .~ .G W1 W1                    Def: 5
(7)    5       ⊢  [G_{oιι} J_ι J] ⊃ .~ .G J J
                                       ∀I: [λW1_{oι} .W1 J_ι ] 6
(8)    5       ⊢  [~ .G_{oιι} J_ι J] ⊃ .~ .~ .G J J
                                       ∀I: [λW1_{oι} .~ .W1 J_ι ] 6
(96)   3,5     ⊢  ⊥                                           RuleP: 7 8
(97)   3       ⊢  ⊥                                           RuleC: 4 96
(98)   1,3     ⊢  ⊥                                           Deduct,RuleP: 97
(99)   1       ⊢  ⊥                                           RuleC: 2 98
(100)          ⊢  ~ .∃G_{oιι} ∀F_{oι} ∃J_ι .[G J] = F         Indirect: 99
----------------------------------------------------------------
There are no plan lines.

(RunTime 31.780359   ConsCount 103655)
```

Since every line of the proof now has a justification, the proof is complete. Note that it is simply a formalized presentation of the traditional diagonal proof of Cantor's Theorem. It was obtained in less than 32 seconds by a purely syntactic analysis of the theorem.

§3 A Metatheorem for Extending the Capabilities of TPS

TPS is able to prove certain theorems of higher-order logic, such as Cantor's Theorem, by combining higher-order unification with a theorem-proving method which was really designed for first-order logic. We're a long way from having a reasonable theorem-proving procedure for higher-order logic which is logically complete even in principle, but in this section we shall present a metatheorem which provides a conceptual framework in which it may be possible to develop such a procedure.

It may be recalled that in his thesis Herbrand discussed properties A, B, and C of wffs, and asserted for each of the properties that a wff of first-order logic is provable iff it has the property. In its present formulation, the metatheorem below is a generalization to higher-order logic of Herbrand's Theorem using Property A, in which skolem functions are not used. (The metatheorem as formulated below was proved by Andrews. Another formulation due to Miller will be presented in [11].)

In order to state the metatheorem precisely enough so that it can be proved in §4, we must next give some technical definitions. However, the reader who wishes to temporarily skip these can get a general idea of what the metatheorem says by looking at the example later in this section.

In §1 we used T as a name for a language of type theory, but we shall now use T in a more precise way as the name of the logical system described in §2 of [1]; T consists of the system originally presented by Church in [6], minus axioms of extensionality, descriptions, choice, and infinity. Thus T simply embodies the logic of propositional connectives, quantifiers, and λ-conversion in the context of type theory, and (as in [2]) we shall refer to T as *elementary type theory*.

The primitive logical constants of T are $\sim_{oo}$ (negation), $\vee_{ooo}$ (disjunction), and $\Pi_{o(o\alpha)}$. $\Pi_{o(o\alpha)}[\lambda x_\alpha C_o]$ may be abbreviated as $\forall x_\alpha C_o$, and $\Pi_{o(o\alpha)} A_{o\alpha}$ is provably equivalent to $\forall x_\alpha [A_{o\alpha} x_\alpha]$ if x_α is not free in $A_{o\alpha}$. $\wedge, \supset, \equiv$, and $\exists$ are defined in familiar ways.

The *well-formed formulas (wffs)* of T are defined in [6] and in [1]. We shall use *occwfp* as an abbreviation for the phrase *occurrence of a well-formed part*, and *occwfps* for *occurrences of well-formed parts*. We say that an occwfp A of a wff$_o$ (wff of type o) W_o is *very accessible* in W_o iff A is only in the scope of propositional connectives in W_o; for greater precision this notion is defined inductively as follows:

1. W_o is very accessible in W_o.
2. If $\sim B$ is very accessible in W_o, so is B.
3. If $[B \lor C]$ is very accessible in W_o, so are B and C.

An occwfp A of W_o is *negative* in W_o iff A is in the scope of an odd number of occurrences of negation signs in W_o, and is *positive* in W_o otherwise. An occwfp of W_o having the form $\Pi_{o(o\alpha)} A_{o\alpha}$ is *essentially universal* iff it is positive in W_o, and *essentially existential* iff it is negative in W_o. (The essentially universal [existential] occwfps of W_o correspond to the universal [existential] quantifiers in prenex normal forms of W_o.) An occwfp A of a wff W_o is *critical in* W_o iff A is very accessible and essentially existential in W_o.

Let W be a wff_o which has a very accessible occwfp M having the form $\Pi_{o(o\alpha)} A_{o\alpha}$. Let U be the result of replacing the given occurrence of M in W by an occurrence of a wff N which we discuss below.

1. If M is positive in W and N is $A_{o\alpha} x_\alpha$, where x_α does not occur free in W, we say that U is obtained from W by *universal quantifier deletion* ($\forall$-deletion).
2. If M is negative in W and N is $[\Pi_{o(o\alpha)} A_{o\alpha} \land \Pi_{o(o\alpha)} A_{o\alpha}]$, we say that U is obtained from W by *existential duplication* ($\exists$-duplication).
3. If M is negative in W and N is $A_{o\alpha} B_\alpha$, where B_α is any wff_α, we say that U is obtained from W by *existential instantiation* ($\exists$-instantiation).

We shall say that we apply λ-*reduction* to a wff when we apply λ-contraction (rule 2.6.2 of [1]) to one of its parts of the form $[[\lambda x_\alpha B_\beta] A_\alpha]$, after making any necessary alphabetic changes of bound variables (rule 2.6.1 of [1]). A wff is in λ-*normal form* if it has no parts of the form $[[\lambda x_\alpha B_\beta] A_\alpha]$.

We call $\forall$-deletion, $\exists$-duplication, $\exists$-instantiation, and λ-reduction the four *basic operations*. A *development* of a sentence W of $\mathcal{T}$ is any wff_o obtained from it by any sequence of applications of the basic operations.

It can be seen that if W_o contains a very accessible positive occwfp $\forall x_\alpha C_o$, where x_α does not occur free in W_o, then this occurrence of $\forall x_\alpha C_o$ can be replaced by C_o by $\forall$-deletion and λ-reduction. Indeed, $\forall x_\alpha C_o$ is an abbreviation for $\Pi_{o(o\alpha)}[\lambda x_\alpha C_o]$, which can be replaced by $[\lambda x_\alpha C_o] x_\alpha$ by $\forall$-deletion, and λ-reduced to C_o. Similarly, a very accessible positive occwfp

$\exists x_\alpha C_o$ is an abbreviation for $\sim \Pi_{o(o\alpha)}[\lambda x_\alpha \sim C_o]$, which can be $\exists$-instantiated to $\sim [\lambda x_\alpha \sim C_o]B_\alpha$, which λ-reduces (modulo a double negation which we often omit writing) to the result of instantiating the quantifier in $\exists x_\alpha C_o$ with the wff B_α. Thus, when we abbreviate wffs so that quantifiers always occur as $\forall$ or $\exists$, and $\Pi_{o(o\alpha)}$ does not explicitly appear, we see that developments of wffs are obtained by deleting essentially universal quantifiers, duplicating or instantiating essentially existential quantifiers, and performing λ-reductions.

We next illustrate these concepts by developing THM5. In abbreviated form, THM5 is:

$\sim .\exists G_{o\iota\iota} \; \forall F_{o\iota} \; \exists J_\iota \; .[G \; J] = F$

We start by instantiating the definition of = (which is really just an application of λ-reduction), and obtain:

$\sim .\exists G_{o\iota\iota} \; \forall F_{o\iota} \; \exists J_\iota \; \forall Q_{o(o\iota)} \; .[Q \; .G \; J] \supset .Q \; F$

Delete $\exists G$:

$\sim .\forall F_{o\iota} \; \exists J_\iota \; \forall Q_{o(o\iota)} \; .[Q \; .G_{o\iota\iota} \; J] \supset .Q \; F$

Instantiate $\forall F$ with $\lambda W_\iota \; .\sim .G_{o\iota\iota} \; W \; W$:

$\sim .\exists J_\iota \; \forall Q_{o(o\iota)} \; .[Q \; .G_{o\iota\iota} \; J] \supset .Q \; .\lambda W_\iota \; .\sim .G \; W \; W$

Delete $\exists J$:

$\sim .\forall Q_{o(o\iota)} \; .[Q \; .G_{o\iota\iota} \; J_\iota] \supset .Q \; .\lambda W_\iota \; .\sim .G \; W \; W$

Duplicate $\forall Q$:

$\sim . \quad [\forall Q_{o(o\iota)} \; .[Q \; .G_{o\iota\iota} \; J_\iota] \supset .Q \; .\lambda W_\iota \; .\sim .G \; W \; W]$
$\quad \wedge .\forall Q \; .[Q \; .G \; J] \supset .Q \; .\lambda W \; .\sim .G \; W \; W$

Instantiate the first $\forall Q$ with $\lambda U_{o\iota} \; .U \; J_\iota$:

$\sim . \quad [[[\lambda U_{o\iota} \; .U \; J_\iota].G_{o\iota\iota} \; J] \supset .[\lambda U \; .U \; J].\lambda W_\iota \; .\sim .G \; W \; W]$
$\quad \wedge .\forall Q_{o(o\iota)} \; .[Q \; .G \; J] \supset .Q \; .\lambda W \; .\sim .G \; W \; W$

λ-reduce:

$\sim\ .\quad [[G_{o\iota\iota}\ J_\iota\ J\ ] \supset .[\lambda U_{o\iota}\ .U\ J\ ].\lambda W_\iota\ .\sim\ .G\ W\ W\ ]$
$\wedge\ .\forall Q_{o(o\iota)}\ .[Q\ .G\ J\ ] \supset .Q\ .\lambda W\ .\sim\ .G\ W\ W$

$\sim\ .\quad [[G_{o\iota\iota}\ J_\iota\ J\ ] \supset .[\lambda W_\iota\ .\sim\ .G\ W\ W\ ]\ J\ ]$
$\wedge\ .\forall Q_{o(o\iota)}\ .[Q\ .G\ J\ ] \supset .Q\ .\lambda W\ .\sim\ .G\ W\ W$

$\sim\ .\quad [[G_{o\iota\iota}\ J_\iota\ J\ ] \supset .\sim\ .G\ J\ J\ ]$
$\wedge\ .\forall Q_{o(o\iota)}\ .[Q\ .G\ J\ ] \supset .Q\ .\lambda W_\iota\ .\sim\ .G\ W\ W$

Instantiate the remaining $\forall Q$ with $\lambda U_{o\iota}\ .\sim\ .U\ J_\iota$:

$\sim\ .\quad [[G_{o\iota\iota}\ J_\iota\ J\ ] \supset .\sim\ .G\ J\ J\ ]$
$\wedge .\quad [[\lambda U_{o\iota}\ .\sim\ .U\ J\ ].G\ J\ ]$
$\supset .[\lambda U\ .\sim\ .U\ J\ ].\lambda W_\iota\ .\sim\ .G\ W\ W$

λ-reduce:

$\sim\ .\quad [[G_{o\iota\iota}\ J_\iota\ J\ ] \supset .\sim\ .G\ J\ J\ ]$
$\wedge\ .[\sim\ .G\ J\ J\ ] \supset .[\lambda U_{o\iota}\ .\sim\ .U\ J\ ].\lambda W_\iota\ .\sim\ .G\ W\ W$

$\sim\ .\quad [[G_{o\iota\iota}\ J_\iota\ J\ ] \supset .\sim\ .G\ J\ J\ ]$
$\wedge\ .[\sim\ .G\ J\ J\ ] \supset .\sim\ .[\lambda W_\iota\ .\sim\ .G\ W\ W\ ]\ J$

$\sim\ .\quad [[G_{o\iota\iota}\ J_\iota\ J\ ] \supset .\sim\ .G\ J\ J\ ]$
$\wedge\ .[\sim\ .G\ J\ J\ ] \supset .\sim\ .\sim\ .G\ J\ J$

Note that we have obtained a development of THM5 which is tautologous (a substitution instance of a tautology of propositional calculus). Our main metatheorem asserts that this can be done for every theorem of T.

Metatheorem. A sentence of T is provable in T iff it has a tautologous development.

It can be seen from the results in [11] that once one knows how to obtain a tautologous development of a sentence, one can construct a natural-deduction proof of it without further search. Thus, the metatheorem above shows that we can focus our research in higher-order theorem proving on the problem of finding tautologous developments. Of course, this is a partial extension to higher-order logic of the basic approach in [4].

Although a tautologous development of THM5 can be found by purely automatic methods, significant new ideas must be found in order to expand the set of sentences for which tautologous developments can be found automatically. Nevertheless, it seems reasonable to hope that the search for tautologous developments provides a context in which further progress on automating higher-order logic can be made.

§4 Proof of the Metatheorem

In this section we prove the metatheorem stated in §3. We shall write $\vdash$ W to indicate that W is a theorem of T, and $\models$ W to indicate that W is tautologous.

By examining the axioms and rules of inference of T, one easily sees that analogues of all theorems and derived rules of inference of first-order logic are provable in T, and we shall use this fact freely. In particular, if $\models$ W, then $\vdash$ W. Also, it is easy to establish the following:

Lemma (Substitutivity of Implication). Let M and N be wffs$_o$ such that $\vdash$ M $\supset$ N, let W and U be wffs$_o$ such that M has an occurrence in W as a very accessible occwfp, and U is obtained from W by replacing the designated occurrence of M in W by an occurrence of N. Then $\vdash$ W $\supset$ U if the designated occurrence of M is positive in W, and $\vdash$ U $\supset$ W if it is negative.

It is easy to see that if a sentence W has a tautologous development, then $\vdash$ W. Indeed, suppose $W^0, W^1, ..., W^n$ is a sequence of wffs$_o$ such that W^0 is W, W^n is tautologous, and W^{i+1} is obtained from W^i by a basic operation for each $i < n$. Since $\vdash W^n$, it suffices to show for each $i < n$ that if $\vdash W^{i+1}$, then $\vdash W^i$. Since $\vdash \Pi_{o(o\alpha)} A_{o\alpha} \supset [\Pi_{o(o\alpha)} A_{o\alpha} \wedge \Pi_{o(o\alpha)} A_{o\alpha}]$ and $\vdash \Pi_{o(o\alpha)} A_{o\alpha} \supset A_{o\alpha} B_\alpha$ and operations of λ-conversion are reversible, it is easy to see with the aid of the Lemma that if W^{i+1} is obtained from W^i by $\exists$-duplication, $\exists$-instantiation, or λ-reduction, then $\vdash W^{i+1} \supset W^i$, which is more than sufficient. Also, if W^{i+1} is obtained from W^i by $\forall$-deletion, then W^i can be inferred from W^{i+1} by universal generalization and anti-prenex rules for pushing in quantifiers. (Here we use the condition on the variable in the

∀-deletion rule). This completes the proof of the theorem in the trivial direction, and shows that finding tautologous developments is a sound method for establishing theorems of T.

Next we must show that the method is complete. Here we shall assume familiarity with §4 of [1] (the cut-elimination theorem for T, which is based on the work of Takahashi [16]). It is interesting to note that just as there is a close relation between Gentzen's cut-elimination theorem and Herbrand's theorem for first-order logic, there is a close relation between cut-elimination for T and our generalization of Herbrand's theorem to T.

We start by introducing some definitions. A wff_o is *open* iff it is in λ-normal form and contains no positive very accessible occwfps of the form $\Pi_{o(o\alpha)} A_{o\alpha}$. We *open* a wff_o W by applying the operations of ∀-deletion and λ-reduction to it repeatedly until it is open. This process is called *opening* W, and the resulting wff is called an *open form* of W. It can be seen (as a minor extension of the Church-Rosser Theorem for typed λ-calculus) that every wff_o W has an open form (i.e., the process terminates), which is unique modulo alphabetic changes of free and bound variables.

If W is a wff_o and U is obtained from W by applying ∃-instantiation (once) to W and opening the resulting wff, then we say that U is obtained from W by *major* ∃-*instantiation*. The sequence $W^o,..., W^i,..., W^n$ (where $n \geq 0$) is a *development sequence* (d-seq) for a wff_o W iff W^o is an open form of W, and for each $i < n$, W^{i+1} is obtained from W^i by ∃-duplication or by major ∃-instantiation. Clearly, each of the wffs W^i in a d-seq for W is an open development of W. If $W^o,..., W^n$ is a d-seq for W and $\models W^n$, we say that the d-seq $W^o,..., W^n$ *verifies* W and is a *verification of* W, and write $\Box W\{W^o,...,W^n\}$. If some d-seq verifies W we say that W is *verified* and write $\Box W$.

Let $W^o,...,W^i,..., W^n$ be any development sequence, and let R be any very accessible occwfp of W^i (where $0 \leq i \leq n$). For each $k \geq i$, we define the *descendant* R^k of R in W^k by induction on k:

1. If $k = i$, then R^k is R.
2. Suppose $k > i$, so W^{k+1} is obtained from W^k by replacing some critical occwfp M of W^k by an occwfp N. M has the form $\Pi_{o(o\alpha)} A_{o\alpha}$, so M is either a subformula of R^k or does not overlap R^k. In the former case R^{k+1} is the subformula of W^{k+1} obtained

when M is replaced by N in R^k. In the latter case R^{k+1} is the ocurrence of R^k in W^{k+1} which corresponds to the occurrence of R^k in W^k.

Now we are ready to start the completeness proof. We shall show that if $\vdash W$, then $\Box\, W$. Suppose that $\vdash W$; then by Theorem 4.10 of [1], W has a proof $P^1,...,P^k,...,P^m$ in the cut-free system $\mathcal{G}$ of [1]. We show by induction on k that $\Box\, P^k$ for each k (k = 1,..., m).

Case 0. P^k is an axiom $A \vee \sim A$ of $\mathcal{G}$, where A is atomic.
Clearly $\Box\, P^k\{A^o \vee \sim A^o\}$, where A^o is a λ-normal form of A.

For the sake of brevity, in all the cases below we shall assume that P^k is inferred from P^i [or from P^i and P^j in Case 5] by the indicated rule of inference of $\mathcal{G}$, where i < k [and j < k]. By inductive hypothesis we are given that $\Box\, P^i\{\mathcal{I}\}$ (where $\mathcal{I}$ is the d-seq $W^o, W^1,..., W^n$), and we must show that $\Box\, P^k$.

Case 1. P^k is inferred by λ-conversion.
Then $\Box\, P^k\{\mathcal{I}\}$.

Case 2. P^k is inferred by disjunction rules.
Clearly one can apply the same disjunction rules as were used to obtain P^k from P^i to each wff in $\mathcal{I}$ to obtain a verification of P^k.

Case 3. P^k is $P^i \vee A$.
Let A^o be an open form of A. $\Box\, P^k\{W^o \vee A^o, W^1 \vee A^o,..., W^n \vee A^o\}$.

Case 4. P^i is $M \vee A$ and P^k is $M \vee \sim\sim A$.
W^o has the form $M^o \vee A^o$. For each p, let U^p be obtained from W^p by replacing the descendant D of A^o in W^p by the wff $\sim\sim D$. It is easy to see that $\Box\, P^k\{U^o,...,U^n\}$.

Case 5. P^i is $M \vee \sim A$ and P^j is $M \vee \sim B$ and P^k is $M \vee \sim[A \vee B]$.
Since M has an essentially unique open form (modulo renaming of variables), we may assume that we are given verifications $\mathcal{I} = (W^o,..., W^n)$ for $M \vee \sim A$ and $\mathcal{J} = (U^o,..., U^q)$ for $M \vee \sim B$, where W^o is $M^o \vee \sim A^o$, U^o is $M^o \vee \sim B^o$, and M^o is an open form of M. Also, we may assume that the only variables which occur free both in wffs of $\mathcal{I}$ and in wffs of $\mathcal{J}$ are those which occur free both in $M^o \vee \sim A$ and in $M^o \vee \sim B$. Clearly W^n has the form $M^a \vee \sim A^1$ and U^q has the form $M^b \vee \sim B^1$, where M^a, M^b, $\sim A^1$, and $\sim B^1$ are descended from M^o, M^o, $\sim A^o$, and $\sim B^o$, respectively. (See Figure 4-1.)

We now describe a verification of $M \vee \sim[A \vee B]$. Its initial wff X^o is $M^o \vee \sim[A^o \vee B^o]$. Next, for each critical occwfp R of M^o, perform an

Figure 4-1: Notations for Case 5

$$P^i = [M \vee \sim A] \qquad P^j = [M \vee \sim B] \qquad P^k = [M \vee \sim [A \vee B]]$$

$$W^o = [M^o \vee \sim A^o] \qquad U^o = [M^o \vee \sim B^o] \qquad X^o = [M^o \vee \sim [A^o \vee B^o]]$$

$$\vdots \qquad\qquad \vdots \qquad\qquad \vdots$$

$$W^n = [M^a \vee \sim A^1] \qquad U^q = [M^b \vee \sim B^1] \qquad X^r = [M^c \vee \sim [A^o \vee B^o]]$$

$$\vdots$$

$$X^{r+n} = [M^d \vee \sim [A^1 \vee B^o]]$$

$$\vdots$$

$$X^{r+n+q} = [M^e \vee \sim [A^1 \vee B^1]]$$

∃-duplication; the two copies of R thus produced will be called the *left* and the *right* copies. This produces a development $X^r = [M^c \vee \sim [A^o \vee B^o]]$ of $M \vee \sim [A \vee B]$, where $\models [M^c \equiv M^o]$. Continue from here by imitating the operations in the verification I; whenever an operation used to construct I was applied to an occwfp of a descendant of a critical occwfp R of M^o, perform that operation within the descendant of the left copy of R in M^c. Whenever an operation in I was applied to an occwfp in a descendant of $\sim A^o$ (in W^o), perform the same operation to the corresponding occwfp in the descendant of the copy of $\sim A^o$ in X^r. One thus obtains a development X^{r+n} of $M \vee \sim [A \vee B]$. Now continue by imitating the operations in J, but use the occwfps in descendants of right copies of occwfps of M^o. One thus obtains a development $X^{r+n+q} = [M^e \vee \sim [A^1 \vee B^1]]$ of $M \vee \sim [A \vee B]$. Note that all this is possible without changing any of the variables which were introduced in I and J.

Lemma. Let C be any very accessible occwfp of M^o, and let C^a, C^b, and C^e be the descendants of C in M^a, M^b, and M^e, respectively. Then $\models [C^a \vee C^b] \supset C^e$ if C is positive in M^o, and $\models C^e \supset [C^a \wedge C^b]$ if C is negative in M^o.

The lemma is proved by induction on the construction of C.

Case a. C is atomic.
Then C^e, C^a, and C^b are all the same as C, so $\models [C^a \vee C^b] \supset C^e$ and $\models C^e \supset [C^a \wedge C^b]$.

Case b. C is critical in M^o.
Then C^e is $[C^a \wedge C^b]$, so $\models C^e \supset [C^a \wedge C^b]$. Since C is negative in M^o, this is

the desired conclusion.

Case c. C has the form $\sim$D.

Then C^e is $\sim D^e$, C^a is $\sim D^a$, and C^b is $\sim D^b$. If C is positive in M^o, then D is negative in M^o, so $\models D^e \supset [D^a \wedge D^b]$ by inductive hypothesis, so $\models [C^a \vee C^b] \supset C^e$. Similarly, if C is negative in M^o, then $\models [D^a \vee D^b] \supset D^e$, so $\models C^e \supset [C^a \wedge C^b]$.

Case d. C has the form $[D \vee E]$.

Then C^i is $[D^i \vee E^i]$ for $i = a, b, e$. If C is positive in M^o, then D and E are too, so by inductive hypothesis $\models [D^a \vee D^b] \supset D^e$ and $\models [E^a \vee E^b] \supset E^e$, so $\models [C^a \vee C^b] \supset C^e$. If C is negative in M^o, then $\models D^e \supset [D^a \wedge D^b]$ and $\models E^e \supset [E^a \wedge E^b]$, so $\models C^e \supset [[D^a \wedge D^b] \vee [E^a \wedge E^b]]$, so $\models C^e \supset [[D^a \vee E^a] \wedge [D^b \vee E^b]]$, so $\models C^e \supset [C^a \wedge C^b]$.

This completes the proof of the lemma, since every very accessible occwfp of the open wff M^o must fall under one of these cases.

We now apply the Lemma to see that $\models M^a \vee M^b \supset M^e$. Since $\models W^n$ and $\models U^q$ we know that $\models M^a \vee \sim A^1$ and $\models M^b \vee \sim B^1$, so $\models M^e \vee \sim [A^1 \vee B^1]$. Thus $\models X^{r+n+q}$, so $\square [M \vee \sim [A \vee B]]\{X^o,..., X^{r+n+q}\}$. This concludes the treatment of Case 5.

Case 6. P^i is $M \vee \sim \Pi_{o(o\alpha)} A_{o\alpha} \vee \sim A_{o\alpha} B_\alpha$ and P^k is $M \vee \sim \Pi_{o(o\alpha)} A_{o\alpha}$. By the argument in Case 4 we see that $\square M \vee \sim\sim [\sim \Pi_{o(o\alpha)} A_{o\alpha} \vee \sim A_{o\alpha} B_\alpha]$, so this wff has a verification $U^o, ..., U^n$, where U^o has the form $M^o \vee \sim\sim [\sim \Pi_{o(o\alpha)} A^o_{o\alpha} \vee \sim C_o]$, and $\sim C_o$ is an open form of $\sim A_{o\alpha} B_\alpha$. Since $A^o_{o\alpha}$ is a λ-normal form of $A_{o\alpha}$, and $\sim C_o$ is an open form of $\sim A^o_{o\alpha} B_{\alpha'}$ it is easy to see that
$\square P^k \{ M^o \vee \sim \Pi_{o(o\alpha)} A^o_{o\alpha'}$
$\quad M^o \vee \sim [\Pi_{o(o\alpha)} A^o_{o\alpha} \wedge \Pi_{o(o\alpha)} A^o_{o\alpha}]$
$\quad$ (which is $M^o \vee \sim\sim [\sim \Pi_{o(o\alpha)} A^o_{o\alpha} \vee \sim \Pi_{o(o\alpha)} A^o_{o\alpha}])$,
$\quad M^o \vee \sim\sim [\sim \Pi_{o(o\alpha)} A^o_{o\alpha} \vee \sim C_o]$,
$\quad U^1, ..., U^n\}$.

Case 7. P^i is $M \vee A_{o\alpha} x_\alpha$ and P^k is $M \vee \Pi_{o(o\alpha)} A_{o\alpha}$, and x_α is not free in M or $A_{o\alpha}$.

Clearly $\square P^k \{ \}$, since any open form of P^i is also an open form of P^k.

Since we have now considered all the rules of inference of $\mathcal{G}$, the proof of the metatheorem is complete.

I. Appendix: Rules of Inference

$\mathcal{H}$ denotes a (possibly empty) set of wffs$_0$, and $\mathcal{H}$,A denotes $\mathcal{H} \cup \{A\}$.

Hypothesis Rule (Hyp): Infer $\mathcal{H}$, A ⊢ A.

Deduction Rule (Deduct): From $\mathcal{H}$, A ⊢ B infer $\mathcal{H}$ ⊢ A ⊃ B.

Rule of Propositional Calculus (RuleP): From $\mathcal{H}_1$ ⊢ A_1, ... and $\mathcal{H}_n$ ⊢ A_n infer $\mathcal{H}_1 \cup \ldots \cup \mathcal{H}_n$ ⊢ B, provided that $[[A_1 \wedge \ldots \wedge A_n] \supset B]$ is tautologous. (For the case that n = 0, this rule takes the form: infer $\mathcal{H}$ ⊢ B, provided that B is tautologous.)

Negation-Quantifier Rule (RuleQ): From $\mathcal{H}$ ⊢ A infer $\mathcal{H}$ ⊢ B, where A is ∼∀xC, ∼∃xC, ∀x∼C, or ∃x∼C, and B is ∃x∼C, ∀x∼C, ∼∃xC, or ∼∀xC, respectively.

Rule of Indirect Proof (Indirect): From $\mathcal{H}$, ∼A ⊢ ⊥ infer $\mathcal{H}$ ⊢ A.

Rule of Cases (Cases): From $\mathcal{H}$ ⊢ A ∨ B and $\mathcal{H}$, A ⊢ C and $\mathcal{H}$, B ⊢ C infer $\mathcal{H}$ ⊢ C.

Universal Generalization (∀G): From $\mathcal{H}$ ⊢ A infer $\mathcal{H}$ ⊢ ∀xA, provided that x is not free in any member of $\mathcal{H}$.

Existential Generalization (∃G): Let A(x) be a wff and let t be a term which is free for x in A(x). (t may occur in A(x).) From $\mathcal{H}$ ⊢ A(t) infer $\mathcal{H}$ ⊢ ∃xA(x).

Universal Instantiation (∀I): From $\mathcal{H}$ ⊢ ∀xA(x) infer $\mathcal{H}$ ⊢ A(t), provided that t is a term free for x in A(x).

When one has inferred a wff of the form ∃xA, one often "existentially instantiates" to obtain the wff A, which asserts that x (a free variable of A) is an entity of the sort whose existence is asserted by ∃xA. We shall simply regard A as an additional hypothesis, which is eventually to be eliminated by Rule C below. (The name Rule C was introduced by Rosser [15], but our formulation of the rule differs from his.)

Rule C: From $\mathcal{H}$ ⊢ ∃xA and $\mathcal{H}$, A ⊢ B infer $\mathcal{H}$ ⊢ B, when x is not free in B or in any member of $\mathcal{H}$.

Rules of Alphabetic Change of Bound Variable ($\alpha\beta$) and λ-conversion (λ). See [6] or [1] for the formal statements of these rules.

Rule of Definition (Def): Eliminate or introduce a definition.

Bibliography

1. Peter B. Andrews, *Resolution in Type Theory*, Journal of Symbolic Logic 36 (1971), 414-432.

2. Peter B. Andrews, *Provability in Elementary Type Theory*, Zeitschrift fur Mathematische Logic und Grundlagen der Mathematik 20 (1974), 411-418.

3. Peter B. Andrews, "Transforming Matings into Natural Deduction Proofs," in *5th Conference on Automated Deduction, Les Arcs, France*, edited by W. Bibel and R. Kowalski, Lecture Notes in Computer Science 87, Springer-Verlag, 1980, 281-292.

4. Peter B. Andrews, *Theorem Proving via General Matings*, Journal of the Association for Computing Machinery 28 (1981), 193-214.

5. W. W. Bledsoe. A Maximal Method for Set Variables in Automatic Theorem-proving. Machine Intelligence 9, 1979, pp. 53-100.

6. Alonzo Church, *A Formulation of the Simple Theory of Types*, Journal of Symbolic Logic 5 (1940), 56-68.

7. Gérard P. Huet. A Mechanization of Type Theory. Proceedings of the Third International Joint Conference on Artificial Intelligence, IJCAI, 1973, pp. 139-146.

8. Gérard P. Huet, *A Unification Algorithm for Typed λ-Calculus*, Theoretical Computer Science 1 (1975), 27-57.

9. D. C. Jensen and T. Pietrzykowski, *Mechanizing ω-Order Type Theory Through Unification*, Theoretical Computer Science 3 (1976), 123-171.

10. Dale A. Miller, Eve Longini Cohen, Peter B. Andrews, "A Look at TPS," in *6th Conference on Automated Deduction, New York*, edited by Donald W. Loveland, Lecture Notes in Computer Science 138, Springer-Verlag, June 1982, 50-69.

11. Dale A. Miller. *Proofs in Higher-Order Logic*, Ph.D. Th., Carnegie-Mellon University, 1983. (to appear)

12. Tomasz Pietrzykowski, *A Complete Mechanization of Second-Order Type Theory*, Journal of the Association for Computing Machinery 20 (1973), 333-364.

13. T. Pietrzykowski and D. C. Jensen. A complete mechanization of (ω)-order type theory. Proceedings of the ACM Annual Conference, Volume I, 1972, pp. 82-92.

14. J. A. Robinson. Mechanizing Higher-Order Logic. In *Machine Intelligence 4*, Edinburgh University Press, 1969, pp. 151-170.

15. J. Barkley Rosser, *Logic for Mathematicians*, McGraw-Hill, 1953.

16. Moto-o-Takahashi, *A proof of cut-elimination theorem in simple type-theory*, Journal of the Mathematical Society of Japan 19 (1967), 399-410.

ABELIAN GROUP UNIFICATION ALGORITHMS FOR ELEMENTARY TERMS

D. Lankford, G. Butler, and B. Brady[1]

ABSTRACT. We describe a method of constructing Abelian group unification algorithms for elementary terms which uses any basis algorithms for the integer solutions of linear equations with integer coefficients. We illustrate the method with computer generated examples using two different basis algorithms.

Unification algorithms are fundamental to computational logic. An ordinary unification algorithm is implicit in the original development of the semi-decidability of first order logic by Herbrand [2] and is explicit in many of the earliest developments of computer oriented first order logic semi-decision procedures inspired by Herbrand's work, such as in Prawitz [17], Robinson [18], and Guard, et al. [1]. Like linear programming algorithms, there seems to be a reversal of theoretical and experimental complexity for ordinary unification algorithms. Patterson and Wegman [15] have developed a linear unification algorithm, but Martelli and Montanari [13] have presented experimental evidence that their n + log(n) unification algorithm is more efficient for current practical applications.

The notion of ordinary unification was generalized to unification in an equational theory E, called E-unification, by Plotkin [16] who showed that equational theories with a complete E-unification algorithm can be entirely built into Herbrand proof procedures so that no extraneous equality inferences are required. E-unification algorithms have important applications to term rewriting system generators, such as Knuth and Bendix [7] for ϕ-unification (ordinary unification), Huet [3] for ϕ-unification, Lankford and Ballantyne [9, 10] for commutative and commutative-associative unification, Stickel and Peterson [22] for commutative and commutative-associative unification, and others. Plotkin's E-unification first order proof procedures have been combined with term rewriting system methods by Lankford and Ballantyne [11],

---
1. Department of Mathematics, Louisiana Tech University, Ruston, Louisiana 71272
   This work was supported in part by NSF Grant MCS-8209143 and a Louisiana Tech University research grant.

© 1984 American Mathematical Society
0271-4132/84 $1.00 + $.25 per page

building on the work of Slagle [19] and Lankford [8]. The most widely applied
E-unification algorithms have been commutative-associative. The completeness
of C + A unification for elementary terms with constants was shown independently
by Stickel [21] and Livesey and Siekmann [12]. The general C + A unification
problem is open.

For an adequate summary of the necessary terminology and notation the
reader should consult Taylor [23]. When an equation t = u is a consequence
of a set E of equations we write $E \mid\!\!- t = u$. When an equation t = u holds in an
algebra A we write $A \models t = u$. When $A \models t = u$ for each t = u in E we say A
is a model of E. When $A \models t = u$ for each model A of E we write $E \models t = u$.
Birkhoff's Theorem states $E \mid\!\!- t = u$ iff $E \models t = u$.

E-unification is a natural generalization in which equality of terms,
t = u, is replaced by proof (or, equivalently, truth) of equality, $E \mid\!\!- t = u$
(or, equivalently, $E \models t = u$). A substitution θ is an E-unifier of terms
t and u in case $E \mid\!\!- t\theta = u\theta$. For any two substitutions θ and ψ we define
$E \mid\!\!- \theta = \psi$ iff for each variable symbol v, $E \mid\!\!- v\theta = v\psi$. For any substitution
θ and any set X of variable symbols, let the restriction of θ to X be denoted
by $\theta|X = \{t/v \in \theta \mid v \in X\}$.

A complete set C of E-unifiers of terms t and u is a set C of substitutions satisfying (1) any $\mu \in C$ is an E-unifier of t and u, and (2) if θ is
any E-unifier of t and u, then there is some $\mu \in C$ and a substitution λ such
that $E \mid\!\!- \theta = \mu\lambda|V$ where V is the set of all variable symbols which occur in
t or u. A complete E-unification algorithm is an algorithm which returns
complete (necessarily finite) sets of E-unifiers for any two E-unifiable terms
t and u, and halts otherwise with an indication that t and u are not E-
unifiable. An E-unification algorithm is called simple in case the condition
$E \mid\!\!- \theta = \mu\lambda|V$ in (2) above may be replaced by $E \mid\!\!- \theta = \mu\lambda$.

The equational theory of Abelian groups is denoted AG and is defined by
the equations $(x + y) + z = x + (y + z)$, $x + y = y + x$, $x + 0 = x$, and
$x + (-x) = 0$. Elementary AG terms are terms consisting entirely of variable
symbols, constants, and the Abelian group function symbols: +, -, and 0.
Elementary AG terms t are frequently denoted by their normal forms $\Sigma t_i^v v_i + \Sigma t_i^c c_i$ or their matrix normal forms $T^v V + T^c C$ where the t_i^v and t_i^c are integers,
the v_i are variable symbols, and the c_i are constants.

LEMMA 1 There is a complete AG-unification algorithm iff there is a
restricted complete AG-unification algorithm which requires the second term of
the input pair to be 0.

PROOF: $AG \mid\!\!- t\theta = y\theta$ iff $AG \mid\!\!- (t-u)\theta = 0$.

We say that a substitution θ is AG-elementary in case for each variable symbol v, $v\theta$ is an elementary AG term. Elementary AG substitutions $\theta = \{\Sigma a_{i1}w_i + \Sigma b_{i1}c_i/v_1, \ldots, \Sigma a_{ij}w_i + \Sigma b_{ij}c_i/v_j\}$ are frequently denoted by their normal matrix forms $AW + BC/V$.

LEMMA 2 If t is AG-elementary, and ψ is an AG unifier of t and 0, then there exist substitutions θ and λ such that θ is AG-elementary, θ is an AG-unifier of t and 0, and AG $|\!\!-\psi = \theta\lambda|V$.

PROOF: AG $|\!\!-\psi = A_1W_1 + BC + A_2T/V$
$= A_1W_1 + BC + A_2W_2(T/W_2)/V$,
where W_2 has no variable symbols in common with W_1,
$= (A_1W_1 + BC + A_2W_2/V)(T/W_2)$
$= ([A_1 \mid A_2][W_1^\tau \mid W_2^\tau] + BC/V)(T/W_2)$
where M^τ denotes the transpose of M,
and $[M_1 \mid M_2]$ denotes the juxtaposition of M_1 and M_2,
$= (AW + BC/V)(T/W_2)$
$= \theta\lambda|V.$

LEMMA 3 If t and θ are AG elementary, then AG $|\!\!- t\theta = 0$ iff $T^V A = 0$ and $T^V B + T^c = 0$.

PROOF: AG $|\!\!- t\theta = 0$ iff $T^V(AW + BC) + T^c C = 0$ iff $(T^V A)W + (T^V B + T^c)C = 0$ iff $T^V A = 0$ and $T^V B + T^c = 0$.

An integer basis of a linear integral equation $PX + [q] = 0$ is a pair of integer matrices R, S such that the solutions of the linear equation are $RY + S$ for arbitrary integer matrices Y (of course, Y must have the correct size).

LEMMA 4 A linear integral equation has an integer basis iff it has an integral solution, and there are algorithms which construct bases when they exist.

PROOF: (=>) Obvious. (<=) See, for example, Hurt and Waid [5], Knuth [6], and Niven and Zuckerman [14].

THEOREM There are complete AG-unification algorithms for elementary AG terms.

PROOF: By Lemma 1 it suffices to consider the AG-unification problem for t and 0. Let a basis algorithm be given. Input terms t are expressed in their matrix normal forms $T^V V + T^C C$. Apply the basis algorithm to the linear equations $T^V X + [t_i^c] = 0$, $i = 1,\ldots,j$. If any basis does not exist, return that t and 0 are not AG-unifiable. If all bases R_i, S_i exist, return the singleton $\{\mu\}$ where $\mu = [R_i Y_i] + SC/V$, the variable symbols of the Y_i are distinct, the columns of S are the S_i, and $[M_i]$ denotes the juxtaposition of $M_1,\ldots,M_j$. The substitutions μ are AG-unifiers because

$$\begin{aligned} AG \vdash t\mu &= T^V([R_i Y_i] + SC) + T^C C \\ &= [(T^V R_i) Y_i] + (T^V S + T^C) C \\ &= [0 Y_i] + 0 C \\ &= 0. \end{aligned}$$

Let θ be an AG unifier of t and 0. By Lemma 2 we may assume θ is AG elementary, so let $AW + BC/V$ be the matrix normal form of θ. By Lemma 3 the columns of A are solutions of $T^V X = 0$ and the jth column of B is a solution of $T^V X + [t_j^c] = 0$. Thus there exist I_i such that $A = [R_i I_i]$ and there exist J_i such that $B = [R_i J_i] + S$. Let λ be the substitution $\{I_i W + J_i C / Y_i\}$. Then

$$\begin{aligned} AG \vdash \mu\lambda &= ([R_i Y_i] + SC/V)\lambda \\ &= ([R_i (I_i W + J_i C)] + SC/V)\lambda \\ &= ([R_i I_i]W + ([R_i J_i] + S)C/V)\lambda \\ &= (AW + BC/V)\lambda, \text{ and clearly} \end{aligned}$$
$$AG \vdash \theta = \mu\lambda | V.$$

The algorithm described above is not the most practical. Let R be a basis of the homogeneous system $T^V X = 0$. Then $[R_i I_i] = RI$ for some I and $[R_i J_i] = RJ$ for some J and we may take $\mu = RY + SC/V$. The number of rows r of R equals the number of variables of V, but the number of columns c of R is not a priori restricted. For the basis algorithm of Hurt and Waid [5], $c \leq r$; while for Knuth [6], $c \leq r - 1$. It is a well-known fact of number theory (see, for example, Niven and Zuckerman [14]) that a linear integral equation has an integer solution iff the greatest common divisor (GCD) of the homogeneous coefficients divides the constant term. Given an algorithm which computes GCD's g and integers y_i such that $\Sigma t_i^V y_i = g$, find all z_i such that $g z_i = t_i^c$. If g fails to divide some t_i^c, then t and 0 are not AG-unifiable. Otherwise take $S_i = -z_i Y$, $S = [S_i]$, and $\mu = RY + SC/V$. This is obviously more efficient than computing the S_i from a given basis algorithm. When $c < r$, as in the basis algorithm of Knuth [6], we may take the variable symbols of Y from among the variable symbols of X. Consequently there is a simple AG-unification algorithm for elementary terms.

EXAMPLE 1   Consider $t = 2v_1 + 3v_2 - v_3 - 4v_4 - 5c$. A basis of the homogeneous equation $t = 0$ is

$$R = \begin{bmatrix} 1 & 0 & 0 \\ 0 & 1 & 0 \\ 2 & 3 & -4 \\ 0 & 0 & 1 \end{bmatrix}.$$

The GCD of the homogeneous coefficients is 1 and hence divides 5, so t and 0 are AG-unifiable. From the solution $(0, 0, -1, 0)$ of $2w + 3x - y - 4z = 1$ we get

$$S = 5 \begin{bmatrix} 0 \\ 0 \\ -1 \\ 0 \end{bmatrix} = \begin{bmatrix} 0 \\ 0 \\ -5 \\ 0 \end{bmatrix}$$

and

$$\mu = RY + SC/V$$
$$= \{y_1/v_1,\ y_2/v_2,\ 2y_1 + 3y_2 - 4y_3 - 5c/v_3,\ y_3/v_4\}.$$

EXAMPLE 2   Larger examples were tried by randomly selecting the last four digits of telephone numbers, such as $t = 3827v_1 - 2223v_2 + 1934v_3 - 3400v_4 + 8418v_5 - 6646v_6 + 7833v_7 - 9433v_8 + 4584v_9 - 4462v_{10}$. With larger examples we were naturally concerned about practicality, so we compared the basis algorithms of Hurt and Waid [5] and Knuth [6] (Niven and Zuckerman's is similar to Knuth's). To conserve space only the most general unifiers are given.

For Hurt and Waid, $\mu = \{y_1/v_1,\ y_2/v_2,\ 2{,}954{,}444\ y_1 - 1{,}716{,}156\ y_2 + 1{,}493{,}049\ y_3 - 2{,}624{,}800\ y_4 + 6{,}498{,}696\ y_5 - 5{,}130{,}712\ y_6, + 6{,}047{,}076\ y_7 - 7{,}289{,}996\ y_8 + 3{,}538{,}848\ y_9 - 3{,}444{,}664\ y_{10}/v_3,\ y_4/v_4,\ 122{,}464\ y_1 - 71{,}136\ y_2 + 61{,}888\ y_3 - 108{,}800\ y_4 + 269{,}377\ y_5 - 212{,}672\ y_6 + 250{,}656\ y_7 - 301{,}856\ y_8 + 146{,}688\ y_9 - 142{,}784\ y_{10}/v_5,\ y_6/v_6,\ -861{,}075\ y_1 + 500{,}175\ y_2 - 435{,}150\ y_3 + 765{,}000\ y_4 - 1{,}894{,}050\ y_5 + 1{,}495{,}350\ y_6 - 1{,}762{,}424\ y_7 + 2{,}124{,}675\ y_8 - 1{,}031{,}400\ y_9 + 1{,}003{,}950\ y_{10}/v_7,\ y_8/v_8,\ y_9/v_9,\ y_{10}/v_{10}\}$.

For Knuth, $\mu = \{y_1/v_1,\ y_2/v_2,\ 475\ y_1 - 47\ y_2 - 251\ y_3 - 771\ y_4 - 380\ y_5 - 285\ y_6 + 1200\ y_7 + 130\ y_8 + 74\ y_9/v_3,\ y_4/v_4,\ 22\ y_1 - 7\ y_2 - 11\ y_3 - 35\ y_4 - 13\ y_5 - 16\ y_6 + 58\ y_7 + 7\ y_8 + 4\ y_9/v_5,\ y_6/v_6,\ -139\ y_1 + 17\ y_2 + 75\ y_3 + 226\ y_4 + 109\ y_5 + 86\ y_6 - 335\ y_7 - 39\ y_8 - 22\ y_9/v_7,\ 2\ y_1 - 2\ y_2 + y_3 - 2\ y_4 + 3\ y_5 - 2\ y_6 + y_7 + y_8/v_8,\ y_8/v_9,\ y_9/v_{10}\}$.

We have described a method of constructing Abelian group unification algorithms for elementary terms which uses any basis algorithm for the integer solutions of linear equations with integer coefficients. The existence of a general Abelian group unification algorithm for arbitrary terms is open. Computational experience suggests that the basis algorithm of Knuth [6] is among the more efficient. The methods we have developed generalize to simultaneous unification where instead of a single equation there is a system of equations. Interestingly, a dual method (with constants in place of variables) solves the uniform word problem for finitely presented Abelian groups, cf. Smith [20].

BIBLIOGRAPHY

1. Guard, J., F. Oglesby, J. Bennett, and L. Settle "Semi-automated mathematics," J. Assoc. Comput. Mach. 18 (1969), 42-62.

2. Herbrand, J. "Recherches sur la théorie de la demonstration," translation in From Frege to Gödel, Harvard University Press, 1967, 525-581.

3. Huet G. "Confluent reductions: abstract properties and applications to term rewriting systems," J. Assoc. Comput. Mach. 27, 4 (Oct. 1980), 797-821.

4. Huet, G. and Oppen, D. "Equations and rewrite rules: a survey," in Formal Languages: Perspectives and Open Problems, R. Book, ed., Academic Press, 1980.

5. Hurt, M. and Waid, C. "A generalized inverse which gives all the integer solutions to a system of linear equations," SIAM J. Appl. Math. 19 (1970), 547-550.

6. Knuth, D. The Art of Computer Programming, Vol. 2, Addison-Wesley Pub. Co., 1969, 303-304.

7. Knuth, D. and Bendix, P. "Simple word problems in universal algebras," Computational Problems in Abstract Algebras, J. Leech, ed., Pergamon Press, 1969, 263-297.

8. Lankford, D. "Canonical inference," Depts. Math. and Comp. Sci., University of Texas, report ATP-32, Dec. 1975

9. Lankford, D. and Ballantyne, A. "Decision procedures for simple equational theories with commutative axioms: complete sets of commutative reductions, Depts. Math. and Comp. Sci., University of Texas, report ATP-35, Mar. 1977.

10. Lankford, D. and Ballantyne, A. "Decision procedures for simple equational theories with commutative-associative axioms: complete sets of commutative-associative reductions," Depts. Math. and Comp. Sci., University of Texas, report ATP-39, Aug. 1977.

11. Lankford, D. and Ballantyne, A. "The refutation completeness of blocked permutative narrowing and resolution," Proc. Fourth Workshop on Automated Deduction, W. Joyner, ed., Feb. 1979, 168-174.

12. Livesey, M. and Siekmann, J. "Unification of A+C-terms (bags) and A+C+I-terms (sets)," Interner Bericht Nr. 3/76, Institut für Informatik I, Universität Karlsruhe, West Germany.

13. Martelli, A. and Montanari, U. "An efficient unification algorithm," Instituto di Elaborazione delle Informazione, C.N.R., Via S. Maria, 46-56100 Pisa, Italy.

14. Niven, I. and Zuckerman, H. An Introduction to the Theory of Numbers, John Wiley & Sons, Inc., 1966, 103-109.

15. Patterson, M. and Wegman, M. "Linear Unifications," J. Comput. System Sci., 16 (1978), 158-167.

16. Plotkin, G. "Building-in equational theories," Machine Intelligence 7, John Wiley and Sons, 1972, 73-89.

17. Prawitz, D. "An improved proof procedure," Theoria 26 (1960), 102-139.

18. Robinson, J. "A machine-oriented logic based on the resolution principle," J. Assoc. Comput. Mach. 12 (1965), 23-41.

19. Slagle, J. "Automated theorem-proving for theories with simplifiers, commutativity, and associativity," J. Assoc. Comput. Mach. 21, 4 (Oct. 1974), 622-642.

20. Smith, D. "A basis algorithm for finitely generated Abelian groups" Math. Algorithms 1, 1 (Jan. 1966), 13-26.

21. Stickel, M. "A complete unification algorithm for associative-commutative functions," J. Assoc. Comput. Mach. 28 (1981), 423-434.

22. Stickel, M. and Peterson, G. "Complete sets of reductions for equational theories with complete unification algorithms," unpublished manuscript, Aug. 1977.

23. Taylor, W. "Equational logic," Houston J. Math., Survey 1979.

COMBINING SATISFIABILITY PROCEDURES BY EQUALITY-SHARING

Greg Nelson[1]

Abstract. Given procedures for determining the satisfiability of conjunctions of literals in two formal theories, the paper describes how to derive a satisfiability procedure for the combination of the two theories, assuming the two theories share no functions or predicates.

Let R be the set $\{+, -, <, \leq, =, \neq, 0, 1\}$ of the common arithmetic functions, predicates, and constants, excluding multiplication and division; and let E be the set $\{=, \neq, f, g, \ldots\}$, containing many "uninterpreted functions" taking various numbers of arguments. Consider the problem of determining the satisfiability over the reals of a conjunction of literals (that is, signed atomic formulas) containing only variables and the symbols in $R \cup E$. This problem is important in mechanical verification, where the formulas that arise often contain both arithmetic and non-arithmetic functions. A typical instance of the problem is to determine the satisfiability of

$$f(f(x) - f(y)) \neq f(z) \wedge x \leq y \wedge y + z \leq x \wedge z \geq 0. \qquad (1)$$

This conjunction is unsatisfiable, since the three inequalities imply that $x = y$ and $z = 0$, hence the distinction is equivalent to $f(0) \neq f(0)$.

In general, the *satisfiability problem* for a set of functions, predicates and constants is the problem of determining the satisfiability of a conjunction of literals containing, in addition to variables, only functions, predicates and constants in the set. Note that the satisfiability problem differs from the *decision problem*, in which quantified statements are allowed. The satisfiability problem for $R \cup E$ is unusual because it allows mixed expressions containing functions and predicates from two natural theories that intuitively have nothing to do with one another: the additive theory of the real numbers and the theory of equality with uninterpreted function

---

[1] Xerox PARC, 3333 Coyote Hill Rd, Palo Alto, CA 94304

symbols. The separate satisfiability problems for R and E were solved long ago—the first by Fourier, the second by Ackermann. But the combined problem was not considered until recently, when it became important in program verification. A solution to the satisfiability problem for $R \cup E$ was first given by Shostak [8]; another solution for a larger class of functions has been given by Suzuki and Jefferson [9].

This section describes a general method for factoring the satisfiability problem for a combined theory $S \cup T$ into the two satisfiability problems for S and T, given certain conditions on S and T. The method gives practical solutions to many satisfiability problems that are important in practice, including that for $R \cup E$. This material represents joint work with Derek Oppen, and was described previously in [6]. We used the method successfully in the theorem-prover of the Stanford Pascal Verifier.

Strictly speaking, the satisfiability problem is not defined for a set of functions and predicates, but for the *theory* of these functions and predicates under some axiomatization. A theory is determined by a set of function and predicate symbols, which will be called the *free functions and predicates* of the theory, together with a set of axioms constraining the interpretation of these functions and predicates. The axioms are written in the first-order predicate calculus with function symbols and equality; thus every theory contains the equality predicate. A constant is regarded as a special case of a function symbol. We now introduce, as examples, some theories that are important in program verification.

Let L be the theory of Lisp list structure with the three axioms

$$(\forall x, y: \text{car}(\text{cons}(x, y)) = x \land \text{cdr}(\text{cons}(x, y)) = y)$$
$$(\forall z: \text{atom}(z) \leftrightarrow (\forall x, y: z \neq \text{cons}(x, y)))$$
$$\text{atom}(\text{nil})$$

These axioms imply that $z = \text{cons}(\text{car}(z), \text{cdr}(z))$ whenever z is non-atomic, since in this case $z = \text{cons}(u, v)$ for some u and v (by the second axiom), and it follows from the first axiom that $u = \text{car}(z)$ and $v = \text{cdr}(z)$. Notice that these axioms do not imply $\text{cdr}(x) \neq x$; the axiomatization allows circular structure. The free functions and predicates of L are car, cdr, cons, atom, and nil.

The theory A of arrays has two free functions, select and store. The intended interpretation is that $\text{select}(a, i)$ is the value of the ith component of the array a, and $\text{store}(a, x, i)$ is the array that agrees with a in every component except the ith, where it has the value x. This interpretation is enforced by the axioms:

$$(\forall x, a, i: \text{select}(\text{store}(a, x, i), i) = x)$$
$$(\forall i, j, x, a: i \neq j \Rightarrow \text{select}(\text{store}(a, x, i), j) = \text{select}(a, j))$$

For brevity's sake, we write $a[i]$ instead of $\text{select}(a, i)$, and $a[i: x]$ instead of $\text{store}(a, x, i)$.

Let $\mathcal{R}$ be the theory of the real numbers under addition, whose free functions and predicates are $+$, $-$, $<$, and the numerals. Let $\mathcal{Z}$ be the theory of the reals under addition with the additional predicate int recognizing integers. Let $\mathcal{R}^\times$ and $\mathcal{Z}^\times$ be the extensions of $\mathcal{R}$ and $\mathcal{Z}$ that include multiplication and division.

It is convenient to define the theory $\mathcal{E}$ of equality with uninterpreted function symbols, whose theorems, like $x = y \Rightarrow f(x) = f(y)$, are valid because of the properties of equality. The function symbols of $\mathcal{E}$ denote arbitrary total functions. It is not necessary for $\mathcal{E}$ to have any axioms, since the semantics of equality are considered part of the underlying logic. We want all uninterpreted function symbols to be free functions of $\mathcal{E}$. In the context of this paper an "uninterpreted" function symbol is one that is not a free function of $\mathcal{Z}^\times$, $\mathcal{L}$, or $\mathcal{A}$, so we define $\mathcal{E}$ to be the theory with no axioms and whose free functions are all function symbols except $+$, $-$, $\times$, $<$, $/$, the numerals, int, cons, car, cdr, atom, nil, store, or select.

If $\mathcal{T}$ is a theory, then a term, literal, or formula will be called an $\mathcal{T}$-term, $\mathcal{T}$-literal, or $\mathcal{T}$-formula respectively if all function and predicate symbols appearing in it are free functions or predicates of $\mathcal{T}$. For example, $x = y$ and $x \leq y + 1$ are $\mathcal{R}$-literals but $x \leq \text{car}(y)$ is not.

The *satisfiability problem* for a theory $\mathcal{T}$ is the problem of determining the satisfiability in $\mathcal{T}$ of conjunctions of $\mathcal{T}$-literals. This makes the earlier notion of a satisfiability problem precise. The general quantifier-free decision problem for $\mathcal{T}$ can be reduced to the satisfiability problem, since a formula is satisfiable if and only if one of the disjuncts of its disjunctive normal form is satisfiable. A *satisfiability procedure* for $\mathcal{T}$ is an algorithm that solves the satisfiability problem for $\mathcal{T}$.

If $\mathcal{T}_1$ and $\mathcal{T}_2$ are two theories with no common free functions or predicates, their *combination* is the theory with all the axioms and free functions and predicates of both $\mathcal{T}_1$ and $\mathcal{T}_2$. Given satisfiability procedures for $\mathcal{T}_1$ and $\mathcal{T}_2$, this paper describes a method for constructing a satisfiability procedure for their combination.

The method works only for theories with no common free functions or predicates. For example, it cannot be used to combine satisfiability procedures for $\mathcal{R}^\times$ and $\mathcal{Z}$ into a satisfiability procedure for $\mathcal{Z}^\times$. This is as it should be, since the satisfiability problems for $\mathcal{R}^\times$ and $\mathcal{Z}$ are recursive, but that for $\mathcal{Z}^\times$ is not.

We begin with an example illustrating how the method determines that the conjunction (1) is unsatisfiable. Assume that we have satisfiability procedures for $\mathcal{R}$ and $\mathcal{E}$. We will use the same names "$\mathcal{R}$" and "$\mathcal{E}$" for the satisfiability procedures that we use for the theories themselves.

The first step is to construct two conjunctions $F_\mathcal{E}$ and $F_\mathcal{R}$ such that $F_\mathcal{E}$ is a conjunction of $\mathcal{E}$-literals, $F_\mathcal{R}$ is a conjunction of $\mathcal{R}$-literals, and $F_\mathcal{E} \wedge F_\mathcal{R}$ is satisfiable if and only if (1) is. This is achieved by introducing new variables to represent terms of the wrong type, and adding equalities defining these new variables. Thus, the first literal is made into an $\mathcal{E}$-literal by replacing the term $\mathrm{f}(x) - \mathrm{f}(y)$ with the new symbol g_1, and the equality $g_1 = \mathrm{f}(x) - \mathrm{f}(y)$ is added to define g_1. The new equality is then made into an $\mathcal{R}$-literal by replacing the $\mathcal{E}$-terms $\mathrm{f}(x)$ and $\mathrm{f}(y)$ by the variables g_2 and g_3, which are defined by $\mathcal{E}$-literals. The result is:

$F_\mathcal{R}$	$F_\mathcal{E}$
$x \leq y$	$\mathrm{f}(g_1) \neq \mathrm{f}(z)$
$y + z \leq x$	$\mathrm{f}(x) = g_2$
$z \geq 0$	$\mathrm{f}(y) = g_3$
$g_2 - g_3 = g_1$	

These two conjunctions are given to $\mathcal{R}$ and $\mathcal{E}$. Each of the conjunctions is satisfiable by itself, so there must be interaction between $\mathcal{R}$ and $\mathcal{E}$ to detect the unsatisfiability. The interaction takes a particular, restricted form: each satisfiability procedure is required to deduce and propagate to the other satisfiability procedure all equalities between variables implied by the conjunction it is considering. For example, if the conjunction $F_\mathcal{R}$ contained the literals $a \leq b$ and $b \leq a$, then $\mathcal{R}$ would be required to deduce and propagate the consequence $a = b$.

In the example, $F_\mathcal{E}$ does not imply any equalities between variables, but $F_\mathcal{R}$ implies $x = y$. (It also implies $z = 0$, but 0 is not a variable, so this equality is not propagated.) $\mathcal{R}$ therefore propagates $x = y$. Given this equality, $\mathcal{E}$ deduces and propagates $g_2 = g_3$. This enables $\mathcal{R}$ to deduce and propagate $z = g_1$, which enables $\mathcal{E}$ to detect the contradiction.

The method illustrated by this example works for any conjunctions $F_\mathcal{R}$ and $F_\mathcal{E}$. Obviously, if one of the conjunctions $F_\mathcal{R}$ or $F_\mathcal{E}$ becomes unsatisfiable as a result of equality propagation, the original conjunction must be unsatisfiable. It is a consequence of the results below that the converse holds as well: if the original conjunction $F_\mathcal{R} \wedge F_\mathcal{E}$ is unsatisfiable, then one of the conjunctions $F_\mathcal{R}$ and $F_\mathcal{E}$ will become unsatisfiable as a result of propagations of equalities between variables.

The method described so far must be extended to handle the combination of $\mathcal{E}$ and $\mathcal{Z}$. Suppose, for example, that the conjunctions are

$$F_Z \qquad\qquad F_\mathcal{E}$$
$$\text{int}(x) \qquad\qquad f(x) \neq f(a)$$
$$1 \leq x \qquad\qquad f(x) \neq f(b)$$
$$x \leq 2$$
$$a = 1$$
$$b = 2$$

Then $F_Z \wedge F_\mathcal{E}$ is unsatisfiable, but neither F_Z nor $F_\mathcal{E}$ imply any equalities between variables. The problem is that F_Z implies the *disjunction* $a = x \vee b = x$, and each case is inconsistent with $F_\mathcal{E}$. To handle this problem, our procedure will consider separate cases. We now consider how the procedure can determine that case-splitting is necessary.

A formula F is *non-convex* in a theory $\mathcal{T}$ if there exist $2n$ variables, $x_1, y_1, \ldots, x_n, y_n$, where $n > 1$ and F implies in $\mathcal{T}$ the disjunction

$$\bigvee_{1 \leq i \leq n} (x_i = y_i), \tag{2}$$

but for no proper subset S of $\{1, 2, \ldots, n\}$ does F imply the disjunction

$$\bigvee_{i \in S} (x_i = y_i).$$

In this case, (2) is a *split implied by F*. If no such disjunction exists, F is *convex*. Note that any unsatisfiable formula is convex.

That is, a formula is non-convex if it implies a proper disjunction of equalities between variables. If $\mathcal{T}$ is the theory of an infinite structure, and F is a non-convex $\mathcal{T}$-formula, then any split implied by F will contain only free variables of F. But if $\mathcal{T}$ is the theory of a finite structure, this is not the case. For example, consider the theory with no free functions or predicates, and the single axiom $(\forall x, y, z\colon x = y \vee y = z \vee x = z)$. In this theory, even the formula *true* is non-convex, and implies infinitely many splits! Therefore, we restrict our attention from now on to theories of infinite structures.

When one of the conjunctions in the equality-sharing procedure described above becomes non-convex, it is necessary to do a case-split; that is, to guess which of the equalities is true, saving the current state so that if the guess is wrong then the state can be restored and another guess tried. This is a straightforward non-deterministic procedure that can be implemented with an extra stack for backtracking. The following algorithm specifies the process precisely. The algorithm uses two stacks to hold the literals and one stack to hold the untried cases. The *size* of a stack is the number of entries in it.

Algorithm E (*Equality Sharing Procedure*). Given two stacks S_1 and S_2 such that the entries in S_1 are T_1-literals, the entries in S_2 are T_2-literals, where T_1 and T_2 are theories with no common free functions or predicates, this algorithm determines the satisfiability in the combination of T_1 and T_2 of the conjunction of all literals in S_1 and S_2. It is assumed that both T_1 and T_2 are theories of infinite structures. The algorithm uses another stack X whose entries are lists of the form $(n_1, n_2, u_1, v_1, \ldots, u_k, v_k)$, where n_1 and n_2 are integers, $k > 0$, and the u_i and v_i are variables.

E1. [Initialize X.] Let X be the empty stack.

E2. [Unsatisfiable?] For $i = 1, 2$, if the conjunction of the literals in S_i is unsatisfiable, then go to step E6.

E3. [Propagate equalities.] For $i = 1, 2$, if the conjunction of literals in S_i implies some equality $u = v$ between variables that is not implied by S_{3-i}, then push the equality $u = v$ on the stack S_{3-i} and go to step E2.

E4. [Case split necessary?] For $i = 1, 2$, if the conjunction of literals in S_i is non-convex, then let $u_1 = v_1 \vee \ldots \vee u_k = v_k$ be a split implied by S_i, n_1 the size of S_1, and n_2 the size of S_2. Push the entry $(n_1, n_2, u_2, v_2, \ldots, u_k, v_k)$ onto the stack X, push the equality $u_1 = v_1$ onto both S_1 and S_2, and go to step E2.

E5. [Satisfiable.] Halt with the answer *satisfiable*.

E6. [Try next case.] If the stack X is empty, then the algorithm halts with result *unsatisfiable*; else let the top entry on X be $(n_1, n_2, u_1, v_1, \ldots, u_k, v_k)$. For $i = 1, 2$, pop S_i as many times as necessary in order to reduce its size to n_i. Push $u_1 = v_1$ on both S_1 and S_2. If $k = 1$ then pop X, else replace the top entry of X with $(n_1, n_2, u_2, v_2, \ldots, u_k, v_k)$. Go to step E2. □

Algorithm E always halts, since there can be no more than $n-1$ non-redundant equalities among n variables. It is not difficult to prove that the algorithm is correct when it answers *unsatisfiable*. Before proving that it is correct when it answers *satisfiable*, we make some observations about how it is used in program verification.

To implement the algorithm efficiently, the satisfiability procedures for T_1 and T_2 must have several properties. They must be *incremental* and *resettable*; that is, it must be possible to add and remove literals from the conjunctions without restarting the procedures. The procedures must also detect the equalities and splits implied by the conjunctions. In theory, any satisfiability procedure can be used to find the equalities and splits, simply by testing each of the finitely many equalities and splits to see if it is implied. In practice, we construct our satisfiability procedures from algorithms that find the equalities and splits for free, or at little extra cost.

Here is good news: *if F is a conjunction of $\mathcal{E}$-literals, $\mathcal{R}$-literals, or $\mathcal{L}$-literals, then F is convex.* Therefore, case-splitting is never caused by these satisfiability procedures. It is easy to see that this is the case for $\mathcal{R}$: the solution set of a conjunction of linear inequalities is a convex set; the solution set of a disjunction of equalities is a finite union of hyperplanes; and a convex set cannot be contained in a finite union of hyperplanes unless it is contained in one of them.

Satisfiability procedures for $\mathcal{E}$ and $\mathcal{L}$, and proofs that conjunctions of $\mathcal{E}$-literals and $\mathcal{L}$-literals are convex, are described in [5] and [4]. As described, these satisfiability procedures take time $O(n^2)$ to process a sequence of instructions of the form "assume the literal ..." and "remove the last-assumed literal", where to assume a literal the procedure determines the satisfiability of the resulting conjunction and propagates any equalities that are implied, and n is the length of the sequence of instructions in characters. The time bounds can be reduced to $O(n \log^2 n)$ by using a different algorithm (see [2]).

A practical incremental and resettable satisfiability procedure for $\mathcal{R}$ can be based on the simplex algorithm, as described in [4]. In the worst case the simplex algorithm takes exponential time. Khachiyan [3] has discovered a polynomial-time linear programming algorithm, which could be used in $\mathcal{R}$, but it is unlikely to be as fast as the simplex algorithm in practice.

The satisfiability problem for the theory $\mathcal{A}$ of arrays is NP-complete. An incremental and resettable satisfiability procedure for $\mathcal{A}$ is described in [4].

It is lucky that $\mathcal{E}$, $\mathcal{R}$, and $\mathcal{L}$ do not require case-splitting, because most theories do. For example, in the theory $\mathcal{R}^\times$, $x \times y = 0 \land z = 0$ implies the split $x = z \lor y = z$. We have already seen that $\mathcal{Z}$ causes case-splitting. (Notice that, since it is only necessary to propagate equalities between variables, not between variables and constants, conjunctions such as $int(z) \land 1 \leq z \leq 100$ will not cause hundred-way splits—unless each of the numbers $1, \ldots, 100$ is constrained equal to a variable!) The theory of sets is non-convex; for example, $\{a, b, c\} \cap \{d, e, f\} \neq \{\}$ forces a nine-way case split. It is also true, unfortunately, that the theory $\mathcal{A}$ can in the worst case cause quadratically wide splits.

We now turn to the problem of proving that Algorithm E is correct if it returns *satisfiable*.

The *residue* of a formula F in the theory $\mathcal{T}$ is the strongest boolean combination of equalities between variables implied by F and the axioms of $\mathcal{T}$. We write $res_\mathcal{T}(F)$ to denote this residue. Here are some examples of formulas and their residues, taken in the obvious theories:

Formula	Residue
$x = \mathrm{f}(a) \wedge y = \mathrm{f}(b)$	$a = b \Rightarrow x = y$
$x \leq y \wedge y \leq x$	$x = y$
$x + y - a - b > 0$	$\neg(x = a \wedge y = b) \wedge \neg(x = b \wedge y = a)$
$x = v[i\mathpunct{:}e][j]$	$i = j \Rightarrow x = e$
$x = v[i\mathpunct{:}e][j] \wedge y = v[j]$	$i = j \wedge x = e \vee i \neq j \wedge x = y$

As another example, here are the residues of the formulas $F_{\mathcal{E}}$ and $F_{\mathcal{R}}$ from the first example in this section:

$$\mathrm{res}_{\mathcal{R}}(F_{\mathcal{R}}): \quad x = y \wedge (g_2 = g_3 \Rightarrow g_1 = z)$$
$$\mathrm{res}_{\mathcal{E}}(F_{\mathcal{E}}): \quad (x = y \Rightarrow g_2 = g_3) \wedge g_1 \neq z$$

Obviously these residues are inconsistent. The reason that Algorithm E works is that the residues are always inconsistent if the conjunctions of literals are. The remainder of the paper states this fact formally as Lemma 1, then proves the correctness of Algorithm E, assuming Lemma 1, and finally proves Lemma 1. In this order the technicalities come last.

Lemma 1. *Let $\mathcal{T}_1$ and $\mathcal{T}_2$ be theories of infinite structures. Let F_1 be a conjunction of $\mathcal{T}_1$-literals and F_2 a conjunction of $\mathcal{T}_2$-literals. Then the conjunction of F_1 and F_2 is satisfiable in the combination of $\mathcal{T}_1$ and $\mathcal{T}_2$ if and only if the conjunction of their residues, $\mathrm{res}_{\mathcal{T}_1}(F_1) \wedge \mathrm{res}_{\mathcal{T}_2}(F_2)$, is not the constant "false".*

Proof of correctness of Algorithm E. Assuming Lemma 1, we can prove the correctness of Algorithm E. Let F_1 be the conjunction of literals in S_1 and F_2 the conjunction of literals in S_2 at the moment that step E5 returns *satisfiable*. By Lemma 1, the step will be justified if we can prove that $\mathrm{res}_{\mathcal{T}_1}(F_1) \wedge \mathrm{res}_{\mathcal{T}_2}(F_2)$ is satisfiable. Let V be the set of all variables appearing in either F_1 or F_2, E the set of equalities $u = v$ such that u and v are in V and $F_1 \Rightarrow u = v$, (equivalently, since step E3 was passed, $F_2 \Rightarrow u = v$), and D the set of equalities $u = v$ such that u and v are in V but $F_1 \not\Rightarrow u = v$ (equivalently, since step E3 was passed, $F_2 \not\Rightarrow u = v$). Let Ψ be any interpretation that satisfies every equality in E and no equality in D. We claim Ψ satisfies $\mathrm{res}_{\mathcal{T}_1}(F_1)$ and $\mathrm{res}_{\mathcal{T}_2}(F_2)$. (Since $\mathrm{res}_{\mathcal{T}_1}(F_1)$ and $\mathrm{res}_{\mathcal{T}_2}(F_2)$ are just boolean combinations of equalities between variables, an interpretation for them can be specified by listing the equalities it satisfies.) Suppose Ψ does not satisfy, say, $\mathrm{res}_{\mathcal{T}_1}(F_1)$. Since, by the construction of Ψ, any interpretation that satisfies F_1 must satisfy every equality that Ψ satisfies, and by supposition, Ψ does not satisfy

F_1, it follows that any interpretation that satisfies F_1 must satisfy some equality that Ψ does not satisfy, that is, some equality in D. Thus F_1 implies the disjunction of the equalities in D. The proof is completed by considering three cases. If D is empty, then F_1 is unsatisfiable, which is impossible since step E2 was passed. If D contains a single equality, then F_1 implies the equality, so the equality would be in E, not D. If D contains more than one equality, then F_1 is non-convex, which is impossible, since step E4 was passed. Thus, Ψ does satisfy $\text{res}_{T_1}(F_1)$. Similarly, it satisfies $\text{res}_{T_2}(F_2)$, and step E5 is correct. □

It remains to prove Lemma 1. This lemma follows in a straightforward way from a well-known result of mathematical logic, Craig's Interpolation Lemma. In our proof, we will assume that T_1 and T_2 are finitely axiomatizable. By using the compactness theorem for first-order logic, the arguments below can be adapted to apply in the case that the theories have infinitely many axioms, but we omit this technicality.

Our first goal is to prove that residues exist. To this end, we define a *parameter* of a formula to be any nonlogical symbol that occurs free in the formula. For example, the parameters of $a = b \vee (\forall x\colon f(x) < c)$ are a, b, f, $<$, and c. A *simple* formula is one whose only parameters are variables. For example, $x \neq y \vee z = y$ and $(\forall x\colon x \neq y)$ are simple, but $x < y$ and $f(x) = y$ are not. Thus an unquantified simple formula is a propositional formula whose atomic formulas are equalities between variables. Lemma 2 characterizes quantified simple formulas.

Lemma 2. *Every quantified simple formula F is equivalent, for infinite interpretations, to some unquantified simple formula G. The formula G can be chosen so that its variables are all free variables of F.*

Proof. Suppose F is of the form $(\exists x\colon \Psi(x))$. Let Ψ_0 be the formula resulting from Ψ by first replacing any occurrences of $x = x$ by *true*, and replacing all equalities between x and any other variable by *false*. Then, if $v_1, \ldots, v_k$ are the parameters of Ψ, F is equivalent to $\Psi_0 \vee \Psi(v_1) \vee \ldots \vee \Psi(v_k)$, since, in any interpretation, x either equals one of the v's or differs from all of them. By repeatedly eliminating quantifiers in this manner, we eventually obtain an equivalent quantifier-free simple formula whose only variables are free variables of F. □

Lemma 3 (Craig's Interpolation Lemma). *If F and G are formulas and F implies G, then there exists a formula H such that F implies H and H implies G, and each parameter of H is a parameter of both F and G.*

Proof. See Craig [1] or Shoenfield [7]. □

Lemma 4. *If F is any formula, then there exists a simple formula $\mathrm{res}(F)$, the residue of F, which is the strongest simple formula that F implies; that is, if H is any simple formula implied by F, then $\mathrm{res}(F)$ implies H. The formula $\mathrm{res}(F)$ can be written so that all its variables are free variables of F. In this lemma, "implies" means "implies for all infinite interpretations".*

Proof. Let S be the set of all simple formulas implied by F. For each G in S, choose H_G so that $F \Rightarrow H_G$, $H_G \Rightarrow G$, the only parameters of H_G are parameters of both F and G, and H_G is unquantified. The existence of H_G follows from Lemmas 2 and 3. Each H_G is a propositional formula whose atomic formulas are equalities between parameters of F. There are only finitely many such equalities, so the infinite conjunction of all H_G for G in S is equivalent to some finite subconjunction H. Any simple formula G implied by F is implied by H_G, and so by H. The only parameters of H are free variables F. Thus H is the residue of F. □

We now define $\mathrm{res}_{\mathcal{T}}(F)$, where $\mathcal{T}$ is a finitely axiomatizable theory, to be $\mathrm{res}(A \wedge F)$, where A is the conjunction of the axioms of $\mathcal{T}$.

Lemma 5. *If A and B are formulas whose only common parameters are variables, then $\mathrm{res}(A \wedge B) \equiv \mathrm{res}(A) \wedge \mathrm{res}(B)$.*

Proof. It is easy to see that the left side of the equivalence implies the right side, since $A \wedge B$ implies $\mathrm{res}(A) \wedge \mathrm{res}(B)$, which is simple. To show the converse, again start from $A \wedge B \Rightarrow \mathrm{res}(A \wedge B)$, and rewrite it as $A \Rightarrow (B \Rightarrow \mathrm{res}(A \wedge B))$. Now Craig's Lemma shows that there is a formula H implied by A which implies $B \Rightarrow \mathrm{res}(A \wedge B)$, and whose only parameters are parameters of A and B. These must be variables, so H is simple. Therefore $\mathrm{res}(A) \Rightarrow (B \Rightarrow \mathrm{res}(A \wedge B))$. Writing this as $B \Rightarrow (\mathrm{res}(A) \Rightarrow \mathrm{res}(A \wedge B))$, and observing that the right hand side is simple, we deduce that $\mathrm{res}(B) \Rightarrow (\mathrm{res}(A) \Rightarrow \mathrm{res}(A \wedge B))$, or, equivalently, that $\mathrm{res}(B) \wedge \mathrm{res}(A) \Rightarrow \mathrm{res}(A \wedge B)$, which proves Lemma 5. □

Proof of Lemma 1. Lemma 1 now follows immediately from Lemma 5. Let A_1 and A_2 be the conjunctions of the axioms of the theories $\mathcal{T}_1$ and $\mathcal{T}_2$. Then

$\qquad\qquad F_1 \wedge F_2$ is satisfiable in the combination of $\mathcal{T}_1$ and $\mathcal{T}_2$
iff $\qquad F_1 \wedge F_2 \wedge A_1 \wedge A_2$ is satisfiable
iff $\qquad \mathrm{res}(F_1 \wedge F_2 \wedge A_1 \wedge A_2) \neq \mathit{false}$
iff $\qquad \mathrm{res}(F_1 \wedge A_1) \wedge \mathrm{res}(F_2 \wedge A_2) \neq \mathit{false}$
iff $\qquad \mathrm{res}_{\mathcal{T}_1}(F_1) \wedge \mathrm{res}_{\mathcal{T}_2}(F_2) \neq \mathit{false}$ □

Bibliography

[1] W. Craig: "Three Uses of the Herbrand-Gentzen Theorem in Relating Model Theory and Proof Theory", *Journal of Symbolic Logic*, vol. 22.

[2] P. J. Downey, R. Sethi, and R. E. Tarjan: "Variations on the common subexpression problem", *JACM* vol. 27 no. 4, October 1980, pp. 758–71.

[3] L. G. Khachiyan: "A polynomial algorithm in linear programming," *Soviet Math. Dokl.* vol. 20, 1979, pp. 191-194.

[4] G. Nelson: "Techniques for program verification", Report CSL-81-10, Xerox PARC, Palo Alto, CA, June 1981.

[5] G. Nelson and D. C. Oppen: "Fast Decision Algorithms based on Congruence Closure", *JACM* vol. 27 no. 2, April 1980, pp. 356–64.

[6] G. Nelson and D. C. Oppen: "Simplification by cooperating decision procedures", *TOPLAS* vol. 1 no. 2, October 1979, pp. 245–57.

[7] J. R. Shoenfield: *Mathematical Logic*, Addison-Wesley, Reading, MA 1967.

[8] R. Shostak: "An efficient decision procedure for arithmetic with function symbols", *JACM* vol. 26 no. 2 April 1979, pp. 351–60.

[9] N. Suzuki and D. Jefferson: "Verification decidability of presburger array programs," *JACM* vol. 27 no. 1 January 1980, pp. 191–205.

ON THE DECISION PROBLEM AND THE MECHANIZATION OF THEOREM-PROVING IN ELEMENTARY GEOMETRY[1]

Wu Wen-Tsün[2]

Abstract

The idea of proving theorems mechanically may be dated back to Leibniz in the 17th century and has been formulated in precise mathematical forms in this century through the school of Hilbert as well as his followers on mathematical logic. The problem consists in essence in replacing qualitative difficulties inherited in usual mathematical proofs by quantitative complexities of calculations on standardizing the proof procedures in an algorithmic manner. Such quantitative complexities of calculations, formerly far beyond the reach of human abilities, have become more and more trivial owing to the occurrence and rapid development of computers. In spite of vigorous efforts, however, researches in this direction give rise quite often to negative results in the form of undecidable mathematical theories. To cite a notable positive result, we may mention Tarski's method of proving theorems mechanically in elementary geometry and elementary algebra. The methods of Tarski as well as later ones are largely based on a generalization of Sturm theorem and are still too complicated to be feasible, even with the use of computers. The present paper, restricted to theorems with betweenness out of consideration and based on an

---

[1] Reprinted from *Scientia Sinica 21* (2), 1978.

[2] Institute of Systems-Sciences, Academia Sinica, Beijing 100080, People's Republic of China

entirely different principle, aims at giving a mechanical procedure which permits to prove quite non-trivial theorems in elementary geometry even by hands.

I. Formulation of the problem

A. Tarski in a classic paper [14] of 1948 has settled the decision problem of real closed field with one of its main aims to give mechanical proofs of theorems in elementary geometry. Alternative proofs of Tarski's result have later been given by Seidenberg, A. Robinson and P.J. Cohen, cf [12,9,2]. These authors have even suggested construction of certain decision machines to carry out such mechanical proofs. However, such a procedure seems to be far from being realized. In fact, only proofs of very trivial theorems in elementary geometry have actually been carried out on computers, cf. e.g. [6,7]. The purpose of the present paper is, leaving aside questions involving betweenness of points, to give an alternative solution of the decision problem of elementary geometry based on a principle entirely different from those employed by the authors above-mentioned. Our method permits to furnish mechanical proofs of quite difficult geometrical theorems which can be practiced even by hands, i.e., by means of papers and pencils only. The programming on a computer, based on such a method, though has not yet been done, will present no actual difficulties at all.

We shall restrict our considerations wholly to plane elementary geometry, though our method may be applied to the consideration of various other kinds of geometry. The first step of our method consists in the algebraic formalization of the geometrical problems involved. Points in the plane are to be defined as ordered pairs of numbers in a fixed field, say the field of rational numbers R. A dictionary is then set up turning geometrical relations into algebraic expressions which may be considered as either definitions or axioms. For example, for points $A_i = (x_i, y_i)$, distinct or not, we shall say:

A_1A_2 is parallel to A_3A_4

if $(x_1-x_2)(y_3-y_4) - (x_3-x_4)(y_1-y_2) = 0$,

A_1A_2 is orthogonal to A_3A_4

if $(x_1-x_2)(x_3-x_4) + (y_1-y_2)(y_3-y_4) = 0$,

the length-square of A_1A_2 is $r^2 = (x_1-x_2)^2 + (y_1-y_2)^2$, etc.

We may replace the basic field R by other fields, make correspondence the points to other kinds of number-sets, or modify the algebraic expressions in the axioms, e.g., instead of the length-square function r^2 given above, we consider the function $r^4 = (x_1-x_2)^4 + (y_1-y_2)^4$. We then go to other realms of geometry, non-euclidean geometry, real or complex projective geometry, finite geometry, etc. We shall however stick ourselves in what follows to plane elementary geometry only which has some representative character.

To illustrate our method of treatment, let us cite first a simple example. Consider the following statement:

(S_n) Let $A_0A_1A_2$ be a right-angled triangle with right angle at A_0. If x_1, x_2 denote the lengths of sides A_0A_1, A_0A_2 and x_3 is the length of the hypotenuse, then

$$x_1^n + x_2^n = x_3^n.$$

The problem is to decide whether the statement (S_n) is true or not and to give an algorithmic procedure of proving or disproving (S_n) which holds good for all statements alike in elementary geometry (with betweenness out of consideration).

To solve this problem, let us remark first that the points, etc., occurring in the statement have a generic character subjected to the conditions implied in the hypothesis of the statement. Thus, if we represent the points in question in coordinates with $A_0 = (x_0, v_0)$, $A_1 = (u_1, v_1)$, $A_2 = (u_2, v_2)$, the coordinates v_0, u_1, v_1, u_2, v_2 can be considered as indeterminates. On the other hand, the other coordinates and geometric entities x_0, x_1, x_2, x_3 are then algebraically dependent on these indeterminates, being restricted by following algebraic equations according to hypothesis of the statement (S_n):

$$f_0 \equiv (u_1-x_0)(u_2-x_0) + (v_1-v_0)(v_2-v_0) = 0,$$

$$f_1 \equiv x_1^2 - (u_1-x_0)^2 - (v_1-v_0)^2 = 0,$$

$$f_2 \equiv x_2^2 - (u_2-x_0)^2 - (v_2-v_0)^2 = 0,$$

$$f_3 \equiv x_3^2 - (u_1-u_2)^2 - (v_1-v_2)^2 = 0.$$

The conclusion in the statement (S_n) is equivalent to

$$g_n \equiv x_3^n - x_1^n - x_2^n = 0.$$

Let us now take once and for all the rational number field R as the base field. Let A^9 be the affine space on R with coordinates $(v_0, u_1, v_1, u_2, v_2, x_0, x_1, x_2, x_3)$ arranged in that definite order. Then the above equations $f_i = 0$ define an algebraic variety V of dimension 5, in the present case irreducible over R, with some generic point $(v_0, u_1, v_1, u_2, v_2, \bar{x}_0, \bar{x}_1, \bar{x}_2, \bar{x}_3)$ of which v_0, u_1, v_1, u_2, v_2 are indeterminates. The truth or untruth of the statement (S_n) amounts to $g_n \equiv 0$ or $g_n \not\equiv 0$ on V respectively.

It turns out that the general decision problem can be formulated in the following manner.

Problem. In a certain affine space A^n of dimension $n = r + d$ over R with coordinates $(u_1, \ldots, u_d, x_1, \ldots, x_r)$, consider an algebraic variety V with defining equations (u_i being independent indeterminates)

(I)
$$\begin{cases} f_1(u_1, \cdots, u_d, x_1) = 0, \\ f_2(u_1, \cdots, u_d, x_1, x_2) = 0, \\ \cdots \\ f_r(u_1, \cdots, u_d, x_1, x_2, \cdots, x_r) = 0. \end{cases}$$

The variety V may eventually split into irreducible components, all of (real) dimension $\leq d$. Those of dimension $= d$, with generic points of the form $(u_1, \ldots, u_d, \bar{x}_1, \ldots, \bar{x}_r)$ for which $\bar{x}_j$ are algebraic over the field $K = R(u_1, \ldots, u_d)$ will have a union V^* usually coincident with V. Let a polynomial $g(u_1, \ldots, u_d, x_1, \ldots, x_r)$ (or a set of such polynomials g_k) in $R[u_1, \ldots, u_d, x_1, \ldots, x_r]$ be given. It is to decide in an algorithmic manner whether

$$g \equiv 0$$

(or all $g_k \equiv 0$) on V^* or not.

In the above formulation the algebraic variety V, or preferably V^* reflects the hypothesis of the geometric statement considered. Either V or V^* will be called the *associated variety* of the statement in question. The variety V^* considered as one defined on the field $K = R(u_1, \ldots, u_d)$ is of dimension 0. The form of equations (I) shows that the algebraically dependent variables $x_1, \ldots, x_r$ are to be adjoined to K *successively* which reflects the geometrical fact that, starting from some generic points on certain generic lines, circles, etc., new points are to be *successively* adjoined in an algebraic manner by various geometric operations of joining points, drawing parallels, perpendiculars or circles, forming intersections of lines and circles, etc. These geometrical constructions give rise to algebraic equations involving x and u which can easily be turned into the form (I) be simple elimination procedure. In fact, the starting equations in x and u are rarely higher than 2. We shall call the variables u as the *parameters* and x as the *dependents*. It is also to be remarked that the condition for all components of V to be of dimension $\leq d$ over R reflects just the *determinate* character of the geometric statements to be considered, and the restriction to V^* reflects the depriving of degeneracies in our consideration. Both of these are, in reality, implicitly implied in the hypothesis of ordinary geometrical theorems. On the other hand, the equation $g \equiv 0$ on V^* (or set of $g_k \equiv 0$ on V^*) is the algebraic equivalent of the conclusion of the statement to be proved or disproved. We shall call g or set of g_k in what follows the *deciding polynomial(s)* of the geometrical statement in question.

Theoretically, the methods given by Hermann in [4] permit already to solve the above decision problem in an algorithmic manner. However, his methods are so complicated to give rise to astronomical expansions that even the simplest geometrical theorems can hardly be proved. On the contrary the decision procedure given below takes advantage of the particular character of the equations (I) and permits to prove mechanically quite non-trivial theorems even by hands, i.e. by means of pencil and paper only.

Our method of decision procedure is based on the following three theorems:

Theorem 1. *There is an algorithmic procedure permitting us to split the associated variety V^* of any determinative geometric statement defined by (I) into subvarieties V' irreducible over R each of which, considered as defined over the field $K = R(u_1, \ldots, u_d)$ has a representative basis (i.e. basis of the* *associated prime ideal) of the form*

$$(p_1, \cdots p_r),$$

having the following properties:

(T1)$_1$. Each p_i is a polynomial in $R[u_1, \ldots, u_d, x_1, \ldots, x_i]$ of some degree $m_i > 0$ in x_i.

(T1)$_2$. The coefficients of p_i, considered as a polynomial in x_i, are polynomials in $R[u_1, \ldots, u_d, x_1, \ldots, x_{i-1}]$ having no common factor and with degree in x_j less than m_j for $j = 1, \ldots, i-1$.

(T1)$_3$. The leading coefficient of p_i, considered as a polynomial in x_i, is a polynomial $\neq 0$ in $R[u_1, \ldots, u_d]$ free of all x.

(T1)$_4$. p_1, as a polynomial in x_1, is irreducible in the field $K = R(u_1, \ldots, u_d)$, and for each $i>1$, p_i, as a polynomial in x_i, is irreducible in the field obtained by adjoining $x_1, \ldots, x_{i-1}$ to K by the algebraic equations $p_1 = 0, \cdots, p_{i-1} = 0$.

It is clear that the polynomials $p_1, \ldots, p_r$ are uniquely determined by V' up to multipliers in R and will be said to form a *privileged basis* of V' (more exactly, of the prime ideal associated to V' over K), with respect to the given order $x_1, \ldots, x_r$ of the dependents. Remark that the notion is in reality due to Gröbner under the name of prime basis, cf. e.g. [3] and cf. also [8] for the intimately related concept of characteristic sets introduced by R. F. Ritt.

Theorem 2. *Let $(p_1, \ldots, p_r)$ be a privileged basis of any irreducible component V' of the associated variety V^*. There is an algorithmic procedure which permits us to determine, for any polynomial h in $R[u_1, \ldots, u_d, x_1, \ldots, x_r]$, an equation of the form*

$$Dh = \sum_{i_s=0}^{m_s-1} h_{i_1 \ldots i_r} x^{i_1} \cdots x^{i_r} + \sum_{i=1}^{r} A_i p_i,$$

verifying the following conditions:

(T2)$_1$. $D, h_{i_1 \ldots i_r}$ are all polynomials in $R[u_1, \ldots, u_d]$ and $D \neq 0$.

(T2)$_2$. A_i are all polynomials in $R[u_1, \ldots, u_d, x_1, \ldots, x_r]$.

The polynomials $h_{i_1, \ldots, i_r}$ which are uniquely determined up to multipliers in R by the algorithmic procedure, will be called the **remainder constituents** of the polynomial h with respect to the privileged basis $(p_1, \ldots, p_r)$ of V', or, by abuse of language, simply the remainder constituents of h with respect to V'.

Theorem 3. *For a geometrical statement with associated variety V^* and deciding polynomial g (or a set g_k of deciding polynomials) to be true, it is necessary and sufficient that for any irreducible component V' of V^*, all remainder constituents of g (or of all g_k) should be identically zero.*

II. Examples

Before giving proofs of these theorems in IV, we shall illustrate their use by some examples below.

Ex. 1. For the geometrical statement (S_n) about right-angled triangles as cited in the beginning of the present paper, we see readily that the associated variety V in parameters v_0, u_1, v_1, u_2, v_2 and dependents x_0, x_1, x_2, x_3, is already irreducible over R and possesses a privileged basis (p_1, p_2, p_3, p_4), where

$$p_1 = x_0^2 - (u_1+u_2)x_0 + u_1 u_2 + (v_1-v_0)(v_2-v_0),$$

$$p_2 = x_1^2 + (u_1-u_2)x_0 - u_1^2 + u_1 u_2 - (v_1-v_0)(v_1-v_2),$$

$$p_3 = x_2^2 - (u_1-u_2)x_0 - u_2^2 + u_1 u_2 + (v_2-v_0)(v_1-v_2),$$

$$p_4 = x_3^2 - (u_1-u_2)^2 - (v_1-v_2)^2.$$

It is readily verified that

$$g_2 = p_4 - p_2 - p_3,$$

while for $n > 2$, we have

$$g_n \neq 0 \mod (p_1, p_2, p_3, p_4).$$

It follows that (S_n) is true only for $n=2$ which corresponds to the Keu-Kou Theorem.

Ex. 2. Our decision procedure can also be applied to give mechanical proofs of trigonometric identities. Consider, e.g. the following statements:

(S) If $A_1 + A_2 + A_3 = 180°$, then

$$\sin 2A_1 + \sin 2A_2 + \sin 2A_3 = 4 \sin A_1 \sin A_2 \sin A_3.$$

(C) If $A_1 + A_2 + A_3 = 180°$, then

$$\cos 2A_1 + \cos 2A_2 + \cos 2A_3 + 4 \cos A_1 \cos A_2 \cos A_3 = 0.$$

To decide whether (S) or (C) is true, let us set

$$\sin A_i = s_i, \quad \cos A_i = c_i, \quad (i = 1,2),$$

$$\sin 2A_i = x_i, \quad \cos 2A_i = y_i, \quad (i = 1,2,3),$$

$$\sin A_3 = z_1, \quad \cos A_3 = z_2.$$

Take c_1 and c_2 as parameters and

$$s_1, s_2, z_1, z_2, x_1, x_2, x_3, y_1, y_2, y_3$$

as dependents (in this order), then the associated variety of (S) or (C) is already irreducible in R and possesses a privileged basis $(p_1, \ldots, p_{10})$ given by

$$p_1 = s_1^2 + c_1^2 - 1,$$

$$p_2 = s_2^2 + c_2^2 - 1,$$

$$p_3 = z_1 - c_2 s_1 - c_1 s_2,$$

$$p_4 = z_2 - s_1 s_2 + c_1 c_2,$$

$$p_5 = x_1 - 2c_1 s_1,$$

$$p_6 = x_2 - 2c_2 s_2,$$

$$p_7 = x_3 - 2z_1 z_2,$$

$$p_8 = y_1 - 2c_1^2 + 1,$$

$$p_9 = y_2 - 2c_2^2 + 1,$$

$$p_{10} = y_3 + 4c_1 c_2 s_1 s_2 - 4c_1^2 c_2^2 + 2c_1^2 + 2c_2^2 - 1.$$

The deciding polynomials of the statements (S) and (C) are given respectively by

$$g_s = x_1 + x_2 + x_3 - 4s_1 s_2 z_1,$$

$$g_c = y_1 + y_2 + y_3 + 4c_1 c_2 z_2.$$

We verify readily that

$$g_s = -2c_2 s_2 p_1 - 2c_1 s_1 p_2 - 2p_3(s_1 s_2 + c_1 c_2)$$

$$+ 2z_1 p_4 + p_5 + p_6 + p_7$$

$$\equiv 0 \mod(p_1, \ldots, p_{10}),$$

$$g_c = -1 + 4c_1 c_2 p_4 + p_8 + p_9 + p_{10}$$

$$\not\equiv 0 \mod(p_1, \ldots, p_{10}).$$

Hence (S) gives a true identity while (C) does not.

Ex. 3. Let us consider the Simson-Line Theorem which corresponds to the following statement:

(S_s). From a point A_4 on the circumscribed circle of a triangle $A_1 A_2 A_3$, perpendiculars are drawn to the sides of the triangle. Then the feet of the perpendiculars are in a line.

To prove this, let us take for simplicity the center of the circumscribed circle as the point (0,0), while the radius is r. Let the points A_i ($i = 1,2,3,4$) be (x_i, u_i) and the feet of perpendiculars A_j be (x_j, y_j), $j = 5,6,7$. Consider r, u_1, u_2, u_3, u_4 as parameters and

$$x_1, x_2, x_3, x_4, y_5, y_6, y_7, x_5, x_6, x_7$$

as the dependents in the order indicated.

A prime basis ($p_1, \ldots, p_{10}$) of the associated Simson variety which is irreducible is readily given as follows:

$$p_1 = x_1^2 + u_1^2 - r^2,$$

$$p_2 = x_2^2 + u_2^2 - r^2,$$

$$p_3 = x_3^2 + u_3^2 - r^2,$$

$$p_4 = x_4^2 + u_4^2 - r^2,$$

$$p_5 = 2r^2 y_5 - h_5,$$

$$p_6 = 2r^2 y_6 - h_6,$$

$$p_7 = 2r^2 y_7 - h_7,$$

$$p_8 = u_{23} x_5 - x_{23} y_5 + u_3 x_2 - u_2 x_3,$$

$$p_9 = u_{31} x_6 - x_{31} y_6 + u_1 x_3 - u_3 x_1,$$

$$p_{10} = u_{12} x_7 - x_{12} y_7 + u_2 x_1 - u_1 x_2.$$

In the above equations we have set for simplicity:

$$u_{ij} = u_i - u_j, \quad x_{ij} = x_i - x_j, \quad (i,j = 1,2,3),$$

$$h_5 = u_4 x_2 x_3 - (u_2 x_3 + u_3 x_2) x_4 + r^2(u_2 + u_3 + u_4) - u_2 u_3 u_4.$$

Similarly for h_6 and h_7.

The deciding polynomial is given by

$$g = x_5(y_6 - y_7) + x_6(y_7 - y_5) + x_7(y_5 - y_6).$$

Straightforward calculation shows again

$$g \equiv 0 \mod(p_1, \ldots, p_{10}),$$

which proves the truth of Simson statement (S_s).

Ex. 4. For a less trivial example, let us consider the Feuerbach theorem which corresponds to the following statement:

(S_F) The 9-point circle of a triangle is tangent to the four inscribed and the escribed circles of the triangle.

Let us take the three vertices of the triangle as $(2u_i, 2v_i)$, the center and radius of the 9-point circle as (x_1, y_1) and r_1, and the center and radius of either the inscribed or any of the escribed circles as (x_2, y_2) and r_2. Introduce also variables z_1, z_2, z_3 corresponding to lengths of the sides of the triangle. Then with $u_1, v_1, u_2, v_2, u_3, v_3$ as the parameters and $z_1, z_2, z_3, x_1, y_1, x_2, y_2, r_1, r_2$ as dependents in this order, the associated Feuerbach variety V_F splits into irreducible ones having privileged basis $(p_1, \ldots, p_9)$ given by:

$$p_1 = z_1^2 - u_{23}^2 - v_{23}^2,$$

$$p_2 = z_2^2 - u_{31}^2 - v_{31}^2,$$

$$p_3 = z_3^2 - u_{12}^2 - v_{12}^2,$$

$$p_4 = 4\Delta x_1 - \alpha_1,$$

$$p_5 = 4\Delta y_1 - \beta_1,$$

$$p_6 = 2\Delta x_2 - \alpha_2,$$

$$p_7 = 2\Delta y_2 - \beta_2,$$

$$p_8 = 4\Delta r_1 - z_1 z_2 z_3,$$

$$p_9 = 2\Delta r_2 + \gamma.$$

In the above formulas we have put for simplicity

$$u_{ij} = u_i - u_j, \quad v_{ij} = v_i - v_j, \quad (i,j = 1,2,3),$$

$$2\Delta = \begin{vmatrix} u_1, & v_1, & 1 \\ u_2, & v_2, & 1 \\ u_3, & v_3, & 1 \end{vmatrix},$$

$$\alpha_1 = v_{23} v_{31} v_{12} + u_1^2 v_{23} + u_2^2 v_{31} + u_3^2 v_{12}$$
$$\quad - 2(u_2 u_3 v_{23} + u_3 u_1 v_{31} + u_1 u_2 v_{12}),$$

$$\beta_1 = -u_{23} u_{31} u_{12} - v_1^2 u_{23} - v_2^2 u_{31} - v_3^2 u_{12}$$
$$\quad + 2(v_2 v_3 u_{23} + v_3 v_1 u_{31} + v_1 v_2 u_{12}),$$

$$\alpha_2 = -[v_{23} v_{31} v_{12} + v_1(u_2^2 - u_3^2) + v_2(u_3^2 - u_1^2) + v_3(u_1^2 - u_2^2)$$
$$\quad + \varepsilon_2 \varepsilon_3 v_{23} z_1 z_3 + \varepsilon_3 \varepsilon_1 v_{31} z_3 z_1 + \varepsilon_1 \varepsilon_2 v_{12} z_1 z_2],$$

$$\beta_2 = u_{23} u_{31} u_{12} + u_1(v_2^2 - v_3^2) + u_2(v_3^2 - v_1^2) + u_3(v_1^2 - v_2^2)$$
$$\quad + \varepsilon_2 \varepsilon_3 u_{23} z_2 z_3 + \varepsilon_3 \varepsilon_1 u_{31} z_3 z_1 + \varepsilon_1 \varepsilon_2 u_{12} z_1 z_2,$$

$$\gamma = \varepsilon_1(u_{31} u_{12} + v_{31} v_{12}) z_1 + \varepsilon_2(u_{12} u_{23} + v_{12} v_{23}) z_2$$
$$\quad + \varepsilon_3(u_{23} u_{31} + v_{23} v_{31}) z_3 + \varepsilon_1 \varepsilon_2 \varepsilon_3 z_1 z_2 z_3.$$

Remark that the choice of $\varepsilon_i = +1$ or -1 corresponds to the 4 inscribed or escribed circles and reflects the reducibility of the

Feuerbach variety.

Now the deciding polynomial of statement (S_F) is of the form

$$g_\eta = (\eta r_1 - r_2)^2 - (x_1 - x_2)^2 - (y_1 - y_2)^2,$$

in which $\eta = +1$ or -1. By a straightforward calculation, however lengthy and tedious, we verify that g_η will be a linear combination of $p_1, \ldots, p_9$ for $\eta = +\varepsilon_1\varepsilon_2\varepsilon_3$ but not so for $\eta = -\varepsilon_1\varepsilon_2\varepsilon_3$. Thus the Feuerbach statement is a true theorem and $\eta = +\varepsilon_1\varepsilon_2\varepsilon_3$ reflects the manner of contact between the 9-point circle and the respective in- or es-cribed circle. Remark that in our formulation we have left aside the question of betweenness and the manner of contact of circles is not of interest to us. In this way we can take $g = g_{+1} \cdot g_{-1}$ as the deciding polynomial if we like. We may also take the Feuerbach variety in the affine space of u_i, v_i, x_j, y_j only without the introduction of z_1, z_2, z_3 so that it is irreducible at the outset.

III. Some lemmas

To make some preparations we shall consider a field $R(u_1, \ldots, u_d, x_1, \ldots, x_r)$ in which $u_1, \ldots, u_d$ are transcendental while $x_1, \ldots, x_r$ are algebraic over the base field R. The algebraic extensions to $x_1, \ldots, x_r$ are defined successively by the following equations

(II)
$$\begin{cases} p_1(x_1) \equiv p_{10}x_1^{m_1} + p_{11}x_1^{m_1-1} + \cdots + p_{1m_1} = 0, \\ p_2(x_1,x_2) \equiv p_{20}x_2^{m_2} + p_{21}x_2^{m_2-1} + \cdots + p_{2m_2} = 0, \\ \quad \cdots \\ p_r(x_1,\cdots,x_r) \equiv p_{r0}x_r^{m_r} + p_{r1}x_r^{m_r-1} + \cdots + p_{rm_r} = 0. \end{cases}$$

It is assumed that for $1 \le i \le r$, the p_{ij}'s are polynomials in the ring $P_{i-1} = R[u_1, \ldots, u_d, x_1, \ldots, x_{i-1}]$, that p_{i0} are polynomials $\neq 0$ in $P_0 = R[u_1, \ldots, u_d]$, and that p_i considered as polynomials in x_i are irreducible over the field $K_{i-1} = R(u_1, \ldots, u_d, x_1, \ldots, x_{i-1})$ defined by the equations $p_1(x_1) = 0, \cdots, p_{i-1}(x_1, \ldots, x_{i-1}) = 0$. We put here

$$K_0 = K = R(u_1, \cdots, u_d).$$

Let Υ be the collection of sets of indices $I = (i_1, \ldots, i_r)$ with $0 \le i_j \le m_j - 1$. For such an I we shall write symbolically

$$x^I = x_1^{i_1} \cdots x_r^{i_r}.$$

Any polynomial of the form
with coefficients a_I in a certain ring or field F will be called then a *normalized* one in $F[x_1, \ldots, x_r]$.

Lemma 1. *There is an algorithmic procedure which permits to determine uniquely for any polynomial A in P_r, a set of integers $s_1, \ldots, s_r \ge 0$ and a set of polynomials A_I in P_0 for $I \in \Upsilon$ verifying the following conditions:*

$(L1)_1$. *Modulo some linear combination of p_i over P_r, we have*

$$p_{10}^{s_1} \cdots p_{r0}^{s_r} A \equiv \sum A_I x^I.$$

$(L1)_2$. *A_I are polynomials in P_0 with coefficients linear in those of A, considered as polynomial in P_r.*

$(L1)_3$. *$s_1, \ldots, s_r$ are the least integers ≥ 0 to make $(L1)_1$ and $(L1)_2$ possible.*

Proof. Considering both A and p_r as polynomials in x_r with coefficient in p_{r-1}, we get by division for some integer $s_r \ge 0$ taken to be least possible

$$p_{r,0}^{s_r} A = Q_r p_r + R_{r-1},$$

with Q_r, R_{r-1} polynomials in P_r for which the degree of R_{r-1} in x_r is $< m_r$. Considering now p_{r-1} and R_{r-1} as polynomials in x_{r-1} with coefficients in $R[u_1, \ldots, u_d, x_1, \ldots, x_{r-2}, x_r]$, we get by division for some integer $s_{r-1} \geq 0$ taken to be least possible

$$p_{r-1,0}^{s_{r-1}} R_{r-1} = Q_{r-1} p_{r-1} + R_{r-2},$$

with Q_{r-1}, R_{r-2} polynomials in P_r for which the degrees of R_{r-2} in x_r and x_{r-1} are $< m_r$ and $< m_{r-1}$ respectively. Proceeding in this manner, we get successively

$$p_{r-2,0}^{s_{r-2}} R_{r-2} = Q_{r-2} p_{r-2} + R_{r-3},$$

$$\cdots$$

$$p_{1,0}^{s_1} R_1 = Q_1 p_1 + R_0,$$

with R_0 as a polynomial in P_r, for which the degree in x_i is $< m_i$, $1 \leq i \leq r$. We may then write R_0 as $\sum A_I x^I$ and get the expression verifying all the conditions in the Lemma as required.

Lemma 2. *There is an algorithmic procedure which permits us to determine for any polynomial in some indeterminate y of the form*

$$A = A_0 y^m + A_1 y^{m-1} + \cdots + A_m$$

with each A_i in P_r and $A_0 \neq 0$ in K_r expressions of the form

$$HA = B + C_1 p_1 + \cdots + C_r p_r,$$

with

$$B = B_0 y^m + B_1 y^{m-1} + \cdots + B_m,$$

verifying the following conditions:

$(L2)_1$. All B_i are normalized polynomials in P_r and B_0 is a polynomial $\neq 0$ in P_0.

$(L2)_2$. C_i are all polynomials in $P_r[y]$ and H is one in P_r.

$(L2)_3$. $H \not\equiv 0 \mod (p_1, \ldots, p_r)$.

Any such polynomial B satisfying $(L2)_1$ will then be said to be a *normalized polynomial over P_r in the indeterminate y.*

Proof. By Lemma 1, we have for some integers $s_1, \ldots, s_r \geq 0$ an expression of the form

$$p_1^{s_1} \cdots p_r^{s_r} A = A'_0 y^m + A'_1 y^{m-1} + \cdots + A'_m \mod (p_1, \cdots, p_r)$$

with all A'_i normalized in P_r and $A'_0 \neq 0$ in K_r. Suppose that A'_0 is free of $x_{i+1}, \ldots, x_r$ but not so for x_i. As A'_0 has a degree $< m_i$ in x_i while p_i is irreducible in K_{i-1} and of degree m_i in x_i, we find by the usual division algorithm polynomials h, k in P_i and A'_{00} in P_{i-1} such that

$$hA'_0 + kp_i = A'_{00},$$

in which h is $\neq 0$ in K_i and $A'_{00} \neq 0$ in K_{i-1}. We have then an expression of the form

$$hp_1^{s_1} \cdots p_r^{s_r} A + kp_i y^m = A'_{00} y^m + hA'_1 y^{m-1} + \cdots + hA'_m$$

$$\mod (p_1, \cdots, p_r).$$

Applying Lemma 1 again to hA'_i and A'_{00}, we get then some expression

$$h''A = A''_0 y^m + A''_1 y^{m-1} + \cdots + A''_m \mod (p_1, \cdots, p_r),$$

for which all A''_i are normalized in P_r with $A''_0 \neq 0$ in K_r and not containing any x_j for $j \geq i$, and h'' is some polynomial in P_r which is $\neq 0$ in K_r. If A''_0 is free of all $x_1, \ldots, x_r$, we may take

$$B = A''_0 y^m + A''_1 y^{m-1} + \cdots + A''_m$$

as the polynomial B required or otherwise we proceed as before.

Lemma 3. *There is an algorithmic procedure which permits to factorize in K_r any polynomial in $P_r[y]$*

$$A = A_0 y^m + A_1 y^{m-1} + \cdots + A_m$$

with A_i in P_r and $A_0 \neq 0$ in K_r, $m \geq 2$. More precisely, it permits to find expression of the form

$$HA = A' + \sum C_i p_i,$$

with

$$A' = \bar{A}_1 \cdots \bar{A}_t,$$

verifying the following conditions:

$(L3)_1$. *Each $\bar{A}_j$ is a normalized polynomial in $P_r[y]$ and is irreducible in K_r.*

$(L3)_2$. *H is a polynomial in P_r which is $\neq 0$ in K_r.*

$(L3)_3$. *C_i are polynomials in $P_r[y]$.*

Proof. The method of Hermann in [4] permits us to give, in an algorithmic manner, a factorization of A into irreducible ones in K_r, so that after clearing of fractions we have an expression of the form

$$DA = B_1 \cdots B_t + \sum B'_i p_i,$$

with B_i, B'_i polynomials in $P_r[y]$, B_i irreducible in K_r and D a polynomial in P_r which is $\neq 0$ in K_r. Applying now Lemma 2 to each B_i, we get then the expression required. Consider also the reference [15], p. 130.

IV. Proofs of the theorems

We are now in a position to give proofs of Theorems 1-3 quite simply as follows:

Proof of Theorem 1. For the defining system of equations (I) of the associated variety V, let us consider $f_1(u_1, \ldots, u_d, x_1)$ as a polynomial in x_1 with coefficients in $R[u_1, \ldots, u_d]$ and factorize f_1 into ones irreducible in $K = R(u_1, \ldots, u_d)$ by applying the algorithm in Lemma 3, with x_1 as y and 0 as r there. Take any such irreducible factor as $p_1(x_1)$. Let K_1 be the field obtained from K by adjoining x_1 defined by the equation

$$p_1(x_1) = 0$$

Now f_2, considered as a polynomial in the indeterminate x_2, cannot be identical with 0 in the field K_1, for otherwise the variety V would be of dimension $> d$, contrary to the determinancy hypothesis of our geometric statement. Applying Lemma 3 to f_2 with x_2 as y and 1 as r there, we get a certain polynomial f'_2 in $P_2 = R[u_1, \ldots, u_d, x_1, x_2]$ with an expression of the form

$$h_2 f_2 = f'_2 + C_{21} p_1,$$

in which f'_2, as a polynomial in the indeterminate x_2, is a product of normalized factors irreducible and $\neq 0$ in the field K_1. Take any such factor as $p_2(x_1, x_2)$. Let K_2 be the field obtained from K_1 by adjoining x_2 defined by the equation

$$p_2(x_1, x_2) = 0.$$

Then f_3 is a polynomial in the indeterminate x_3 over P_2 which cannot be identical with 0. Applying Lemma 3, we get an expression of the form

$$h_3 f_3 = f'_3 + C_{31} p_1 + C_{32} p_2,$$

in which the polynomial f'_3 in x_3 is a product of normalized factors irreducible and $\neq 0$ in K_2. Take any factor as $p_3(x_1,x_2,x_3)$, adjoin x_3 to K_2 by the equation

$$p_3(x_1,x_2,x_3) = 0,$$

and proceed further as before. In this manner, we get finally system of subvarieties V' irreducible over R, each defined on K by systems of equations of the type

(III) $\begin{cases} p_1(x_1) = 0, \\ p_2(x_1, x_2) = 0, \\ \cdots \\ p_r(x_1, x_2, \cdots, x_r) = 0, \end{cases}$

verifying some obvious conditions as described in III.

It is easy to see that the collection of all such subvarieties exhausts the given variety V^*. In fact, consider an irreducible component in R of V^* with a generic point, say $(u_1, \ldots, u_d, \bar{x}_1, \ldots, \bar{x}_r)$, in which $u_1, \ldots, u_d$, are independent indeterminates while $\bar{x}_1, \ldots, \bar{x}_r$ depend algebraically on them. As $(u_1, \ldots, u_d, \bar{x}_1)$ should satisfy the system of equations (I), in particular the equation $f_1 = 0$, they should annul some one of the irreducible factors of f_1, say p_1 before. Now, $(u_1, \ldots, u_d, \bar{x}_1, \bar{x}_2)$ should satisfy the other equations in the system (I) as well as the equation $p_1(x_1) = 0$. It follows from the expression about f'_2 given above that they should also satisfy the equation $f'_2 = 0$ and hence should annul one of the irreducible factors of f'_2, say $p_2(x_1,x_2)$ above. Proceeding in the same manner we see that $(u_l, \ldots, u_d, \bar{x}_1, \ldots, \bar{x}_r)$ should satisfy a system of equations of the type (III) and hence is a generic point of an irreducible subvariety among the collection found above. This completes the proof of the theorem.

Proofs of Theorems 2 and 3.

Theorem 2 follows immediately from the algorithm in Lemma 1. The truth of Theorem 3 is quite clear from Theorems 1 and 2.

Final Remark. Our algorithm for the mechanization of theorem-proving in elementary geometry involves mainly such polynomial manipulations as arithmetic operations and simple eliminations, which were all originated and quite developed in 12-14c. Chinese mathematics, cf. e.g. the book of late Chien [1] for the explanations. In fact, the algebrization of geometrical problems and systematic method of their solutions by algebraic tools were some of the main achievements of Chinese mathematicians at that time, much earlier than the appearance of analytic geometry in 17c.

Added in Proof. (Dec. 1977).

The same principle has been applied to the mechanization of theorem-proving in elementary differential geometry with the aid of Ritt's theory of differential algebra, of which the details will be given later.

Bibliography

[1] Chien, Baozong. *The history of Chinese mathematics*, Science Press, 1964.

[2] Cohen, P.J. Decision procedures for real and p-adic fields, *Comm. Pure Appl. Math*, **22** (1969), 131-152.

[3] Gröbner, W. *Moderne algebraische Geometrie* (1949), Wien.

[4] Hermann, G. Die Frage der endlich vielen Schritte in der Theorie der Polynomideale, *Math. Ann.*, **95** (1926), 736-788.

[5] Hilbert, D. *Grundlagen der Geometrie.*

[6] McCharen, J.D. *et al.* Problems and experiments for and with automated theorem-proving programs, *IEEE Trans. on Computers*, **C-25** (1976), 773-782.

[7] Reiter, R. A semantically guided deductive system for automatic theorem proving, *IEEE Trans. on Computers*, **C-25** (1976), 328-334.

[8] Ritt, R.F. *Differential Algebra* (1950).

[9] Robinson, A. *Introduction to Model Theory and to the Metamathematics of Algebra* (1963), Amsterdam.

[10] Robinson, A. Algorithms in algebra, in model theory and algebra, *Lect. Notes in Math.*, No. 498 (1975), 14-40.

[11] Robinson, A. A decision method for elementary algebra and geometry-revisited, *Proc. of Tarski Symposium* (1974), 139-152.

[12] Seidenberg, A. A new decision method for elementary algebra, *Annals of Math.*, **60** (1954), 365-374.

[13] Tarski, A. What is elementary geometry? *The Axiomatic Method with Special Reference to Geometry and Physics* (1959), 16-29, Amsterdam.

[14] Tarski, A. and McKinsey, J.C.C. *A Decision Method for Elementary Algebra and Geometry*, 2nd ed., Berkeley and Los Angeles (1948-1951).

[15] Waerden, Van der. *Moderne Algebra*, Bd. 1 (1930).

[16] Waerden, Van der. *Einführung in die algebraischen Geometrie* (1945).

SOME RECENT ADVANCES IN MECHANICAL THEOREM-PROVING OF GEOMETRIES

Wu Wen-tsün[1]

The method described in [3] has been extended in elementary differential geometry, cf. [4] and [5]. It has been programmed and practiced on some microcomputers, viz. HP9835A and HP1000. The whole program, yet to be completed and improved, is named the <u>China-Prover</u>. It consists of three parts, to be called C-P1, C-P2, and C-P3, respectively. The program C-P1 includes subroutines of polynomial manipulations and a set of dictionaries, one for each special kind of geometry, permitting translation of geometric-entities into algebraic ones. The geometries in question may be either planar or solid, metric, affine or projective, euclidean or non-euclidean, etc. For the moment only two small dictionaries, not quite complete, have been done, viz. the planar metric and affine ones. We remark that it is not self-evident that the proof of theorems of, say, the non-euclidean hyperbolic geometry can be mechanized. That this is indeed possible requires itself a mathematical proof which is not a trivial one, cf. the forthcoming book [7] for all theoretical questions.

Program CP-2 is based on the method of R. F. Ritt as described in his books [1] and [2] which permits determination from a given set of polynomials (corresponding to the hypothesis of a geometric theorem) of an equivalent set in standard form as follows (cf. (I) in [3]):

$$f_j(x_1, x_2, \ldots, x_j), \quad j = j_1, j_2, \ldots j_r,$$

with $0 < j_1 < j_2 < \ldots < j_r$. If the degree of x_j in f_j is d_j, then the coefficient of $x_j^{d_j}$ in f_j is called the <u>initial</u> of f_j which is itself a

---

[1] Institute of Systems-Sciences, Academia Sinica

polynomial in $x_1, x_2, \ldots, x_{j-1}$. The two sets of polynomials are said to be equivalent here in that they possess the same zeros in any extended field under the subsidiary conditions that all the initials are kept non-zero. These conditions correspond to the non-degeneracy conditions under which the theorem would be true. In fact a geometric theorem can usually only be true in a generic sense which is almost always the case, cf. e.g. [6].

Program C-P3 is the proper proof itself of geometrical theorems. It consists of determining by successive reductions the remainder of the conclusion polynomial with respect to the hypothesis-polynomial set already in standard form by program C-P2. If the remainder vanished then the theorem is true under the non-degeneracy conditions furnished by the initials of the hypothesis polynomials and the running of the program is itself a proof of the theorem in question. If it is not so then the theorem is in doubt and further investigations, e.g., the reducibility of polynomials, should be made, cf. the work of S. C. Chou [8], (this volume).

For each theorem to be verified we just write down a short program stating the geometrical conditions of the theorem, after the specification of coordinates of the various points involved. Using the C-P1 these hypothesis and conclusion polynomials are then converted into algebraic entities in the form of matrices, C-P2 and C-P3 are then used in turn to decide whether the theorem is true or in doubt. The C-P2 is usually unnecessary in case of elementary geometries but is indispensable in case of differential geometries.

Using the China-Prover we have proved a number of famous theorems in ordinary elementary geometry including the celebrated Morley theorem: Among the 27 triangles formed by the intersections of neighboring trisectors of angles of a triangle 18 are equilateral. During the proof procedure a polynomial of 1960 terms in 12 variables occurs. Our proof shows also that the Morley theorem remains true in more general geometries. We have also proved some theorems of differential geometry, e.g., the affine analogue of the Bertrand curve-pair theorem and the Backlund theorem of surface pair. In forming conjectures and then verifying on the computer the China-Prover gives us an efficient tool in discovering new theorems in geometries. Several theorems have in fact been discovered in this way. As an illustrative example, we shall give below in some detail the discovery and proof of a Pascal-Conic Theorem which was first done on a HP9835A in 1980 and then repeated in 1982 on a HP1000 by means of the China-Prover.

SOME RECENT ADVANCES IN MECHANICAL THEOREM-PROVING OF GEOMETRIES

Pascal-Conic Theorem. Let A_i ($i = 1,2,\ldots,6$) be six points on the same conic (co-conic for short). Call each point of intersection $P = A_i A_j \wedge A_k A_l$ a Pascal point. Let s be the permutation (1 2 3 4 5 6). Then for the point $P_0 = A_1 A_3 \wedge A_2 A_5$ the six Pascal-points $P_i = s^i P_0$ ($i = 0,1,\ldots,5$) will also be co-conic.

Of course, a direct proof of the theorem in the usual manner can easily be found, which will not concern us. For the mechanical proof let us choose coordinates with:

Conic: $x_1 U^2 + x_2 UV + x_3 V^2 + x_4 U + x_5 V + 1 = 0$

(U, V = running coordinates)

$A_1 = (x_{11}, 0)$, $A_2 = (x_6, x_7)$, $A_3 = (0, x_9)$,
$A_4 = (x_{12}, 0)$, $A_5 = (x_{14}, x_{15})$, $A_6 = (0, x_8)$,
$P_0 = A_1 A_3 \wedge A_2 A_5 = (x_{22}, x_{23})$, $P_1 = A_2 A_4 \wedge A_3 A_6 = (0, x_{13})$,
$P_2 = A_3 A_5 \wedge A_4 A_1 = (x_{21}, 0)$, $P_3 = A_4 A_6 \wedge A_5 A_2 = (x_{16}, x_{17})$,
$P_4 = A_5 A_1 \wedge A_6 A_3 = (0, x_{18})$, $P_5 = A_6 A_2 \wedge A_1 A_4 = (x_{10}, 0)$,
$Q_1 = P_4 P_5 \wedge P_0 P_3 = (x_{19}, x_{20})$,
$Q_2 = P_0 P_1 \wedge P_2 P_5 = (x_{24}, 0)$,
$Q_3 = P_2 P_3 \wedge P_1 P_4 = (0, x_{25})$.

In assuming the well-known Pascal theorem it is clear it will be sufficient to prove that the hexagon $P_0 P_1 P_4 P_5 P_2 P_3$ is Pascalian, or that the points Q_1, Q_2, Q_3 are collinear. Using program C-P1 we get then the set of hypothesis-polynomials and the conclusion polynomial as follows.

$H_1 = x_1 x_6^2 + x_2 x_6 x_7 + x_3 x_7^2 + x_4 x_6 + x_5 x_7 + 1$,
$H_2 = x_3 x_8^2 + x_5 x_8 + 1$,
$H_3 = x_3 x_8 + x_3 x_9 + x_5$,
$H_4 = x_6 x_8 + x_7 x_{10} - x_8 x_{10}$,
$H_5 = x_1 x_{11}^2 + x_4 x_{11} + 1$,
$H_6 = x_1 x_{11} + x_1 x_{12} + x_4$,
$H_7 = -x_6 x_{13} - x_7 x_{12} + x_{12} x_{13}$,

$$H_8 = x_1 x_{14}^2 + x_2 x_{14} x_{15} + x_3 x_{15}^2 + x_4 x_{14} + x_5 x_{15} + 1,$$

$$H_9 = -x_6 x_{12} x_{15} + x_6 x_8 x_{12} - x_6 x_8 x_{16} - x_7 x_{12} x_{16}$$
$$+ x_7 x_{12} x_{14} - x_8 x_{12} x_{14} + x_8 x_{14} x_{16} + x_{12} x_{15} x_{16},$$

$$H_{10} = x_8 x_{12} - x_{12} x_{17} - x_8 x_{16},$$

$$H_{11} = x_{11} x_{15} - x_{11} x_{18} + x_{14} x_{18},$$

$$H_{12} = x_6 x_{10} x_{15} + x_6 x_{18} x_{19} - x_6 x_{10} x_{18} + x_7 x_{10} x_{19}$$
$$- x_7 x_{10} x_{14} - x_{14} x_{18} x_{19} + x_{10} x_{14} x_{18} - x_{10} x_{15} x_{19},$$

$$H_{13} = x_{18} x_{19} - x_{10} x_{18} + x_{10} x_{20},$$

$$H_{14} = x_9 x_{21} - x_9 x_{14} - x_{15} x_{21},$$

$$H_{15} = x_6 x_{11} x_{15} + x_6 x_9 x_{11} - x_6 x_9 x_{22} - x_7 x_{11} x_{22}$$
$$+ x_7 x_{11} x_{14} - x_9 x_{11} x_{14} + x_9 x_{14} x_{22} + x_{11} x_{15} x_{22},$$

$$H_{16} = x_9 x_{11} - x_{11} x_{23} - x_9 x_{22},$$

$$H_{17} = x_{13} x_{22} + x_{23} x_{24} - x_{13} x_{24},$$

$$H_{18} = x_{17} x_{21} - x_{21} x_{25} + x_{16} x_{25}.$$

The conclusion polynomial is

$$C = -x_{19} x_{25} - x_{20} x_{24} + x_{24} x_{25}.$$

The hypothesis-polynomial set is already in the standard form so that we can directly apply our program C-P3 without a preliminary use of C-P2. For each polynomial h let us define an <u>index set</u> $\text{Ind}(h) = [T,C,D]$ with T - number of terms in h, C = the largest subscript j for which x_j actually occurs in H and D = highest degree of x_C occurring in h. Thus the conclusion polynomial has the index set [3 25 1] and the successive hypothesis polynomials the index sets:

SOME RECENT ADVANCES IN MECHANICAL THEOREM-PROVING OF GEOMETRIES

[6 7 2],   [3 8 2],   [3 9 1],
[3 10 1],   [3 11 2],   [3 12 1],
[3 13 1],   [6 15 2],   [8 16 1],
[3 17 1],   [3 18 1],   [8 19 1],
[3 20 1],   [3 21 1],   [8 22 1],
[3 23 1],   [3 24 1],   [3 25 1].

The proof procedure of the theorem in using C-P3 may now be sketched in the form of the scheme below, showing the index sets of the successive polynomials appearing in the reductions:

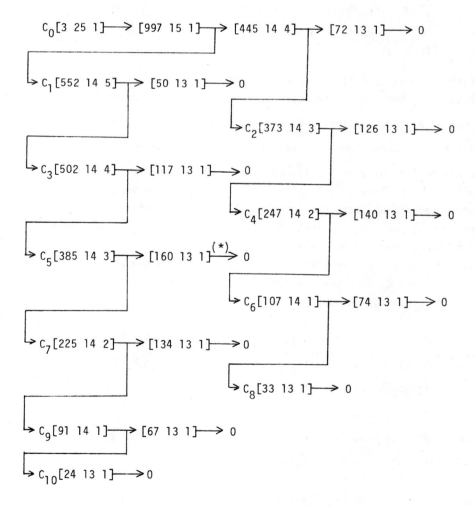

From the above scheme we see that all the remainders in the reductions (11 in number) are finally all 0. It follows that the theorem in question is true under the non-degeneracy conditions furnished by the initials I_i of the hypothesis set, viz.

$$x_3 \neq 0,\ x_7 - x_8 \neq 0,\ x_1 \neq 0,\ x_{12} - x_6 \neq 0,$$

$$-x_6 x_8 - x_7 x_{12} + x_8 x_{14} + x_{12} x_{15} \neq 0,\ x_{12} \neq 0,\ x_{14} - x_{11} \neq 0,$$

$$x_6 x_{18} + x_7 x_{10} - x_{14} x_{18} - x_{10} x_{15} \neq 0,\ x_{10} \neq 0,\ x_{15} - x_9 \neq 0,$$

$$-x_6 x_9 - x_7 x_{11} + x_9 x_{14} + x_{11} x_{15} \neq 0,\ x_{11} \neq 0,\ x_{23} - x_{13} \neq 0,$$

$$x_{16} - x_{21} \neq 0.$$

Each of these conditions has an evident geometric meaning. For example, the negation of the 5th of the above conditions, $I_5 = 0$, means that the line $A_4 A_6$ and $A_2 A_5$ are parallel. We may either verify by our program C-P3 that $I_5 = 0$ is not a consequence of the hypothesis set or we may adjoin $I_5 = 0$ to the original hypothesis set to get a new set and then verify whether the conclusion remains true or not. Clearly we may treat the various non-degeneracy conditions in turn in an algorithmic manner which is trivial for a computer. We may also use preliminarily the program C-P2 to get the initials already reduced before we start to use C-P3.

We add finally some remarks. First, each arrow in the above scheme consists of a series of successive reductions. Thus, the arrow marked (*) would be in details as follows.

$$[160\ 13\ 1] \longrightarrow [156\ 12\ 3] \longrightarrow [311\ 12\ 2] \longrightarrow [431\ 12\ 1]$$
$$\longrightarrow [537\ 11\ 6] \longrightarrow [1016\ 11\ 5] \longrightarrow [1346\ 11\ 4] \longrightarrow [976\ 11\ 3]$$
$$\longrightarrow [513\ 11\ 2] \longrightarrow [141\ 10\ 2] \longrightarrow [141\ 10\ 1] \longrightarrow [94\ 9\ 2]$$
$$\longrightarrow [105\ 9\ 1] \longrightarrow [105\ 8\ 5] \longrightarrow [178\ 8\ 4]$$
$$\longrightarrow [201\ 8\ 3] \longrightarrow [113\ 8\ 2] \longrightarrow [47\ 8\ 1] \longrightarrow [47\ 7\ 5]$$
$$\longrightarrow [81\ 7\ 4] \longrightarrow [64\ 7\ 3] \longrightarrow [21\ 7\ 2] \longrightarrow 0.$$

We see that a polynomial of 1346 terms in 11 variables occurs during this part of the reductions. Secondly we have taken advantage of some separation techniques in the programming when it comes to the reduction with respect to hypothesis H_7, for example. Without this there may occur polynomials of a too high number of terms to be admitted by even a big computer. Finally, the choice of the coordinate system and coordinates of points are somewhat arbitrary. With some different choices the reduction procedures would be different, which furnish us in fact alternative proofs of the same theorem.

The trials made here have proved that our method is a very efficient one. In fact up to now only a very small computer of memory 256K or 512K bytes have been used and the memory actually used is only a little more than 100K. It seems that the method can be applied with success at least in the two following directions. The first one is of academic character: Try to discover and prove new theorems in various kinds of geometries, e.g., non-euclidean geometry in which known interesting theorems are so rare. We are now doing this here in China, mainly in our Institute of Systems-Science of Academia Sinica. The second one is of pedagogic character: Teach students in high schools to apply the method in proving theorems of elementary geometry. For this purpose the work of S. C. Chou [8] is exceedingly valuable.

Bibliography

1. Ritt, R.F., Differential equations from the algebraic standpoint, Amer. Math. Soc. (1932).
2. Ritt, R. F., Differential algebra, Amer. Math. Soc. (1950).
3. Wu Wen-Tsün, On the decision problem and the mechanization of theorem proving in elementary geometry, Scientia Sinica 21 (1978), 150-172.
4. Wu Wen-Tsün, Mechanical theorem proving in elementary differential geometry, Scientia Sinica, Mathematics Supplement (I) (in Chinese) (1979), 94-102.
5. Wu Wen-Tsün, Mechanical theorem proving in elementary geometry and differential geometry, in Proc. 1980 Beijing Symposium on Differential Geometry and Differential Equations, Vol. 2, Science Press, (1982) 1073-1092.
6. Wu Wen-Tsün, Toward mechanization of geometry -- Some comments on Hilbert's "Grundlagen der Geometrie", Acta Math. Scientia 2 (1982), 125-128.
7. Wu Wen-Tsün, Basic principles of mechanical theorem proving in geometries (Part on elementary geometries), (in Chinese), Science Press, to appear.
8. Chou, S.C., Proving elementary geometry theorems using Wu's algorithm, this volume.

Proving Elementary Geometry Theorems Using Wu's Algorithm

SHANG-CHING CHOU[1]

ABSTRACT. In elementary geometry A. Tarski's classic algorithm [9] gives a decidable procedure for proving theorems. But the algorithm is not efficient enough to prove non-trivial theorems in geometry. In 1977, Wu Wen-tsün, a leading mathematician in China, discovered an algorithm which can prove certain kinds of quite non-trivial elementary geometry theorems [12, 13].

The first two sections of this article describe the implementation of a prover based on Wu's work. The prover is written in PASCAL. More than 130 theorems have been proved by the prover. Among those are: Simson's theorem, Pappus' theorem, Desargues' theorem, Euler's theorem, Pascal's theorem, the Butterfly theorem and the nine-point circle theorem. These two sections are self-contained and accessible even to high school students.

Sections 3 and 4 assume the reader is familiar with Wu's original work.

1. The Method.

1.1. The Scope of the Prover.
As we know in elementary analytic geometry, a geometry problem can be reduced to an equivalent algebra problem. This is the first step of Wu's method.

---

[1] Department of Mathematics, The University of Texas at Austin, Austin, Texas 78712.
The work reported here was supported in part by NSF Grants MCS 80-11417 and MCS 81-22039.

© 1984 American Mathematical Society
0271-4132/84 $1.00 + $.25 per page

In his classic book *The Foundations of Geometry* [4], Hilbert proposed five groups of axioms for plane (Euclidean) geometry:

Group 1: Axioms of Incidence.
Group 2: Axioms of Order.
Group 3: Axioms of Congruence.
Group 4: Axioms of Parallelism.
Group 5: Axioms of Continuity.

Wu's method can not prove all theorems in elementary geometry (in their equivalent algebraic form). Crudely speaking, Wu's method can only prove certain kinds of geometry theorems involving only incidence, congruence and parallelism. To be more precise, stated in algebraic terms, Wu's method can address those questions for which the hypotheses and the conclusion can be expressed by (algebraic) equations (not inequations or inequalities).

1.2. A Simple Example.

Before formulating the method, let us first look at a simple example to illustrate the method.

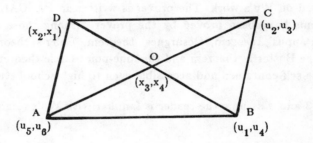

Figure 1

Example: Let ABCD be a parallelogram, O be the intersection of its diagonals AC and BD. Show $|AO| = |CO|$. (Figure 1).

First let us assign to each point a pair of variables (x, y), different variables for different points. For example, we can let $A = (u_5, u_6)$, $B = (u_1, u_4)$, $C = (u_2, u_3)$, $D = (x_2, x_1)$, $O = (x_3, x_4)$. By the hypotheses of the problem these coordinates satisfy the the following equations:

$$(2.1) \begin{array}{ll} f_1: (x_1 - u_3)(u_5 - u_1) - (x_2 - u_2)(u_6 - u_4) = 0 & \{\text{AB is parallel to CD}\} \\ f_2: (x_1 - u_6)(u_2 - u_1) - (x_2 - u_5)(u_3 - u_4) = 0 & \{\text{AD is parallel to BC}\} \\ f_3: (x_1 - u_4)(x_3 - u_1) - (x_2 - u_1)(x_4 - u_4) = 0 & \{\text{O is on BD}\}. \end{array}$$

f_4: $(u_3 - u_6)(x_3 - u_5) - (u_2 - u_5)(x_4 - u_6) = 0$ \{O is on AC\}

Note that the equation $f_1 = 0$, which roughly corresponds to AB $\parallel$ CD, is actually equivalent to the following condition: A = B or C = D or AB $\parallel$ CD. Note also that the condition AB $\parallel$ CD includes the case that AB and CD are on the same line. The conclusion $|AO| = |CO|$ becomes

(2.2) $\quad$ g: $(x_3 - u_5)^2 + (x_4 - u_6)^2 - (u_2 - x_3)^2 - (u_3 - x_4)^2 = 0$.

Now our geometry problem is reduced to whether we can infer (2.2) from (2.1).

In order to simplify the computation we can take advantage of the following two principles:

(1) Geometrical properties are invariant under translations.
(2) Geometrical properties are invariant under rotations.

By (1) we can choose A as the origin of the coordinates (i. e. let $u_5 = 0$, $u_6 = 0$), and by (2) we can choose AB as the x-axis (i.e. let $u_4 = 0$). So hypotheses (2.1) become:

(2.3)
f_1: $-u_1 x_1 + u_1 u_3 = 0$

f_2: $-u_3 x_2 + u_2 x_1 - u_1 x_1 = 0$

f_3: $(u_1 - x_2) x_4 + x_1 x_3 - u_1 x_1 = 0$

f_4: $-u_2 x_4 + u_3 x_3 = 0$

and conclusion (2.2) becomes:

(2.4) $\quad$ g: $2 u_3 x_4 + 2 u_2 x_3 - u_2^2 - u_3^2 = 0$.

It should be emphasized that using (1) and (2) to simplify the computation is vitally important for practical use of the prover. It will lead to reducing the time and memory significantly.[2] In order to infer (2.4) from (2.3), first we transform (2.3) into "triangular" form, i.e. for each hypothesis polynomial f_i we only introduce one new variable (i.e. x_i). Thus (2.3) is not in triangular

---

[2] Another principle often used in practice is:
(3) Geometry properties involving incidence and parallelism only are invariant under linear (affine) transformations (Rotation is a special kind of linear transformation which preserves the distance.) Thus, we can choose any two lines as x-axis and y-axis, provided they intersect in exactly one point.

form because f_3 introduces two new variables x_3 and x_4 at the same time. We can use a simple elimination procedure to eliminate x_4 in f_3: replace f_3 by $-u_2 f_3 - (u_1 - x_2) f_4$. So we have transformed (2.3) into triangular form:

(2.3)'
$$f_1: -u_1 x_1 + u_1 u_3 = 0$$
$$f_2: -u_3 x_2 + u_2 x_1 - u_1 x_1 = 0$$
$$f_3: (u_3 x_2 - u_2 x_1 - u_1 u_3) x_3 + u_1 u_2 x_1 = 0$$
$$f_4: -u_2 x_4 + u_3 x_3 = 0$$

The key step in Wu's method is to divide g successively by f_4, f_3, f_2 and f_1, considered successively as polynomials in x_4, x_3, x_2 and x_1.

Dividing g by f_4 (in x_4)[3] we get

(2.5) $\quad -u_2 g = 2 u_3 f_4 + R_4$

where $\quad R_4 = -2 u_3^2 x_3 - 2 u_2^2 x_3 + u_2 u_3^2 + u_2^3$.

Dividing R_4 by f_3 (in x_3), we get

(2.6) $\quad (u_3 x_2 - u_2 x_1 - u_1 u_3) R_4 = (-2 u_2^2 - 2 u_3^2) f_3 + R_3$

where $\quad R_3 = u_2 u_3^3 x_2 + u_2^3 u_3 x_2 - u_2^2 u_3^2 x_1 + 2 u_1 u_2 u_3^2 x_1 - u_2^4 x_1$
$\quad + 2 u_1 u_2^3 x_1 - u_1 u_2 u_3^3 - u_1 u_2^3 u_3$.

Dividing R_3 by f_2 (in x_2), we get

(2.7) $\quad -u_3 R_3 = (u_2 u_3^3 + u_2^3 u_3) f_2 + R_2$

where $\quad R_2 = -u_1 u_2 u_3^3 x_1 - u_1 u_2^3 u_3 x_1 + u_1 u_2 u_3^4 + u_1 u_2^3 u_3^2$.

Dividing R_2 by f_1 (in x_1) we get

(2.8) $\quad -u_1 R_2 = (-u_1 u_2 u_3^3 - u_1 u_2^3 u_3) f_1 + R$

where $\quad R \equiv 0$.

R is the final remainder; $R \equiv 0$ means that the theorem is true. To see this,

---

[3] For the division algorithm see see section 2.5.

multiplying g by c_1, c_2, c_3 and c_4 and using the identities (2.5) - (2.8) we get:

(2.9) $\quad c_1 c_2 c_3 c_4 g = A_1 f_1 + A_2 f_2 + A_3 f_3 + A_4 f_4 + R.$

where A_1, A_2, A_3, and A_4 are some polynomials, c_1, c_2, c_3 and c_4 are the leading coefficients of f_1, f_2, f_3 and f_4 respectively:

$$c_1 = -u_1$$
$$c_2 = -u_3$$
$$c_3 = u_3 x_2 - u_2 x_1 - u_1 u_3$$
$$c_4 = -u_2.$$

Now $R \equiv 0$, so from the hypotheses $f_1 = 0$, $f_2 = 0$, $f_3 = 0$ and $f_4 = 0$ we can infer $g = 0$ immediately if $c_i \neq 0$. $c_i \neq 0$ ($i = 1, ..., 4$) are called subsidiary conditions. Some of them are essential for the theorem to be true. Let us explain the geometrical meaning of all these conditions.

c_1 (i.e. $-u_1$) $= 0$ is the degenerate case in which AD and BC are on the same line. $c_2 = 0$ means that AB and CD are on the same line. In these two cases the conclusion does not follow. c_3 is the leading coefficient of f_3 (in (2.3)') which comes from the elimination procedure and is the determinant of the coefficient of x_3 and x_4 in f_3 and f_4 in (2.3). Thus $c_3 = 0$ means that the two diagonals AC and BD are parallel. This only happens when $u_1 = 0$ or $u_3 = 0$. $c_4 = 0$ means that C is on the y-axis; in this case the theorem is still true. In fact, c_4 comes by chance: if we swap f_3 and f_4 in (2.3) and eliminate x_4 from f_3, then c_3 is unchanged, but c_4 now is $u_1 - x_2$.[4] In order to infer $g = 0$ from (2.9), $c_4 \neq 0$ is necessary.

So the correct statement of the theorem should be: except for some special cases the theorem is true. It was only after realizing that the degenerate cases must be taken as exceptions, did Wu discover a mechanical procedure for proving geometry theorems.

1.3. The Formulation of the Problem.

Suppose the hypotheses of the geometry theorem in question are given in the form:

---

[4] By a general theorem [2], if (2.3)' is irreducible and $R \equiv 0$ with respect to (2.3)' then $R \equiv 0$ with respect to (2.3)'' obtained from swapped (2.3) by eliminating x_3 from f_3. It follows that $c_4 \neq 0$ is not necessary.

$$f_1(u_1, \ldots, u_d, x_1) = 0$$

$$f_2(u_1, \ldots, u_d, x_1, x_2) = 0$$

(3.1) ...

$$f_r(u_1, \ldots, u_d, x_1, \ldots, x_r) = 0.$$

and the conclusion to be proved is given as

(3.2) $\quad g(u_1, \ldots, u_d, x_1, \ldots, x_r) = 0$

where $f_1, \ldots, f_r$ and g are polynomials in u_j and x_j with integer coefficients. The geometry theorem to be proved is equivalent to saying that

(3.3) For all u_i in $\mathbf{R}$ (real field) and all x_j in $\mathbf{R}$, $f_1 = 0$ & ... & $f_r = 0$ implies $g = 0$.

1.4. The Main Result.

How do we infer (3.2) from (3.1)? Rewrite (3.1) in the form:

$$f_1: \quad c_{1,0} x_1^{m_1} + c_{1,1} x_1^{m_1 - 1} + \ldots + c_{1, m_1}$$

$$f_2: \quad c_{2,0} x_2^{m_2} + c_{2,1} x_2^{m_2 - 1} + \ldots + c_{2, m_2}$$

(4.1) ...

$$f_r: \quad c_{r,0} x_r^{m_r} + c_{r,1} x_r^{m_r - 1} + \ldots + c_{r, m_r}$$

where each coefficient $c_{i,j}$ is a polynomial in $u_1, \ldots, u_d$ and $x_1, \ldots, x_{i-1}$.

The key step in Wu's method is to divide g successively by $f_r, f_{r-1}, \ldots, f_1$ considered successively as polynomials in $x_r, x_{r-1}, \ldots, x_1$. Let the final remainder we get be R, then we have an identity of the form (for the proof see section 2.7):

(4.2) $\quad c_{1,0}^{s_1} c_{2,0}^{s_2} \ldots c_{r,0}^{s_r} g = A_1 f_1 + \ldots + A_r f_r + R$

where $A_1, \ldots, A_r$ are some polynomials and $s_1, \ldots, s_r$ are non-negative integers. If the final remainder R is identical to zero then we can infer $g = 0$ from $f_1 = 0, \ldots, f_r = 0$ immediately under subsidiary conditions $c_{i,0} \neq 0$. The following theorem is a simple consequence of the important identity (4.2), which ensures the soundness of Wu's method:

Theorem 1. $R \equiv 0$ is a sufficient condition for the theorem to be true under the subsidiary conditions $c_{i,0} \neq 0$.

Proof. Immediate from (4.2).

Remark 1. In (3.1), we implicitly consider $u_1, ..., u_d$ as indeterminates or independent variables. The notion that $u_1, ..., u_d$ are independent (with respect to a given geometry problem and a given choice of coordinates) is equivalent to our intuitive concept that $u_1, ..., u_d$ can be chosen arbitrarily in some intervals in **R** (real field) under given geometrical conditions or equations (3.1).[5] For example, in the example in section 2.2, u_1, u_2 and u_3 can be chosen as independent variables. For most geometry problems the choice of independents are obvious if the user has some intuition in geometry. Later, we will give several examples to explain how to chose coordinates for preparing the input to the prover.

Remark 2. Soundness. In our starting hypothesis equations (3.1), we implicitly assume two things: first, $u_1, ..., u_d$ are independent variables, while $x_1, ..., x_r$ are dependent on them; second, the equations (3.1) is already in triangular form, i.e. each equation f_i introduces only one new variable x_i. Though these assumptions are very important for proving geometry theorems successfully, the soundness is independent of them: if the final remainder is identical to zero, then theorem to be proved is true, no matter what coordinates are chosen or whether (3.1) is in triangular form.

Remark 3. Usually the equations corresponding to geometric hypotheses of a given problem are not in triangular form. It is a very important observation by Wu that for most geometry problems, starting from some fixed (generic) points, new points are to be successively adjoined <u>one by one in a constructive way</u> by various geometric constructions, such as joining points, drawing parallels, perpendiculars or circles, forming intersections of lines and circles. Thus new dependents are to be added at most two by two. So it is very easy to transform the equations corresponding to original hypotheses of the problem into triangular form (3.1), as we have shown in the example in Section 2. If we call such problems constructive, then most geometry theorems can be stated as constructive problems.

---

[5] The precise definition: $u_1, ..., u_d$ are independent with respect to (3.1) and x_i is dependent on them if and only if there exist intervals $I_1, ..., I_d$, in **R**, such that for all u_i in I_i, there exist reals $x_1, ..., x_{i-1}, x_{i+1}, ..., x_r$ and only finite x_i such that $u_1, ..., u_d, x_1, ..., x_r$ satisfy the equations (3.1).

Remark 4. Though the triangular form and the choice of independent variables do not affect soundness, for many theorems they are essential in order to succeed. Incorrect choice of independent variables usually leads to failure (i.e. the final remainder is not identical to zero).

Remark 5. Stated in the form of (3.3), the theorem usually is not true. As we have seen in the example in section 2.2, there are some degenerate cases for which the theorem may be not true. Wu's method can detect those degenerate cases.[6] We also cite the following theorem without proving it (cf. [12]).

Theorem 2. If $u_1, ..., u_d$ are independent in **R** (with respect to (4.1)) and (4.1) is irreducible[7] then $R \equiv 0$ is also a necessary condition for the theorem to be true.

Remark 1. Correct choice of independent variables and the triangular form in (4.1) is essential for the necessary condition.

If (4.1) is reducible and $R \not\equiv 0$ we have to decompose (4.1) into a set of irreducibles, then check g with each of them. The prover described here has not included this step yet. The main difficulty is to factor polynomials over algebraic extensions of the fields of rational functions. Most programs which have been implemented so far simply do successive division to see whether the final remainder is identical to zero; if it is, then the theorem has been proved. If it is not, and if we are sure that (4.1) is irreducible (e.g. $\deg(f_i, x_i) = 1$ for i = 1, ..., r), then it is not a theorem. It turns out that many theorems can be proved without factoring. We will discuss the factoring problem in section 4.

1.5. The Division Algorithm.

The basic stone of the subsequent algorithms is division of two polynomials (in [5], it is called pseudo-division of polynomials). In order to understand the latter parts of this paper more clearly, we emphasize the division algorithm.

Let g and h be polynomials in $P = Z[x_1, ..., x_k, y] = Z[x_1, ..., x_k][y]$, where Z is the ring of integers. We can always write g and h in the following form:

$$g = a_n y^n + ... + a_0$$

---

[6]See examples in section 3.3. As Wu has pointed out that theorems of elementary geometry are usually true only in the generic or non-degenerate case which are implicitly assumed as hypothesis but usually not clearly expressed in the statements of the theorems. (cf. [14])

[7]For the definition of irreducibility cf. [12].

$h = b_m y^m + \ldots + b_0$.

Suppose that h is not identical to zero, and that $m = \deg(h, y)$ and b_m is the leading coefficient of h. Then there exists a non-negative integer k and polynomials q and r in P with $\deg(r, y) < \deg(h, y)$ such that

$b_m^k g = q h + r$.

Note that we have the following convention: $\deg(0, y) = -\infty$.

Proof. If $\deg(g, y) < \deg(h, y)$ the result is clear on writing $g = 0 h + g$. Hence suppose $\deg(h, y) \leq \deg(g, y)$. Let n be $\deg(g, y)$ and a_n be the leading coefficient of g, then we can write

(5.1) $\qquad b_m g - a_n y^{n-m} h = g_1$.

Since the coefficients of y^n in $b_m g$ and $a_n y^{n-m} h$ are both $a_n b_m$, it is clear that $\deg(g_1, y) < \deg(g, y)$. So we can use induction on the $\deg(g, y)$ to obtain a k - 1 (non-negative integer) and q_1, r in P such that

(5.2) $\qquad b_m^{k-1} g_1 = q_1 h + r$.

Then by (5.1) and (5.2), $b_m^k g = b_m^{k-1} a_n y^{n-m} h + q_1 h + r = q h + r$

where $q = b_m^{k-1} a_n y^{n-m} + q_1$.

This proof actually provides an algorithm for calculating q and r. For the purpose of Wu's method we are mainly interested in calculating r, so let us denote it by $r(g, h, y)$.

1.6. Triangulation Procedure.

Generally, the polynomials corresponding to hypotheses are not in triangular form. One of Wu's key observations is that for most geometry problems new dependents are added at most two by two. So it is very easy to transform the hypothesis polynomials into triangular form. Our prover provides a procedure for transforming an arbitrary set of polynomials into triangular form. Suppose we have a set of hypothesis polynomials:

(6.1)
$$f_1(u_1, \ldots, u_d, x_1, \ldots, x_r) = 0$$
$$f_2(u_1, \ldots, u_d, x_1, \ldots, x_r) = 0$$
$$\ldots$$
$$f_r(u_1, \ldots, u_d, x_1, \ldots, x_r) = 0.$$

In order to transform it into triangular form, we first eliminate x_r from $r-1$ polynomials, obtaining

$$f_1'(u_1, ..., u_d, x_1, ..., x_{r-1}) = 0$$

$$f_2'(u_1, ..., u_d, x_1, ..., x_{r-1}) = 0$$

(6.2)
$$...$$

$$f_r'(u_1, ..., u_d, x_1, ..., x_{r-1}, x_r) = 0.$$

we then apply the same procedure to x_{r-1} to eliminate it from $r-2$ polynomials, and so on. The program eliminates x_r in the following way.

Case 1. No x_r appears in any f_i, stop. Give the user a warning: something was wrong in the choice of coordinates.

Case 2. x_r only appears in one f_i, take this f_i as f_r. We then have (6.2).

Case 3. One of the f, say f_r, has degree 1 in x_r: then let $f_i' = r(f_i, f_r, x_r)$, $i = 1, ..., r-1$. f_i' are free of x_r.

Case 4. Take two f_i which have the minimal positive degrees in x_r, say f_r, f_{r-1}. Suppose that $\deg(f_{r-1}, x_r) \leq \deg(f_r, x_r)$. Let $r_1 = r(f_r, f_{r-1}, x_r)$, then $\deg(r_1, x_r) < \deg(f_{r-1}, x_r)$. Usually $\deg(r_1, x_r) = \deg(f_{r-1}, x_r) - 1$. If $\deg(r_1, x_r) > 1$, then let $r_2 = r(f_{r-1}, r_1, x_r)$, and $r_3 = r(r_1, r_2, x_r)$, so on, until for some integer k, we have the following three possibilities:

1. $\deg(r_k, x_r) = 1$. Take r_k and r_{k-1} as new f_r and f_{r-1}, we return to case 3.
2. $r_k = 0$. This means that f_r, f_{r-1} have a common factor. So we can factor f_r and f_{r-1}. Since we do not intend to do factoring in this implementation, we simply stop the program. This situation has never happened in practice so far.
3. $\deg(r_{k-1}, x_r) > 1$, but $\deg(r_k, x_r) = 0$ (i.e. r_k is free of x_r). Take r_k and r_{k-1} as new f_r and f_{r-1}, then repeat the process beginning from case 2.

In this way we can transform (6.1) into triangular form (3.1). In practice, the new dependents are introduced at most two by two, and $\deg(f_i, x_i) \leq 2$. So no large polynomials will be generated by the triangulation procedure.

(6.1) implies (6.2), so (6.1) implies (3.1). The converse is true only under some additional conditions. This does not affect the soundness of the prover, because, if (3.1) implies the conclusion $g = 0$, so does (6.1).

This triangulation procedure is sufficient for practical use. For more satisfactory procedure see [7]. Again the factoring of polynomials over algebraic extensions of the fields of rational functions is necessary for that procedure.

1.7. Successive Division.

Now that we have the hypothesis polynomials in triangular form (4.1), we are ready to do successive divisions, which are the core of Wu's method:

Let $R_1 = r(g, f_r, x_r)$, then $c_{r,0}^{s_r} g = q_1 f_r + R_1$

Let $R_2 = r(R_1, f_{r-1}, x_{r-1})$, then $c_{r-1,0}^{s_{r-1}} R_1 = q_2 f_{r-1} + R_2$

(7.1) ...

Let $R_r = r(R_{r-1}, f_1, x_1)$, then $c_{1,0}^{s_1} R_{r-1} = q_r f_1 + R_r$.

R_r is the final remainder R. From the above algorithm and identities we have the following important theorem:

Theorem 3. There are some polynomials $A_1, ..., A_r$ in $Z[u_1,...,u_d, x_1,...,x_r]$ such that the following identity is true:

(7.2) $\quad c_{1,0}^{s_1} c_{2,0}^{s_2} \cdots c_{r,0}^{s_r} g = A_1 f_1 + ... + A_r f_r + R$

where $c_{1,0}, ..., c_{r,0}$ are the leading coefficients of $f_1, ..., f_r$ respectively (see (4.1)). $s_1, ..., s_r$ are non-negative integers, and R is the final remainder having the following properties:

(7.3) $\quad \deg(R, x_i) < \deg(f_i, x_i), \quad i = 1, ..., r.$

Proof. induction on r. r = 1. In the first identity of (7.1), let $A_1 = q_1$ and $R = R_1$. Then we have $c_{r,0}^{s_r} g = A_1 f_1 + R$, and by division algorithm $\deg(R, x_1) < \deg(f_1, x_1)$.

Suppose that $r > 1$ and that the theorem is true for r - 1. We want to prove it is also true for r. First we have (7.1). The last r - 1 lines in (7.1) are just the conditions for the case of r - 1 with g replaced by R_1. So by induction hypothesis we have

(7.4) $\quad c_{1,0}^{s_1} \cdots c_{r-1,0}^{s_{r-1}} R_1 = A_1 f_1 + ... + A_{r-1} f_{r-1} + R_r$

(7.5) $\quad \deg(R_r, x_i) < \deg(f_i, x_i), \quad i = 1, ..., r-1.$

(7.5) is also true for i = r, because $\deg(R_1, x_r) < \deg(f_r, x_r)$. From the first line of (7.1) we have $R_1 = c_{r,0}^{s_r} g - q_1 f_r$, and substituting it in (7.4) we have

$$(7.6) \quad c_{1,0}^{s_1} \ldots c_{r,0}^{s_r} g = A_1 f_1 + \ldots + A_{r-1} f_r + A_r f_r + R_r$$

where $A_r = c_{1,0}^{s_1} \ldots c_{r-1,0}^{s_{r-1}} q_1$.

2. The Prover.

2.1. The Main Structure of the Prover.

The prover is written in PASCAL. Recently, a version in MACSYMA has been written. The main structures are essentially the same, as shown in the following diagram:

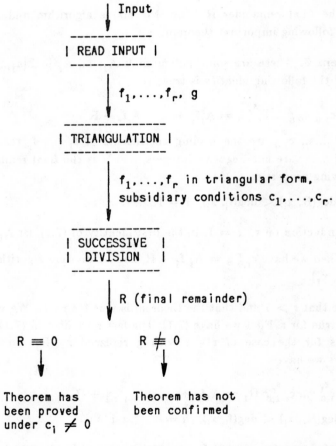

The input, which represents the theorem to be proved, is to be prepared by the user. First, the program reads input commands and transforms them into corresponding polynomials. After generating the polynomials for hypotheses and conclusion, the program transforms the polynomials for hypotheses into triangular form. Then the program does successive division, the crucial step of the prover.

The prover provides a set of input commands which can express most of the common geometric conditions. For most geometry problems these commands are sufficient. So once the choice of the coordinates has been made, it is a simple job to translate each geometric condition to a corresponding input command.

2.2. The Input Commands.

The formats of the input representing the theorem to be proved are essentially the same for both PASCAL and MACSYMA versions. Here we explain the PASCAL input.

The input is put in some file. It consists of a set of commands, one line for each command. The first line is an integer r which is the number of dependents or equivalently the number of hypothesis polynomials. Next should follow r lines corresponding to r hypotheses. Following these lines should be a line corresponding to the conclusion. The last line is symbol E which ends the input.

Each command corresponding to a geometry statement begins with a letter (corresponding to a geometric condition) followed by integers representing coordinates of points. 0 denotes the coordinate zero. The positive integer i denotes the variable x_i. For example, 0 1 represents the point $(0, x_1)$. Below we list the complete set of commands and illustrate each of them.

Notation Convention: If the coordinates of point A and point B are (x_1, y_1) and (x_2, y_2) respectively, then we denote the slope of AB, $(y_1 - y_2)/(x_1 - x_2)$, by S(AB). |AB| denotes the length of segment AB. AB sometimes also denotes the vector from A to B. X(AB) and Y(AB) denote the x-component and the y-component of vector AB respectively, which are $x_2 - x_1$ and $y_2 - y_1$ in this case. So we have: $|AB|^2 = X(AB)^2 + Y(AB)^2$.

Command A (followed by 12 numbers): the tangents of two angles are equal.

Example:

A 1 2 3 4 5 6 7 8 9 10 11 12

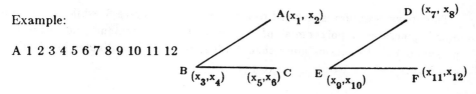

means that $\tan(\angle ABC) = \tan(\angle DEF)$. The corresponding (polynomial) equation is equivalent to

$$\frac{S(BA) - S(BC)}{1 + S(BA)\,S(BC)} = \frac{S(ED) - S(EF)}{1 + S(ED)\,S(EF)}.$$

The exact equation is:

$(Y(BA)\,X(BC) - Y(BC)\,X(BA))\,(X(ED)\,X(EF) + Y(ED)\,Y(EF)) -$
$(Y(ED)\,X(EF) - Y(EF)\,X(ED))\,(X(BA)\,X(BC) + Y(BA)\,Y(BC)) = 0$

As we know in elementary analytic geometry, if the tangents of two angles are equal, then either the two angles are equal if they are in the same direction (clockwise or counterclockwise), or they are complementary if they are in different directions. The exact equivalent geometrical condition corresponding to the above equation is:

$A = B$ or $B = C$ or $D = E$ or $F = E$
or $\angle ABC$ and $\angle DEF$ are congruent and in the same direction
or $\angle ABC$ and $\angle DEF$ are complementary and in the opposite direction.

Note: $\tan(\angle ABC) = -\tan(\angle CBA)$.

Command B (followed by 12 numbers): two angles are congruent.

Example (The same diagram as for command A):

B 1 2 3 4 5 6 7 8 9 10 11 12

means that $\angle ABC \equiv \angle DEF$. The corresponding equation is equivalent to $\cos(\angle ABC) = \cos(\angle DEF)$, i. e.

$$\frac{X(BA)\,X(BC) + Y(BA)\,Y(BC)}{|BA|\,|BC|} = \frac{X(ED)\,X(EF) + Y(ED)\,Y(EF)}{|ED|\,|EF|}$$

The exact geometrical condition corresponding to the above equation (in its equivalent polynomial form) is:

$A = B$ or $C = B$ or $D = E$ or $F = E$ or $\angle ABC$ and $\angle DEF$ are congruent.

The above equation is reduced to a polynomial of degree 8, while the A command generates a polynomial of degree 4. The B command can generate a polynomial that contains more than 1000 terms if all coordinates involved are non-zero.

PROVING GEOMETRY THEOREMS

Command C (followed by 8 numbers): four points are on the same circle.

Example:

C 1 2 3 4 5 6 7 8

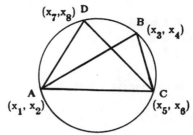

means that A, B, C and D are on the same circle. The order of the four points can be arbitrary. For the order above, the program generates a polynomial equivalent to $\tan(\angle ABC) = \tan(\angle ADC)$.

Note: if B and D are on the same side of line AC then angles ABC and ADC are in the same direction, otherwise they are in opposite directions.

Command L (followed by six numbers): three points are collinear.

Example:

L 1 2 3 4 5 6

means that A, B and C are collinear. The corresponding equation is equivalent to $S(AB) = S(BC)$, or in its equivalent polynomial form:

$Y(AB) * X(BC) - Y(BC) * X(AB) = 0$.

Command M (Followed by six numbers): midpoints.

Example:

M 1 2 3 4 5 6

means that B is the midpoint of segment AC. The corresponding equations are $x_3 - (x_1 + x_5) / 2 = 0$ and $x_4 - (x_2 + x_6) / 2 = 0$. Note: this command usually generates two equations; if the segment is parallel to the x-axis or to the y-axis then only one equation is generated. The exact geometrical condition corresponding to the above equations is:

($A = C$ and $A = B$) or B is the midpoint of AC.

Command O (followed by 8 numbers): ratio of two vectors with the same origin and on the same line is equal to some rational number.

Example:

```
O 1 2 3 4 5 6 2 -1
```

C   A   B
(x_5, x_6) (x_1, x_2) (x_3, x_4)

means that $AB : AC = 2 : -1$. This command usually also generates two equations.[8]

The corresponding equations for the above example are

$X(AB)/X(AC) = 2/-1$ and $Y(AB)/Y(AC) = 2/-1$.

The exact geometrical condition corresponding to this command is (assuming the ratio of the last two number is r):

      $A = B$ and $A = C$
or $r > 0$ and $|AB| : |AC| = r$ and (B is between AC or C is between AB)
or $r < 0$ and $|AB| : |AC| = -r$ and A is between BC.

Command P (followed by 8 numbers): two lines are parallel.

Example:

```
P 1 2 3 4 5 6 7 8
```

A —————— B (x_1,x_2) (x_3,x_4)

C —————— D (x_5, x_6) (x_7, x_8)

means that $AB \parallel CD$. The corresponding equation is equivalent to $S(AB) = S(CD)$, or in equivalent polynomial form

$Y(AB) * X(CD) - Y(CD) * X(AB) = 0$.

The exact geometrical condition corresponding to the above equation is:

$A = B$ or $C = D$ or AB is parallel to CD.

Command R (followed by 5 numbers): the length of a segment (represented by the first 4 numbers) is equal to a certain value (represented by the last number).

Example:

```
R 4 5 2 3 1
```

C •———————————• D
(x_2, x_3) (x_4, x_5)

with length x_1

---

[8]Unlike the other commands, the last two numbers of this command do not represent the coordinates. Command Q has the same situation.

PROVING GEOMETRY THEOREMS   259

means that $|DC| = |x_1|$, i.e. $X(DC)^2 + Y(DC)^2 = x_1^2$.

command T (followed by 8 numbers): two lines are perpendicular.

Example:

T 1 2 3 4 5 6 7 8

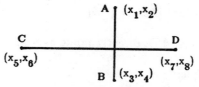

means that AB $\perp$ CD. The corresponding equation is $S(AB) * S(CD) = -1$, or in its polynomial form $Y(AB) * Y(CD) + X(AB) * X(CD) = 0$. The exact geometrical condition corresponding to the above equation is:

A = B or C = D or AB is perpendicular to CD.

Command Q (Followed by 10 numbers): the ratio of two segments.

Example (The same diagram as for command T):

Q 1 2 3 4 5 6 7 8 1 2

means that $|AB| : |CD| = 1 : 2$. The corresponding equation is equivalent to

$2 * (X(AB)^2 + Y(AB)^2) = 1 * (X(CD)^2 + Y(CD)^2)$.

Command Z (followed by 16 numbers): products of two pairs of segments are equal.

Example:

Z 1 2 7 8 5 6 7 8 1 2 5 6 3 4 7 8

means that $|AC| * |BC| = |AB| * |DC|$.

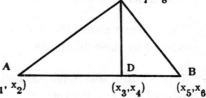

The Command : (followed by 16 numbers): ratios of two pairs of segments are equal.

Example (The same diagram as for command Z):

: 1 2 7 8 1 2 5 6 3 4 7 8 5 6 7 8

means that $|AC| : |AB| = |DC| : |BC|$. The corresponding equation is equivalent to $|AC| * |BC| = |AB| * |DC|$.

The Command = (followed by 8 numbers): the lengths of two segments are equal.

Example:

= 1 2 3 4 5 6 7 8

means that $|AB| = |CD|$.

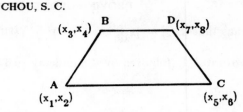

Command D: directly input hypothesis polynomial.

Example:

D 2 2 3 * 4 5 - 2 2 2 * 4 3 * 3 5 - 1 1 1 D

means $2 x_2^3 x_4^5 - 2 x_2^2 x_4^3 x_3^5 - x_1$.

The D command is necessary if the other commands can not express the geometrical conditions for some particular problem.

Command E: end of input.

Unlike the commands above, the following three commands regard the sequence of commands as some sort of stack, providing a mechanism to produce the polynomials at will. So the rule that each command corresponds to an equation is not applicable if the following three commands appear in the sequence of input commands.

The Command + (no symbol following): replace the last two polynomials produced by their sum. After executing this command, the number of polynomials produced decreases by 1.

The Command - (no symbol following): replace the last two polynomials produced by their difference.

The Command * (no symbol following): replace the last two polynomials produced by their product.

2.3. Examples.

Example 1. The example in Sec. 2.2. We choose the coordinates as shown in Figure 2. x_1, x_2 and x_3, are independent and x_4, x_5, x_6 and x_7 are dependent on them. So the input is:

```
4
P 0 0 5 4 1 0 2 3
P 5 4 2 3 0 0 1 0
L 0 0 6 7 2 3
L 5 4 1 0 6 7
= 0 0 6 7 6 7 2 3
E
```

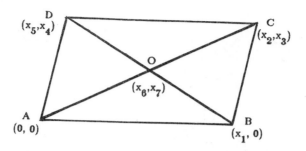

Figure 2

Line 1 tells the program that there are 4 hypothesis polynomials to be generated or equivalently there are 4 dependent variables. Caution: our program always regards the last 4 (or generally r) as dependent, so the order of the variables is important. Line 2, 3, 4 and 5 correspond to the hypotheses that AD ∥ BC, DC ∥ AB, O is on AC, and O is on BD[9] respectively. Line 6 is the conclusion: $|AO| = |OC|$. Line 7: E is the end of the input. This theorem has been proved under the following subsidiary conditions:[10]

$c_1: -x_1 \neq 0$ {AD and BC are not on the same line}

$c_2: -x_3 \neq 0$ {AB and CD are not on the same line}

$c_3: -x_3 x_5 + x_2 x_4 + x_1 x_3 \neq 0$ {AC is not parallel to BD}

$c_4: x_2 \neq 0$

Example 2. Simson's Theorem. Let D be a point on the circumscribed circle of triangle ABC. From D, perpendiculars are drawn to the sides of the triangle. Prove that the feet of the perpendiculars are collinear.

Let the three feet be E, F and G, and let O be the center of the circle. We give three choices of coordinates to illustrate how to choose independents and dependents and to write input commands.

Choice 1. Choose A as origin and OA as x-axis. So $A = (0, 0)$, $O = (x_1, 0)$.

---

[9] The order of O, B and D can be arbitrary e.g. we can write L 5 4 6 7 1 0. They possibly differ by a sign.

[10] The conditions without comments are either trivially satisfied, or not necessary by a general theorem. See footnote 4.

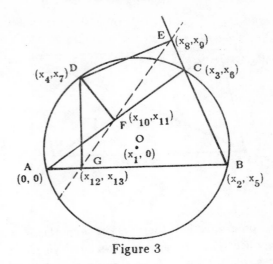

Figure 3

$|x_1|$ = radius of the circle. Once A and O are fixed, the circle is fixed. B can only move along the circle, so one of its two coordinates is independent, and the other is dependent. Points C and D have the same situation. Let B = (x_2, x_5), C = (x_3, x_6), D = (x_4, x_7), let x_2, x_3 and x_4 be independent; then x_5, x_6 and x_7 are dependent. Once A, B, C and D are fixed then the feet E, F and G can be constructed, so their coordinates are all dependent. Summing up, x_1, x_2, x_3 and x_4 are independent, and x_5, x_6, x_7, x_8, x_9, x_{10}, x_{11}, x_{12} and x_{13} are dependent (see figure 3). The complete input is:

```
9
R 2 5 1 0 1
R 3 6 1 0 1
R 4 7 1 0 1
L 2 5 3 6 8 9
T 4 7 8 9 3 6 2 5
L 0 0 3 6 10 11
T 4 7 10 11 0 0 3 6
L 0 0 2 5 12 13
T 4 7 12 13 0 0 2 5
L 8 9 10 11 12 13
E
```

Line 1 tells the program that there are 9 hypothesis polynomials to be generated. The program regards the last 9 variables as dependent. There are 13 variables in all, so the first 4 variables are independent. Lines 2, 3, 4 tell the program that B, C and D are on the circle whose center is O = $(x_1, 0)$, and whose radius is $|x_1|$. Lines 5 and 6 are conditions for point E = (x_8, x_9). (Note: we introduce two new dependents x_8, x_9 at one time in the equations corresponding to line 5 and 6, so the program will use the triangulation procedure to eliminate x_9 from one of the two equations.) Lines 7 and 8 are conditions for point $F = (x_{10}, x_{11})$. Lines 9 and 10 are conditions for point $G = (x_{12}, x_{13})$. Line 11 corresponds to the conclusion, which says E, F and G are collinear. This theorem has been proved under the following subsidiary conditions:[11]

c_1: $1 \neq 0$

c_2: $1 \neq 0$

c_3: $1 \neq 0$

c_4: $-x_6^2 + 2 x_5 x_6 - x_5^2 - x_3^2 + 2 x_2 x_3 - x_2^2 \neq 0$
{DE is not parallel to BC}

c_5: $x_2 - x_3 \neq 0$

c_6: $x_6^2 + x_3^2 \neq 0$ {DF is not parallel to AC}

c_7: $-x_3 \neq 0$

c_8: $x_5^2 + x_2^2 \neq 0$ {DG is not parallel to AB}

c_9: $-x_2 \neq 0$

Remark for Choice 1. The reader may wonder why we did not choose the coordinates in a more natural way: choose the center O as the origin, and OA as X-axis. Of course we can choose in this way. But the surprising fact is that these two choices lead to significant differences in sizes: the maximal intermediate remainder for the former has 56 terms, while it has 285 terms for the latter. Though there are some guidelines for reducing the sizes, sometimes it is unpredictable, such as in this case.

Choice 2. We can regard A, B and C as arbitrarily fixed (generic) points. Let $A = (0, 0)$, $B = (x_1, 0)$, $C = (x_2, x_3)$. Once A, B and C are fixed, then O, the center of the circumscribed circle, can be determined (constructed), let $O = (x_5, x_6)$. So x_5 and x_6 are dependent if we choose x_1, x_2, x_3 as independent. D

---

[11] The ordinary proofs of Simson's theorem assume implicitly some aditional facts besides hypotheses. E.g. the most common proof in textbooks is to reduce to the proof of congruence of ∠AFG and ∠CFE, assuming implicitly the fact that E and F are on either sides of line AC. This fact is by no means trivial. So, besides discovery of degenerate cases, another merit of Wu's mechanical procedure is to give strict proofs of theorems in elementary geometry.

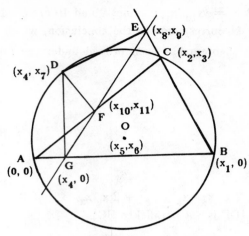

Figure 4

can only move along the circle. So one of its coordinates is independent, the other is dependent. Let $D = (x_4, x_7)$. Because $DG \perp AB$, $G = (x_4, 0)$. The choice of coordinates is shown in Figure 4 and the complete input for this choice is:

```
7
= 0 0 5 6 5 6 1 0
= 0 0 5 6 5 6 2 3
= 5 6 4 7 0 0 5 6
L 1 0 2 3 8 9
T 4 7 8 9 1 0 2 3
L 0 0 10 11 2 3
T 4 7 10 11 0 0 2 3
L 4 0 10 11 8 9
E
```

This theorem has been proved under the following subsidiary conditions:

c_1: $2 x_1 \neq 0$ {B is not A}

c_2: $2 x_3 \neq 0$ {C is not on AB}

c_3: $1 \neq 0$

c_4: $x_3^2 + x_2^2 - 2 x_1 x_2 + x_1^2 \neq 0$ {DE is not parallel to BC}

c_5: $x_1 - x_2 \neq 0$

c_6: $- x_3^2 - x_2^2 \neq 0$ {DF is not parallel to AC}

$c_7: x_2 \neq 0$

This choice is better than choice 1: the largest remainder has only 11 terms.

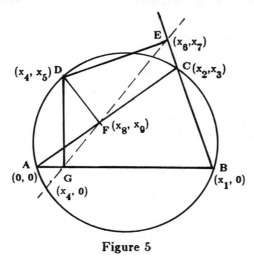

Figure 5

Choice 3. As in choice 2, let $A = (0, 0)$, $B = (x_1, 0)$, $C = (x_2, x_3)$. Now we avoid mentioning the center O explicitly, by using command C. The choice of the coordinates is shown as in Figure 5. The complete input for this choice is:

```
5
C 0 0 1 0 2 3 4 5
L 1 0 2 3 6 7
T 4 5 6 7 1 0 2 3
L 0 0 2 3 8 9
T 0 0 2 3 4 5 8 9
L 4 0 8 9 6 7
E
```

From the input above we know that the first four variables (x_1, x_2, x_3 and x_4) are independent and the last 5 are dependent. This theorem has been proved under the following subsidiary conditions:

$c_1: x_1 x_3 \neq 0$ {C is not on AB}

$c_2: x_3^2 + x_2^2 - 2 x_1 x_2 + x_1^2 \neq 0$ {DE is not parallel to BC}

$c_3: x_1 - x_2 \neq 0$

$c_4: x_3^2 + x_2^2 \neq 0$ {DF is not parallel to AC}

$c_5: -x_2 \neq 0$

Example 3. Pappus' Theorem. Let A_1, A_2, A_3 be on the same line, and B_1, B_2, B_3 be on the same line. Let $D = A_1B_2 \cap A_2B_1$, $E = A_2B_3 \cap A_3B_2$, $F = A_1B_3 \cap A_3B_1$. Show D, E and F are collinear. For the convenience of explaining the choice of coordinates we add another hypothesis: the two lines have exactly one common point, which is not necessary.

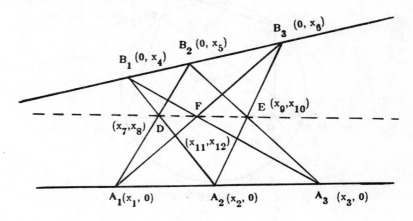

Figure 6

This is a famous theorem in projective geometry. Only incidence is involved in the statement of the theorem. So we can take advantage of the principle (3) in Section 2.2. Thus we can choose line $A_1A_2A_3$ as X-axis, and $B_1B_2B_3$ as Y-axis Let $A_1 = (x_1, 0)$, $A_2 = (x_2, 0)$, $A_3 = (x_3, 0)$, $B_1 = (0, x_4)$, $B_2 = (0, x_5)$, $B_3 = (0, x_6)$. So x_1, x_2, x_3, x_4, x_5 and x_6 are independent. Once A_1, A_2, A_3, B_1, B_2 and B_3 are fixed, then D, E and F can be constructed from them. So their coordinates x_7, x_8, x_9, x_{10}, x_{11} and x_{12} are dependent (Figure 6). The input for this choice of coordinates is:

```
6
L 1 0 0 5 7 8
L 2 0 0 4 7 8
L 2 0 0 6 9 10
L 3 0 0 5 9 10
L 1 0 0 6 11 12
L 3 0 0 4 11 12
L 7 8 9 10 11 12
E
```

Remark. The line DEF is called a Pappus' line. Because the hypotheses do not involve the order, B_2 is not necessarily between B_1 and B_3. So there are six possible combinations, hence there are six Pappus' lines. Our prover has proved all six combinations for the single input. Among six Pappus' lines 3 are concurrent at one point, and the other three are concurrent at another point.

PROVING GEOMETRY THEOREMS

This fact was first discovered and proved by Wu. Our prover also proved it. Pappus' theorem has been proved under the following subsidiary conditions:

$c_1: x_2 x_5 - x_1 x_4 \neq 0$ {A_1B_2 is not parallel to A_2B_1}

$c_2: x_1 \neq 0$

$c_3: x_3 x_6 - x_2 x_5 \neq 0$ {A_2B_3 is not parallel to A_3B_2}

$c_4: x_2 \neq 0$

$c_5: x_3 x_6 - x_1 x_4 \neq 0$ {A_1B_3 is not parallel to A_3B_1}

$c_6: x_1 \neq 0$

Example 4. Let ABCD be a square. CG is parallel to the diagonal BD. E is a point on CG such that $|BE| = |BD|$. F is the intersection of BE and DC. Show $|DF| = |DE|$.

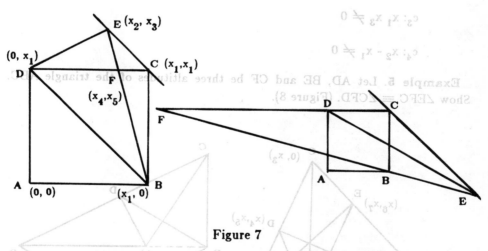

Figure 7

If we let $A = (0, 0)$, $B = (x_1, 0)$, $C = (x_1, x_1)$, $D = (0, x_1)$, then the condition that ABCD is a square is implicit in the above choice of the coordinates. Once x_1 is fixed, E and F can be determined in a constructive way. So their coordinates x_2, x_3, x_4 and x_5 are dependent (Figure 7). The complete input for the prover is:

```
 4
 P 1 1 2 3 1 0 0 1
 = 1 0 2 3 1 0 0 1
 L 2 3 1 0 4 5
 L 0 1 1 1 4 5
 = 0 1 4 5 0 1 2 3
 E
```

It is interesting to point out that there are two points which satisfy the condition for E (see the two diagram of figure 7). Our prover has proved both of them for the above input. This problem is considered fairly difficult in high school geometry. We haven't see the second variant before. It was only after our prover proved it that we realized that there is another variant for the same problem. This theorem has been proved under the following subsidiary conditions (all these conditions are equivalent to $x_1 \neq 0$):

$c_1: 2 x_1^2 \neq 0$ {A, B, C and D are not the same point}

$c_2: - x_1 \neq 0$

$c_3: x_1 x_3 \neq 0$

$c_4: x_2 - x_1 \neq 0$

Example 5. Let AD, BE and CF be three altitudes of the triangle ABC. Show $\angle EFC \equiv \angle CFD$. (Figure 8).

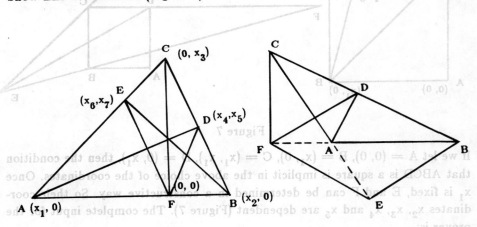

Figure 8

Perhaps the best choice of coordinates for the problems involving altitudes or the orthocenter of a triangle is: $A = (x_1, 0)$, $B = (0, x_2)$, $C = (0, x_3)$. Obviously, x_1, x_2 and x_3 are independent. Once A, B and C are fixed, then D, E and F can be determined in a constructive way. In fact, under this particular choice of coordinates, F = (0, 0), the origin of the coordinates. Let D = (x_4, x_5), E = (x_6, x_7). The input for this theorem is:

```
4
T 1 0 4 5 0 3 2 0
L 0 3 2 0 4 5
T 1 0 0 3 2 0 6 7
L 1 0 0 3 6 7
A 6 7 0 0 0 3 0 3 0 0 4 5
E
```

Note: in the conclusion we use command A, i.e. our prover has proved tan(∠EFC) = tan(∠CFD). What will happen if we use command B (the congruence of two angles) instead of command A in the conclusion? Our prover failed to prove it! The real situation is that the theorem is true only for acute triangles (but in some text books there is no explicit statement about this), otherwise the theorem may not be true. In the right diagram of figure 8 angle EFC and angle CFD are complementary and in opposite directions. Our prover only proved either ∠EFC and ∠CFD are equal, or they are complementary. This theorem has been proved under the following subsidiary conditions:

$c_1: -x_3^2 - x_2^2 \neq 0$ {AD is not parallel to BC}

$c_2: -x_3 \neq 0$

$c_3: -x_3^2 - x_1^2 \neq 0$ {BE is not parallel to AC}

$c_4: x_3 \neq 0$

Example 6. The Butterfly Theorem. A, B, C and D are four points on circle O. E is the intersection of AC and BD. Through E draw a line perpendicular to OE, meeting AD at F and BC at G. Show |FE| = |GE|. (Figure 9).

First let O = (0, 0), A = $(x_1, 0)$; then the circle is fixed. Being on the circle, each of B, C and D has only one independent coordinate. Let B = (x_2, x_5), C = (x_3, x_6), D = (x_4, x_7), regarding x_2, x_3 and x_4 as independent and x_5, x_6 and x_7 as dependent. Once A, B, C and D are fixed, then E, F and G can be determined in a constructive way. The input for this problem is:

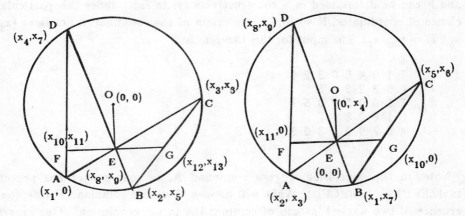

Figure 9

```
9
R 2 5 0 0 1
R 3 6 0 0 1
R 4 7 0 0 1
L 1 0 8 9 3 6
L 4 7 8 9 2 5
T 8 9 10 11 0 0 8 9
L 1 0 10 11 4 7
L 10 11 8 9 12 13
L 2 5 3 6 12 13
M 10 11 8 9 12 13
E
```

Note: in the conclusion we use the midpoint command, which produces two polynomials. We had the prover only pick one and ignore the other. Because F, E and G are collinear (hypothesis), if we have proved one, we actually have proved the other also. We may use the command = in the conclusion instead. But it will lead to increasing the memory and time greatly. Even though this problem is one of the "hardest" problems for our prover: the maximal remainder has 7599 terms, the aggregate number of terms used is 188000, and CPU time is 4 minutes 13.9 seconds on a DEC 2060. This theorem has been proved under the following subsidiary conditions:

c_1: $1 \neq 0$

c_2: $1 \neq 0$

c_3: $1 \neq 0$

c_4: $-x_3 x_7 + x_1 x_7 + x_4 x_6 - x_2 x_6 + x_3 x_5 - x_1 x_5 \neq 0$

{BD is not parallel to AC}

$c_5: x_3 - x_1 \neq 0$

$c_6: x_7 x_9 + x_4 x_8 - x_1 x_8 \neq 0$ {EF is not parallel to AD}

$c_7: x_9 \neq 0$

$c_8: -x_3 x_9 + x_2 x_9 + x_6 x_8 - x_5 x_8 - x_{10} x_6 + x_{10} x_5 + x_{11} x_3 - x_{11} x_2 \neq 0$
{FE is not parallel to BC}

$c_9: x_{10} - x_8 \neq 0$

Perhaps some better choices of coordinates exist. One possibility is shown in the right diagram of figure 9 and the input is:

```
7
L 2 3 0 0 5 6
= 0 4 2 3 0 4 5 6
= 0 4 1 7 0 4 2 3
L 1 7 0 0 8 9
= 0 4 2 3 0 4 8 9
L 1 7 10 0 5 6
L 2 3 11 0 8 9
M 11 0 0 0 10 0
E
```

Indeed, this choice does lead to reducing the memory and time required greatly: the maximal remainder has 16 terms. (This "better" choice was generated by another program for selecting coordinates automatically).

Unfortunately, there are 16 terms left in the final remainder. This is one of the reducible cases. The geometric meaning is as follows. Under this choice, the conditions for determining $C = (x_5, x_6)$ are equations corresponding to the commands "L 2 3 0 0 5 6" and "= 0 4 2 3 0 4 5 6". But $A = (x_2, x_3)$ also satisfies these conditions. So we have two solutions for x_5 and x_6: one is the non-degenerate case as shown the right diagram of figure 9, the other is $x_5 = x_2$ and $x_6 = x_3$. So the corresponding polynomials are reducible (can be factored) in some sense. The situation is similar with point $D = (x_8, x_9)$.

Example 7. Example using input command sequence as stack. Let S be the area of the triangle ABC; a, b and c be the lengths of its three sides BC, AC and AB respectively; r be the radius of its circumscribed circle. Show a * b * c = 4 * r * S.

There is no direct command for expressing the area of a triangle, but we can use command L for this purpose. Let $A = (x_1, x_2)$, $B = (x_3, x_4)$, $C = (x_5, x_6)$. The command "L 1 2 3 4 5 6" generates a polynomial equivalent to $S(AB) = S(BC)$, which is

$$(x_2 - x_4)(x_3 - x_5) - (x_1 - x_3)(x_4 - x_6) \quad \{ = 0\}.$$

This polynomial equals twice the signed area of the oriented triangle ABC, which is positive if A, B and C are clockwise or negative if A, B and C are counterclockwise. If our problem concerns the absolute value of the area, we can take the square of the polynomial as we will do in this example.

For our problem, let $A = (0, 0)$, $B = (x_1, 0)$, $C = (x_2, x_3)$, and $O = (x_4, x_5)$ be the center of the circumscribed circle of $\triangle ABC$. So $c = |x_1|$. The input is:

```
5                        ;only 5 equations for hypotheses
= 0 0 4 5 4 5 1 0        ;condition for determining O
= 0 0 4 5 4 5 2 3        ;condition for determining O
R 4 5 0 0 6              ;let r = |x_6|
R 2 3 0 0 7              ;let b = |x_7|
R 2 3 1 0 8              ;let a = |x_8|
L 0 0 1 0 2 3            ;twice the signed area of △ ABC
L 0 0 1 0 2 3            ;twice the signed area of △ ABC
*                        ;replace above two by its square
D 4 6 2 D                ;4 r^2
*                        ;16 r^2 S^2
D 1 1 2 * 7 2 * 8 2 D    ;a^2 b^2 c^2
-                        ;a^2 b^2 c^2 - 16 r^2 S^2, the conclusion
E
```

This theorem has been proved under the following subsidiary conditions:

c_1: $2 x_1 \neq 0$ {B is not identical to A}
c_2: $2 x_3 \neq 0$ {C is not on AB}

2.4. Some Remarks on Implementation.

The implementation of the version in MACSYMA is easy because in MACSYMA [6] there are many excellent packages for manipulating polynomials. In particular, r(p1, p2, v) can be expressed by two function calls (in fact r(p1, p2, v) and RREM(p1, p2, v) are slightly different):

$$\text{RREM}(p1, p2, v) := \text{NUM}(\text{REMAINDER}(p1, p2, v));$$

Now suppose that $f_1, ..., f_r$, g (after triangulation) are stored in array elements $F[1], ..., F[r], F[r+1]$ respectively, and that variables $x_1, ..., x_{mx}$ are stored in array elements $X[1], ..., X[mx]$ respectively, where mx is the maximal index of the x for a given problem. Then the prover can be as simple as follows:

```
SUCCESSIVE DIVISION( ) :=
   BLOCK([I, M, J, REM],
      I: r,
      M: mx,
      REM: F[r + 1],
      FOR J: 1 THRU r DO
      (REM: RREM(REM, F[I], X[M]),
       M: M - 1,
       I: I - 1),
      IF REM = 0
        THEN PRINT("The theorem is proved. QED.")
        ELSE PRINT("The theorem is not confirmed."));
```

In the process of successive division the remainder grows very rapidly, sometimes as large as several thousand terms. Increasing the speed of division is very important. In our PASCAL version, we reduce the time by hashing, so the time the prover takes is, on average, linear in the sizes of polynomials. As soon as a term is generated, we put it in a hashing table. If there is a like term in the hashing table already, then the two terms are combined or canceled.

3. Discovering New Theorems.

Several "new" theorems[12] have been discovered and proved by Wu's provers. Since the provers are so powerful (within this domain), once we have correct statements of the proposed new theorems it is a simple thing to use the provers to check them. We have three ways to discover new theorems: ingenious guessing, numerical searching and using a modification of Wu's method.

3.1. Ingenious Guessing.

Several theorems have been discovered by Wu in this way. [13, 15]. Among them we cite the Pappus-Point theorem. As we mentioned in example 3 of 2.3, there are six Pappus' lines for one *Pappus' configuration* $A_1A_2A_3$ and $B_1B_2B_3$: taking any permutation [ijk] of 1, 2 and 3 we determine a Pappus' line passing through the three points of intersection $A_1B_j \cap A_2B_i$, $A_2B_k \cap A_3B_j$ and $A_3B_i \cap A_1B_k$; we denote this Pappus' line by [ijk], so there are six Pappus' lines [123], [312], [231], [213], [321] and [132]. Now the Pappus-Point theorem states that [123], [231] and [213] intersect at the same point (point W in figure 10). So do

---

[12]Elementary geometry is one of the oldest branches in mathematics. When we declare some theorem to be "new", it may not be new at all. We will appreciate it if experts in geometry can confirm whether it is new or not.

the other three lines. We call the two points Pappus' points for the given Pappus' configuration.

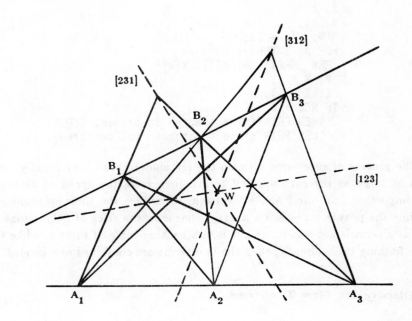

Figure 10

Based on the above theorem we have found another "new" theorem. In figure 11 lines $A_1A_2A_3$, $B_1B_2B_3$ and $C_1C_2C_3$ intersect at point O and lines $A_1B_1C_1$, $A_2B_2C_2$ and $A_3B_3C_3$ intersect at point P. Now consider three Pappus' configurations $(A_1A_2A_3, B_1B_2B_3)$, $(B_1B_2B_3, C_1C_2C_3)$ and $(C_1C_2C_3, A_1A_2A_3)$. For the first configuration we have two Pappus' points: one is P, the other is W_1. For the second we also have two Pappus' points: one is P, the other is W_2. For the third we have: P and W_3. Our version of Wu's prover proved W_1, W_2, W_3 and P are collinear.

3.2. Numerical Searching by Examples.

It is a very important technique to use examples or counterexamples in automated theorem proving [1]. From a given configuration of points (e.g. Pappus' configuration or Pascal's configuration) we form new lines by joining those points. Then we obtain new points by taking intersections of those lines. Then form new lines again and so on. At a certain stage some relations between points or lines appear, e.g. three points are collinear or three lines are concurrent. Some configurations were extensively studied, and many interesting properties were found, e.g. Pascal's configuration [8].

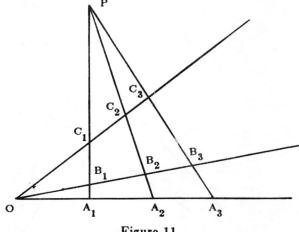

Figure 11

With a modern computer we can use numerical examples to search for new properties of a given configuration. First we assign numerical coordinates (floating point numbers) to each point of the starting configuration. Then we calculate new lines and new points, searching whether there are concurrent lines or collinear points. Once the computer outputs some possible properties, we can use Wu's prover to check whether they are theorems. The searching is usually very fast.

Two sets of possibly new theorems have been found. The following is the simpler one of the two.

Starting from a configuration of five points A, B, C, D and E, where A, B and C are on the same line (figure 12), we form new lines AD, AE, BD, BE, CD, CE and DE (figure 12). Then take all points of intersection of the previous lines: $F = AB \cap ED$, $G = AD \cap BE$, $H = AD \cap CE$, $I = AE \cap BD$, $J = AE \cap CD$, $K = BD \cap CE$, $L = BE \cap CD$. Then we have the following properties:

(2.1) ED, IG, LK and JH are concurrent.
(2.2) AB, GJ and HI are concurrent.
(2.3) AB, GK and IL are concurrent.
(2.4) AB, HL and JK are concurrent.

Our prover has proved the above conclusions.

3.3. Using a modification of Wu's Method.[13]

Suppose that we are trying to prove a theorem with conclusion $g = 0$ and

---

[13]The idea for finding new geometry theorems by computer was first proposed by P. J. Davis [3].

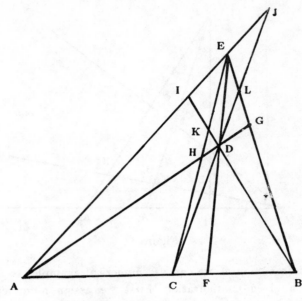

Figure 12

under hypotheses $f_1 = 0, ..., f_r = 0$ and that after successive divisions the final remainder R is not identical to zero. If we add $R = 0$ as a new hypothesis then we have a new theorem: $g = 0$ under hypotheses $f_1 = 0, ..., f_r = 0$, $R = 0$ and some subsidiary conditions. If $R = 0$ does not have a clear meaning in geometry, then we can not think it as a geometry theorem, at least it is not an interesting theorem in geometry. To make the discussion concrete, let us look at the following problem:

The converse of Simson's theorem. Suppose that triangle ABC is fixed (figure 13). Let D be a point satisfying the following condition:

(3.1)   From D three perpendiculars are drawn to the sides of the triangle. The feet of the perpendiculars are collinear.

Now our question is: under what condition on D is (3.1) satisfied ? One trivial answer is: condition (3.1) itself. Our aim is to find a condition involving points A, B, C and D only (not auxiliary points E, F and G). We can put the question in the other form: to find the locus of D satisfying condition (3.1).

The complete geometrical conditions for the problem are:

(3.2) $DE \perp BC$ and E is on BC
(3.3) $DF \perp AC$ and F is on AC
(3.4) $DG \perp AB$ and G is on AB
(3.1)   E, F and G are collinear.

PROVING GEOMETRY THEOREMS

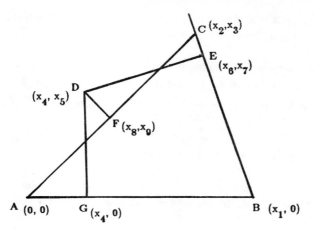

Figure 13

Suppose that now we are going to prove the theorem whose hypotheses are (3.2) - (3.4) and whose conclusion is (3.1). If we chose the coordinates as shown in Figure 13, then the hypotheses (in triangular form) are:

(3.5)

$f_1: (x_1^2 + x_2^2 + x_3^2 - 2 x_1 x_2) x_6 + (x_1 - x_2) x_3 x_5$
$\quad - (x_1^2 + x_2^2 - 2 x_1 x_2) x_4 - x_1 x_3^2 = 0$

$f_2: (x_1 - x_2) x_7 + x_3 x_6 - x_1 x_3 = 0$

$f_3: (x_2^2 + x_3^2) x_8 - x_2 x_3 x_5 - x_2^2 x_4 = 0$

$f_4: x_3 x_9 + x_2 x_8 - x_3 x_5 - x_2 x_4 = 0.$

and the conclusion is

$g: (x_6 - x_4) x_9 + x_4 x_7 - x_7 x_8 = 0.$

After successive divisions we have the final reminder[14]

$R = x_1 x_3^2 (x_3 x_4^2 + x_3 x_5^2 - x_3^2 x_5 - x_2^2 x_5 + x_1 x_2 x_5 - x_1 x_3 x_4).$

Except for some degenerate cases ($x_1 = 0$ or $x_3 = 0$) the equation R = 0 is the circumscribed circle of △ ABC. Thus we "discovered" Simson's theorem (or its converse).

---

[14] As the leading coefficients of f_4 and f_2, x_3 and $x_1 - x_2$, always appear as factors in the final remainder. We have already divided out these two factors in the final remainder below. For the related problem see footnote 4.

To make a non-trivial discovery, let us put another question: what is the locus of points D that keep the area of the oriented $\triangle$ EFG constant?

Let the constant be $A/2$. The hypothesis equations (3.2) - (3.4) are unchanged. But g is replaced by $(x_6 - x_4) x_9 + x_4 x_7 - x_7 x_8 - A$. Now dividing the new g by (3.5) successively, we have the final remainder[15]

$$R = x_1 x_3^2 (x_3 x_4^2 + x_3 x_5^2 - x_3^2 x_5 - x_2^2 x_5 + x_1 x_2 x_5 - x_1 x_3 x_4) + A (x_2^4 + x_3^4 + 2 x_2^2 x_3^2 - 2 x_1 x_2 x_3^2 + x_1^2 x_3^2 - x_1 x_3^3 + x_1^2 x_2^2).$$

The equation $R = 0$ is still a circle with the same center as circumcenter of $\triangle$ ABC (under some non-degenerate conditions). Thus we have discovered a "new" theorem:

Generalized Simson's Theorem. If D is on a circle with the circumcenter of $\triangle$ ABC as its center, then the area of the triangle formed by the three feet of perpendiculars from D to the three sides of $\triangle$ ABC keeps constant; or the locus of points D that keep the area of oriented $\triangle$ ABC constant is a circle with the circumcenter of $\triangle$ ABC as its center.

The situation of general cases is not so simple as in the above example. However, due to Ritt's method ([7], page 95) we can develop a systematic method for finding new theorems (especially theorems of locus type) automatically [2]. Though the complete program has not been finished yet, we use a program modified by Wu's prover to experiment with finding theorems (or loci). Most loci we have met so far are general conics or even the curves of degree greater than 2. It seems to me that the most interesting loci in elementary geometry are circles or lines. So the generalized Simson's theorem is perhaps the most interesting theorem among the loci we have found so far.

4. On the Complete Wu's Prover.

As we mentioned at the end of Section 2.4, if the final remainder $R \not\equiv 0$ then we have to decompose hypothesis polynomials into irreducibles, then check the conclusion polynomial g with each of the irreducibles. The decomposition involves factoring polynomials. As we know, the problem of factoring polynomials over successive algebraic extensions of the fields of rational functions is very difficult, though there exist such algorithms in principle. (cf. [10], page 134). In MACSYMA there is an excellent program for factoring polynomials over simple extensions of $\mathbf{Q}$, cf. [11]. But the base fields we have

---

[15] See the previous footnote.

to use are the fields of rational functions: $\mathbf{Q}(u_1, ..., u_d)$. Besides, generally we want successive extensions. (Theoretically, we can always reduce the successive extensions to a simple extension, but practically it is not so easy to do).

It turns out that almost all problems we have met so far have hypothesis polynomials f_i with $\deg(f_i, x_i) \leq 2$. So it quite desirable to find a fast algorithm just for the case that $\deg(f_i, x_i) \leq 2$. Essentially, we want solve the following problem:

Let K be an algebraic closed field containing $K_0 = \mathbf{Q}(u_1, ..., u_d)$. All fields and elements mentioned below are in K. Let

(I)
$$f_1: a_1 x_1^2 + b_1 x_1 + c_1$$
$$f_2: a_2 x_2^2 + b_2 x_2 + c_2$$
$$...$$
$$f_n: a_n x_n^2 + b_n x_n + c_n$$

be polynomials, where a_i, b_i and c_i are in $K_0[x_1, ..., x_{i-1}]$, and a_i is not zero in the field $K_{i-1} = K_0(x_1, ..., x_{i-1})$ defined by the equations $f_1 = 0, ..., f_{i-1} = 0$ for $i = 1, ..., n$; f_i is irreducible over K_{i-1}, for $i = 1, ..., n-1$. We want to find an algorithm to decide whether $f_n(x_n)$ is reducible over K_{n-1}, and if it is, to factor it.

4.1. Factoring a Sequence of Quadratic Polynomials.

Proposition 1. Let F be a field, and $f(x) = a x^2 + b x + c$ be in F[x], where x is indeterminate and $a \neq 0$. $f(x)$ is reducible iff $D = b^2 - 4ac$ is a perfect square in K, i.e. there is an s in F such that $D = s^2$. If $s^2 = D$ and r is a root for $f(x)$ then $F(r) = F(s)$.

Proof. If $D = s^2$ with s in F, then $f(x) = a (x - (-b + s)/2a) (x - (-b - s)/2a)$. So $f(x)$ is reducible in F. Conversely, suppose $f(x)$ is reducible in F. Let $f(x) = a (x - r_1) (x - r_2)$, with r_1 and r_2 in F. Then $D = s^2$ with $s = a (r_1 - r_2)$ in F, so D is a perfect square in F. If $s^2 = D$ then $r = (-b + s)/2a$ or $r = (-b - s)/2a$. So $F(r) = F(s)$.

Now we want find an algorithm to decide whether an element in F is a perfect square, and if it is, then to find its square root. We call this problem "PSQ in F". Proposition 1 reduces factoring quadratic polynomials over F to deciding PSQ in F.

Proposition 2. Suppose that s^2 is in F and s is not in F. If there is an algorithm for deciding PSQ in F, then there is an algorithm for deciding PSQ in

F(s).

Proof. Let $e = a + bs$ be an element in $F(s)$, where a and b are in F. Suppose that e is a perfect square in $F(s)$, then $e = (n + ms)^2$, where n and m are in F. So we have

(1.1) $\quad a = n^2 + m^2 s^2$ and $b = 2 n m$.

Case 1. If $b = 0$ then $m = 0$ or $n = 0$ by (1.1). If $m = 0$ then $e = a = n^2$. So a is a perfect square in F. If $n = 0$ then $e = a = m^2 s^2$. So a/s^2 is a perfect square in F. Conversely, if a or a/s^2 is a perfect square in F, then e is a perfect square in $F(s)$.

Case 2. If $b \neq 0$ then $n \neq 0$. Substituting $m = b/2n$ in the first equation of (1.1), we have $4n^4 - 4an^2 + b^2s^2 = 0$. So $n^2 = (a + \mathrm{sqrt}(a^2 - b^2 s^2))/2$. Thus

(1.2) $\quad a^2 - b^2 s^2$ is perfect square in F (let its square root be r) and one of $(a + r)/2$ or $(a - r)/2$ is a perfect square in F.

Conversely, if condition (1.2) holds then we can find n and m in F satisfying $(n + ms)^2 = a + bs$. So e is a perfect square in $F(s)$.

Theorem 1. Suppose that $f_1, ..., f_n$ and $K_0, ..., K_n$ are the same as in (I). There is an algorithm for factoring $f_n(x_n)$ over K_{n-1}.

Proof. By proposition 1 factoring $f_n(x_n)$ over K_{n-1} is equivalent to deciding PSQ in K_{n-1}. Now we use induction on n to show there is an algorithm for deciding PSQ in K_{n-1}. If $n = 0$ then by the lemma 1 below there is an algorithm to decide PSQ in $K_0 = Q(u_1, ..., u_d)$. Now suppose there is an algorithm for deciding PSQ in K_{n-2}. Let $r^2 = b_{n-1}^2 - 4a_{n-1}c_{n-1}$, then $K_{n-1} = K_{n-2}(r)$ by proposition 1. So by proposition 2 there is an algorithm for deciding PSQ in K_{n-1}.

Lemma 1. There is an algorithm for deciding PSQ in K_0.

Proof. Let g be an element in K_0. So $g = p_1/p_2$, where p_1 and p_2 are in $Z[u_1, ..., u_d]$. By the GCD algorithm [5, 6], we can assume p_1 and p_2 are relative prime. Suppose that g is a perfect square: $g = (h_1/h_2)^2$, where h_1 and h_2 are in $Z[u_1, ..., u_d]$ and relatively prime. So $p_1 h_2^2 = p_2 h_1^2$. Because $GCD(p_1, p_2) = 1$, $p_1 \mid h_1^2$. Because $GCD(h_1^2, h_2^2) = 1$, $h_1^2 \mid p_1$. Hence $p_1 = h_1^2$ or $p_1 = -h_1^2$. If $p_1 = h_1^2$ then $p_2 = h_2^2$. If $p_1 = -h_1^2$ then $p_2 = -h_2^2$. Thus we reduce deciding PSQ in $Q(u_1, ..., u_d)$ to that in $Z[u_1, ..., u_d]$. Then the lemma follows from the algorithm below.

Algorithm for Deciding PSQ in $A = \mathbf{Z}[u_1, ..., u_d]$.

We define an order in the set of all monomials of A as follows: $u_1^{i_1} ... u_d^{i_d} < u_1^{j_1} ... u_d^{j_d}$ iff there is $k \leq d$ such that $i_h = j_h$ for $h = 1, ..., k-1$ and $i_k < j_k$. Let $t_1 = a\, u_1^{i_1} ... u_d^{i_d}$ and $t_2 = b\, u_1^{j_1} ... u_d^{j_d}$ be two terms in A, $t_1 < t_2$ iff $u_1^{i_1} ... u_d^{i_d} < u_1^{j_1} ... u_d^{j_d}$. By definition it follows that if $t_1 < s_1$ and $t_2 \leq s_2$ then $t_1 t_2 < s_1 s_2$. Let $p = t_1 + t_2 + ... + t_n$ be a polynomials in A, where $t_1, ..., t_n$ are terms in A. If no two terms t_i and t_j ($i \neq j$) have the same monomial, we call $t_1 + ... + t_n$ the reduced form of p and term t_i *is in* p.

Now suppose that p is a perfect square in A. Let $p = g^2$, where $g = a_1 + ... + a_n$ is in the reduced form. We can assume $a_1 > ... > a_n$. So $p =$

(S)
$$a_1^2 +$$
$$2a_1a_2 + a_2^2 +$$
$$2a_1a_3 + 2a_2a_3 + a_3^2$$
$$2a_1a_4 + 2a_2a_4 + 2a_3a_4 + a_4^2$$
$$...$$
$$2a_1a_n + 2a_2a_n + ... + 2a_{n-1}a_n + a_n^2.$$

The sum (S) is not necessarily the reduced form of p. But among all terms in in (S) a_1^2 is the greatest, so it must be in p. The algorithm proceeds as follows:

Step 1. Take the greatest term in p to see whether it is a perfect square. If it is, then take its square root as a_1, otherwise p is not a perfect square.

Step 2. As the greatest, $2a_1a_2$ must be in $p - a_1^2$, so we can compute a_2: $a_2 = (2a_1a_2) / (2a_1)$. If a_2 is not in A, then p is not a perfect square.

Step 3. As the greatest, $2a_1a_3$ must be in $p - a_1^2 - (2a_1a_2 + a_2^2)$. Thus we can compute a_3. ...

Repeating this procedure, we can get a descending sequence $a_1 > a_2 > a_3 > ...$. Because $>$ is well ordered, this sequence can not be infinite. So there is some positive integer k such that either $p - (a_1 + ... + a_k)^2 = 0$, so p is a perfect square in A; or a_k is not in A, so p is not perfect square.

4.2. Applications and Examples.

The major application of the above algorithm is to make Ritt's method practically available to Wu's method in elementary geometry. As we mentioned earlier, one application is in the triangulation procedure, and the other is in the method for finding new theorems (especially theorems of locus type) automatically.

We have already implemented a Wu's prover with the above algorithm. For details of theoretical discussion cf. [12, 7], also cf. [2]. Here we give two examples proved by our Wu's prover with factoring.

Example 1. The Butterfly Theorem. (section 3.3 example 6). As we mentioned earlier, if we choose the coordinates as shown in the right diagram of figure 9, then our previous version of Wu's prover failed to prove it: there are 16 terms left in the final remainder R. The hypothesis polynomials (in triangular form) are:

$$f_1 = -(x_3^2 + x_2^2)x_5^2 + 2x_2x_3x_4x_5 - 2x_2^2x_3x_4 + x_2^2x_3^2 + x_2^4$$

$$f_2 = x_2x_6 - x_3x_5$$

$$f_3 = x_7^2 - 2x_4x_7 + 2x_3x_4 - x_3^2 - x_2^2 + x_1^2$$

$$f_4 = (x_7^2 + x_1^2)x_8^2 - 2x_1x_4x_7x_8 + 2x_1^2x_3x_4 - x_1^2x_3^2 - x_1^2x_2^2$$

$$f_5 = x_1x_9 - x_7x_8$$

$$f_6 = x_5x_7 - x_{10}x_7 + x_{10}x_6 - x_1x_6$$

$$f_7 = -x_2x_9 + x_{11}x_9 + x_3x_8 - x_{11}x_3$$

Our program decomposes it into 4 irreducibles:

(2.1.1) f_1', f_2, f_3, f_4', f_5, f_6, f_7

(2.1.2) f_1', f_2, f_3, $x_8 - x_1$, f_5, f_6, f_7

(2.1.3) $x_5 - x_2$, f_2, f_3, f_4', f_5, f_6, f_7

(2.1.4) $x_5 - x_2$, f_2, f_3, $x_8 - x_1$, f_5, f_6, f_7

where

$$f_1' = x_3^2x_5 + x_2^2x_5 - 2x_2x_3x_4 + x_2x_3^2 + x_2^3$$

$$f_4' = 2x_4x_7x_8 - 2x_3x_4x_8 + x_3^2x_8 + x_2^2x_8 - 2x_1x_3x_4 + x_1x_3^2 + x_1x_2^2.$$

Now check g (the conclusion polynomial) with each of the above irreducibles. Only the final remainder with respect to (2.1.1) is zero. Note that (2.1.2) - (2.1.4) are degenerate cases in which points A and C are the same, or points

B and D are the same. So we have proved the theorem under the non-degenerate conditions $x_5 - x_1 \neq 0$ and $x_8 - x_1 \neq 0$, and the ordinary subsidiary conditions (the leading coefficients of the f are not zero).

Example 2. Let AB be tangent to circle O at point A. M is the midpoint of AB. D is a point on the circle. C and E are the other points of intersection of lines MD and BD with the circle respectively. F is the the other point of intersection of line CB with the circle. Show EF ∥ AB.

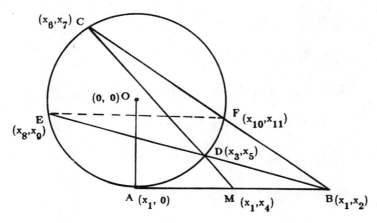

Figure 14

If we choose the coordinates as shown in Figure 14, then this is an example involving factoring of polynomials over quadratic extension fields. The hypothesis polynomials (in triangular form) are:

$f_1 = x_2 - 2 x_4$

$f_2 = x_5^2 + x_3^2 - x_1^2$

$f_3 = (x_5^2 - 2 x_4 x_5 + x_4^2 + x_3^2 - 2 x_1 x_3 + x_1^2) x_6^2 + (x_1(2 x_4 x_5 - 2 x_5^2) + 2 x_3 x_4 x_5 - 2 x_3 x_4^2) x_6 + x_1^2(x_5^2 - x_3^2) - 2 x_1 x_3 x_4 x_5 + x_3^2 x_4^2 + 2 x_1^3 x_3 - x_1^4$

$f_4 = x_3 x_7 - x_1 x_7 - x_5 x_6 + x_4 x_6 + x_1 x_5 - x_3 x_4$

$f_5 = (x_5^2 - 2 x_2 x_5 + x_3^2 - 2 x_1 x_3 + x_2^2 + x_1^2) x_8^2 + (x_1(2 x_2 x_5 - 2 x_5^2) + 2 x_2 x_3 x_5 - 2 x_2^2 x_3) x_8 + x_1^2 (x_5^2 - x_3^2) - 2 x_1 x_2 x_3 x_5 + x_2^2 x_3^2 + 2 x_1^3 x_3 - x_1^4$

$f_6 = x_3 x_9 - x_1 x_9 - x_5 x_8 + x_2 x_8 + x_1 x_5 - x_2 x_3$

$f_7 = x_{10}^2(x_7^2 - 2 x_2 x_7 + x_6^2 - 2 x_1 x_6 + x_2^2 + x_1^2) + x_1^2(x_7^2 - x_6^2) + x_{10}(x_1(2 x_2 x_7$

$- 2 x_7^2) + 2 x_2 x_6 x_7 - 2 x_2^2 x_6) - 2 x_1 x_2 x_6 x_7 + x_2^2 x_6^2 + 2 x_1^3 x_6 - x_1^4$

$f_8 = - x_{10} x_7 + x_1 x_7 - x_2 x_8 + x_{11} x_6 + x_{10} x_2 - x_1 x_{11}$

The program decomposes it into 8 irreducibles:

(2.2.1) $\quad f_1, f_2, f_3', f_4, f_5', f_6, f_7', f_8$

(2.2.2) $\quad f_1, f_2, f_3'\ f_4, f_5', f_6, f_7'', f_8$

(2.2.3) $\quad f_1, f_2, f_3'\ f_4, x_8 - x_3, f_6, f_7', f_8$

(2.2.4) $\quad f_1, f_2, f_3', f_4, x_8 - x_3, f_6, f_7'', f_8$

(2.2.5) $\quad f_1, f_2, x_6 - x_3, f_4, f_5', f_6, f_7''', f_8$

(2.2.6) $\quad f_1, f_2, x_6 - x_3, f_4, f_5', f_6, x_{10} - x_3, f_8$

(2.2.7) $\quad f_1, f_2, x_6 - x_3, f_4, x_8 - x_3, f_6, f_7''', f_8$

(2.2.8) $\quad f_1, f_2, x_6 - x_3, f_4, x_8 - x_3, f_6, x_{10} - x_3, f_8$

where

$f_3' = 4 x_2 x_5 x_6 + 8 x_1 x_3 x_6 - x_2^2 x_6 - 8 x_1^2 x_6 - 4 x_1 x_2 x_5 + x_2^2 x_3 - 8 x_1^2 x_3 + 8 x_1^3$

$f_5' = 2 x_2 x_5 x_8 + 2 x_1 x_3 x_8 - x_2^2 x_8 - 2 x_1^2 x_8 - 2 x_1 x_2 x_5 + x_2^2 x_3 - 2 x_1^2 x_3 + 2 x_1^3$

$f_7' = x_{10}(24 x_1 x_2 x_3 x_5 - 6 x_2^3 x_5 - 24 x_1^2 x_2 x_5 - 8 x_2^2 x_3^2 + 16 x_1^2 x_3^2 - 10 x_1 x_2^2 x_3 - 32 x_1^3 x_3 + x_2^4 + 18 x_1^2 x_2^2 + 16 x_1^4) + 4 x_2^3 x_3 x_5 - 24 x_1^2 x_2 x_3 x_5 + 2 x_1 x_2^3 x_5 + 24 x_1^3 x_2 x_5 + 16 x_1 x_2^2 x_3^2 - 16 x_1^3 x_3^2 - x_2^4 x_3 - 6 x_1^2 x_2^2 x_3 + 32 x_1^4 x_3 - 10 x_1^3 x_2^2 - 16 x_1^5$

$f_7'' = x_{10}(24 x_1 x_2 x_3 x_5 - 6 x_2^3 x_5 - 24 x_1^2 x_2 x_5 - 8 x_2^2 x_3^2 + 16 x_1^2 x_3^2 - 10 x_1 x_2^2 x_3 - 32 x_1^3 x_3 + x_2^4 + 18 x_1^2 x_2^2 + 16 x_1^4) + 2 x_2^3 x_3 x_5 - 24 x_1^2 x_2 x_3 x_5 + 4 x_1 x_2^3 x_5 + 24 x_1^3 x_2 x_5 + 10 x_1 x_2^2 x_3^2 - 16 x_1^3 x_3^2 - x_2^4 x_3 + 6 x_1^2 x_2^2 x_3 + 32 x_1^4 x_3 - 16 x_1^3 x_2^2 - 16 x_1^5$

$f_7''' = 2 x_{10} x_2 x_5 - 2 x_1 x_2 x_5 + x_2^2 x_3 + 2 x_1 x_{10} x_3 - 2 x_1^2 x_3 - x_{10} x_2^2 - 2 x_1^2 x_{10} + 2 x_1^3$

$x_6 - x_3 = 0$ or $x_8 - x_3 = 0$ correspond to the degenerate cases in which points D and C are the same, or points D and E are the same, respectively. If we exclude these cases, only (2.2.1) and (2.2.2) are of interest. Under conditions $f_1 = 0$, $f_2 = 0$, $f_3' = 0$, $f_4 = 0$, $f_5' = 0$, and $f_6 = 0$, f_7 can be factored into two factors f_7' and f_7'':

$$I f_7 = f_7' f_7''$$

where I is the product of powers of some leading coefficients of f_1, f_2, f_3', f_4, f_5', f_6. This factoring is done in a quadratic extension of $Q(x_1, x_2, x_3, x_4)$. The program also checked that $f_7'' = 0$ is in fact equivalent to $x_{10} - x_6 = 0$ (under some subsidiary conditions). So (2.2.2) is the degenerate case in which points F and C are the same. The prover checked that the final remainder is zero with respect to (2.2.1) but not zero with respect to (2.2.2) - (2.2.8). Summing up, the theorem has been proved under the non-degenerate cases $x_6 - x_3 \neq 0$, $x_8 - x_3 \neq 0$ and $x_{10} - x_6 \neq 0$ and the ordinary subsidiary conditions (the leading coefficients of the f are not zero).[16]

Acknowledgments. I would like to acknowledge Dr. W. W. Bledsoe, Dr. R. S. Boyer, Dr. J S. Moore and Dr. Wu Wen-tsün for their guidance, encouragement and help. Dr. Bledsoe's initial encouragement was crucial for me to study Wu's method. Dr. Boyer and Dr. Moore spent hours on helping me to write the article, proposing basic principles, clarifying vague points, correcting numerous mistakes both in English and in mathematics, even teaching me how to use SCRIBE and the laser printer. The correspondence with Dr. Wu was most helpful to me in understanding his elegant method and encouraging me to investigate further problems in this important direction. I also thank Dr. F. E. Gerth for his help in factoring problem and T. C. Wang for helpful discussion.

Bibliography

1. W. W. Bledsoe. "Non-resolution Theorem Proving." *Artificial Intelligence 9* (1977).

---

[16] The reducibility of the hypotheses comes from degenerate cases $x_6 - x_3 = 0$, $x_8 - x_3 = 0$ and $x_{10} - x_6 = 0$, which can be easily seen from the choice of the coordinates. If we can predict these cases in advance then there is a simpler way to solve the problem. Dividing f_3 by $x_6 - x_3$, f_5 by $x_8 - x_3$ and f_7 by $x_{10} - x_6$, we have $f_3 = (x_6 - x_3)q_3 + r_3$, $f_5 = (x_8 - x_3)q_5 + r_5$ and $f_7 = (x_{10} - x_6)q_7 + r_7$. Replacing f_3 by q_3, f_5 by q_5 and f_7 by q_7 in original hypotheses, we have a set of new hypotheses. Under the new hypotheses and subsidiary conditions we have $r_3 = 0$, $r_5 = 0$, $r_7 = 0$ and $g = 0$ (conclusion). The new hypotheses are actually equivalent to (2.2.1). For details cf. [2].

2. Shang-ching Chou. "Topics in Wu's Method." *Automatic Theorem Proving Project, Departments of Mathematics and Computer Sciences, University of Texas, Austin, ATP 76* (1983).

3. P. J. Davis and E. Cerutti. "FORMAC meets Pappus." *The American Mathematical Monthly 76* (1969).

4. D. Hilbert. *Foundations of Geometry.* Open Court Publishing Company, La Salla, Illinois, 1971.

5. D. E. Knuth. *The Art of Computer Programming.* Volume II: *Seminumerical Algorithms.* Addison-Wesley, 1981.

6. R. Bogen et al.. *MACSYMA Reference Manual, Version 10.* the Mathlab Group, Laboratory for Computer Science MIT, 1983.

7. R. F. Ritt. *Differential Algebra.* AMS Colloquium Publications, New York, 1950.

8. George Salmon. *A Treatise on Conic Sections* . Longmans, Green, and Co., London, New York and Bombay, 1900.

9. A. Tarski. *A decision method for elementary algebra and geometry, 2nd ed.* Berkley and Los Angeles, 1951.

10. B. L. van der Waerden. *Modern Algebra, Volume I.* Frederick Ungar Publishing Co., New York, 1949.

11. P. Wang. "Multivariate Polynomials over Algebraic Number Fields." *Mathematics of Computation 30* (1975).

12. Wu Wen-tsün. "On the Decision Problem and the Mechanization of Theorem Proving in Elementary Geometry." *Scientia Sinica 21* (1978), 159-172.

13. Wu Wen-tsün. "Mechanical Theorem proving in Elementary Geometry and Differential Geometry." *Proc. 1980 Beijing Symposium on Differential Geometry and Differential Equations 2* (1982), 1073-1092.

14. Wu Wen-tsün. "Toward Mechanization of Geometry — Some Comments on Hilbert's "Grundlagen der Geometrie"." *Acta Mathematica Scientia 2* (1982), 125-138.

15. Wu Wen-tsün. "Some Recent Advances in Mechanical Theorem-Proving of Geometries." *American Mathematics Society, Contemporiry Mathematics, in this volume* (1984).

AUTOMATED THEORY FORMATION IN MATHEMATICS

Douglas B. Lenat[1]

ABSTRACT. A program called "AM" is described which carries on simple mathematics research: defining and studying new concepts under the guidance of a large body of heuristic rules. The 250 heuristics communicate via an agenda mechanism, a global priority queue of small tasks for the program to perform and reasons why each task is plausible (e.g., "Find generalizations of 'primes', because 'primes' turned out to be so useful a concept"). Each concept is an active, structured knowledge module. One hundred very incomplete modules are initially supplied, each one corresponding to an elementary set-theoretic concept (e.g., union). This provides a definite but immense space which AM begins to explore. In one hour, AM rediscovers hundreds of common concepts (including singleton sets, natural numbers, arithmetic) and theorems (e.g., unique factorization).

1. INTRODUCTION

1.1 Historical Motivation

Scientists often face the difficult task of formulating nontrivial research problems which are soluble. In most branches of science, it is usually easier to tackle a specific given problem than to propose interesting yet managable new questions to investigate. For example, contrast *solving* the Missionaries and Cannibals problem with the more ill-defined reasoning which led to *inventing* it. The first type of activity is formalizable and admits a deductive solution; the second is inductive and judgmental. As another example, contrast *proving* a given theorem versus *proposing* it in the first place.

---
[1] Computer Science Department, Carnegie-Mellon University, Pittsburgh, PA. 15213

This work was supported in part by the Defense Advanced Research Projects Agency (F44620-73-C-0074) and monitored by the Air Force Office of Scientific Research.

This paper appeared in Proc. of the 5th International Joint Conference on Artificial Intelligence, MIT, Aug. 22-25, 1977, pp. 833-842. It is used by permission of International Joint Conferences on Artificial Intelligence, Inc.; copies of the Proceedings are available from William Kaufmann, Inc., 95 First St., Los Altos, CA 94022, USA.

A wealth of AI research has been focussed upon the former type of activity: deductive problem solving (see, e.g., [Bledsoe 71], [Nilsson 71], [Newell & Simon 72]). Approaches to *inductive* inference have also been made. Some researchers have tried to attack the problem in a completely domain-independent way (see, e.g., [Winston 70]). Other AI researchers believe that "expert knowledge" must be present if inductive reasoning is to be done at the level which humans are capable of. Indeed, a few recent AI programs have incorporated such knowledge (in the form of judgmental rules gleaned from human experts) and successfully carried out quite complex inductive tasks: medical diagnosis [Shortliffe 74], mass spectra identification [Feigenbaum 71], clinical dialogue [Davis 76], discovery of new mass spectroscopy rules [Buchanan 75].

The "next step" in this progression of tasks would be that of fully automatic theory formation in some scientific field. This includes two activities: (i) discovering relationships among known concepts (e.g., by formal manipulations, or by noticing regularities in empirical data), and (ii) defining new concepts for investigation. Meta-Dendral [Buchanan 75] performs only the first of these; most domain-independent concept learning programs [Winston 70] perform only the latter of these: while they do create new concepts, the initiative is not theirs but rather is that of a human "teacher" who already has the concepts in mind.

What we are describing is a computer program which defines new concepts, investigates them, notices regularities in the data about them, and conjectures relationships between them. This new information is used by the program to evaluate the newly-defined concepts, concentrate upon the most interesting ones, and iterate the entire process. This paper describes such a program: AM.

1.2 Choice of Domain

Research in distinct fields of science and mathematics often proceeds slightly differently. Not only are the concepts different, so are most of the powerful heuristics. So it was reasonable that this first attempt should be limited to one narrow domain. Elementary mathematics was chosen, because:

1. There are no uncertainties in the raw data (e.g., arising from errorful measuring devices).
2. Reliance on experts' introspection is a powerful technique for codifying the judgmental rules needed to work effectively in a field. By choosing a familiar field, it was possible for the author to rely primarily on personal introspection for such heuristics.
3. The more formal a science is, the easier it is to automate (e.g., the less one needs to use natural language to communicate information.

4. A mathematician has the freedom to explore -- or to give up on -- whatever he wants to. There is no specific problem to solve, no fixed "goal".
5. Unlike some fields (e.g., propositional logic), elementary math research has an abundance (many hundreds) of powerful heuristic rules available.

The limitations of math as a domain are closely intertwined with its advantages. Having no ties to real-world data can be viewed as a liability, as can having no clear "right" or "wrong" behavior. Since math has been worked on for millenia by some of each culture's greatest minds, it is unlikely that a small effort like AM would make many startling new discoveries. Nevertheless, it was decided that the advantages outweighed the limitations.

1.3 Initial Assumptions and Hypotheses

The AM program "got off the ground" only because a number of sweeping assumptions were made about how math research could be performed by a computer program:

1. Very little natural language processing capabilities are required. As it runs, AM is monitored by a human "*user*". AM keeps the user informed by instantiating English sentence templates. The user's input is rare and can be successfully stereotyped.
2. Formal reasoning (including proof) is not indispensable when doing theory formation in elementary mathematics. In the same spirit, we need not worry in advance about the occurence of contradictions.
3. Each mathematical concept can be represented as a list of facets (aspects, slots, parts, property/value pairs). For each new piece of knowledge gained, there will be no trouble in finding which facet of which concept it should be stored in.
4. The basic activity is to choose some facet of some concept, and then try to fill in new entries to store there; this will occasionally cause new concepts to be defined. The high-level decision about which facet of which concept to work on next can be handled by maintaining an "*ordered agenda*" of such tasks. The techniques for actually carrying out a task are contained within a large collection of heuristics.
5. Each heuristic has a well-defined domain of applicability, which coincides perfectly with one of AM's concepts. We say the heuristic "belongs to" that concept.
6. Heuristics superimpose; they never interact strongly with each other. If one concept C1 is a specialization of concept C2, then C1's

heuristics are more powerful and should be tried first.
7. The reasons supporting a task (on the agenda of facet/concept tasks to be carried out) superimpose perfectly. They never change with time, and it makes no difference in what order they were noticed. It suffices to have a single, positive number which characterizes the value of the reason.
8. The tasks on the agenda are completely independent. No task "wakes up" another. Only the general position (near the top, near the bottom) is of any significance.
9. The set of heuristics need not grow, as new concepts are discovered. All common-sense knowledge required is assumed to be already present within the initially-given body of heuristic rules.

It is worth repeating that all the above points are merely convenient falsehoods. Their combined presence made AM doable (by one person, in one year).

One point of agreement between Weizenbaum and Lederberg [Lederberg 76] is that AI can succeed in automating only those activities for which there exists a "strong theory" of how that activity is done by people. Point #4 above is a claim that such a clean, simple model exists for math research: a search process governed by a large collection of heuristic rules. Here is a simplified summary of that model:

1. The order in which a math textbook presents a theory is almost the exact opposite of the order in which it was actually developed. In a text, definitions and lemmata are given with no motivation, and they turn out to be just the ones required for the next big theorem, whose proof magically follows. But in real life, a mathematician would begin by examining some already-known concepts, trying to find some regularity involving them, formulating those as conjectures to investigate further, and using them to motivate some simplifying new definitions.
2. Each step the researcher takes (see #1) involves choosing from a huge set of alternatives -- that is, searching. He uses judgmental criteria (heuristics) to choose the "best" alternative. This saves his search from the combinatorial explosion.
3. Non-formal criteria (aesthetic interestingness, empirical induction, analogy, utility estimates) are much more important than formal methods.
4. All such heuristics can be viewed as situation/action (IF/THEN) rules. There is a common core of (a few hundred) heuristics, basic to all fields of math at all levels. In addition to these, each

field has several specific, powerful rules.
5. Nature is metaphysically pleasant: It is fair, uniform, regular. Statistical considerations are valid and valuable when trying to find regularity in math data. Simplicity and synergy and symmetry abound.

2. DESIGN OF THE 'AM' PROGRAM

A pure production system may be considered to consist of three components: data memory, a set of rules, and an interpreter. Since AM is more or less a rule-based system, it too can be considered as having three main design components: how it represents math knowledge (its frame-like concept/facets scheme), how it enlarges its knowledge base (its collection of heuristic rules), and how it controls the firing of these rules (via the agenda mechanism). These form the subjects of the following three subsections.

2.1 Representation of Concepts

The task of the AM program is to define plausible new mathematical concepts, and investigate them. Each concept is represented internally as a bundle of slots or "facets". Each facet corresponds to some aspect of a concept, to some question we might want to ask about the concept. Since each concept is a mathematical entity, the kinds of questions one might ask are fairly constant from concept to concept. A set of 25 facets was therefore fixed once and for all. Below is that list of facets which a concept C may have. For each facet, we give a typical question about C which it answers.

>Name: What shall we call C when talking with the user?
>Generalizations: Which other concepts have less restrictive (i.e., weaker) definitions than C?
>Specializations: Which concepts satisfy C's definition plus some additional constraints?
>Examples: What things that satisfy C's definition?
>Isa's: Which concepts' definitions does C itself satisfy?
>In-domain-of: Which operations can operate on C's?
>In-range-of: Which operations result in C's when run?
>Views: How can we view some X as if it were a C?
>Intuitions: What abstract, analogic representations are known for C?
>Analogies: Are there any similar concepts?
>Conjec's: What are some potential theorems involving C?
>Definitions: How can we tell if x is an example of C?

Algorithms: What exactly do we do if we want to execute the operation
C on a given argument?

Domain/Range: What kinds of arguments can operation C be executed on?
What kinds of values will it return?

Worth: How valuable is C? (overall, aesthetic, utility, etc.)

Interestingness: What special features can make a C especially
interesting? Especially boring?

In addition, each facet F of concept C can possess a few little subfacets which contain heuristics for dealing with that facet of C's:

F.Fillin: What are some methods for filling in new entries for facet
F of a concept which is a C?

F.Check: How do we verify/debug potential entries?

F.Suggest: If AM bogs down, what are some new tasks (related to
facet F of concept C) to consider doing?

In the Lisp implementation of AM, each concept is maintained as an atom with an attribute/value list (property list). Each facet, and its list of entries, is just a property and its associated value. As an example, here is a rendition of the Sets concept. It is meant to correspond to the notion of a collection of elements.

```
Name(s): Set, Class, Collection
Definitions:
    Recursive: λ(S)
        [S={} or Set.Definition (Remove(Any-member(S),S))]
    Recursive quick: λ(S)[S={} or Set.Definition (CDR(S))]
    Quick: λ(S)[Match S with {...}]
Specializations: Empty-set, nonempty-set, Singleton
Generalizations: Unordered-Structure, Collection,
        Structure-with-no-multiple-elements-allowed
Examples:
    Typical:   {{}}, {A}, {A,B}, {3}
    Barely:    {},    {A,B,{C,{{{A,C,(3,3,9),<4,{B},A>}}}}}
    Non-quite: {A,A}, (), {B,A}
    Foible:    <4,1,A,1>
Conjec's: All unordered-structures are sets.
Intuitions:     Geometric: Venn diagram.
Analogies: {set,set operations} ≡ {list,list operations}
Worth:  600     [on a scale of 0 - 1000]
View:
    Predicate: λ(P) {x(Domainε(P)| P(x)}
    Structure: λ(S)
        Enclose-in-braces(Sort(Remove-multiple-elements(S)))
Suggest: If P is an interesting predicate over X,
            Then consider {xε X| P(x)}.
In-domain-of: Union, Intersection, Set-difference, Subset
            Member, Cartesian-prod, Set-equality
In-range-of: Union, Intersect, Set-difference, Satisfying
```

To decipher the Definitions facet, there are a few things you must know.
Facet F of concept C will occasionally be abbreviated as C.F. In those
cases where F is "executable", the notation C.F will refer to applying the
corresponding function. So the first entry in the Definitions facet is recursive because it contains an embedded call on the function Set.Definition.
Since there are three separate but equivalent definitions, AM may choose whichever one it wants when it recurs. AM can choose one via a random selection
scheme, or always try to recur into the same definition as it was just in, or
perhaps suit its choice to the form of the argument at the moment. All concepts possess executable definitions (Lisp predicates), though not necessarily
effective ones. When given an argument x, Set.definition will return True,
False, or will eventually be interrupted by a timer (indicating that no conclusion was reached about whether or not x is a set).

The Views, Intuitions, and Analogies facets must be distinguished from
each other. <u>Views</u> is concerned with transformations between two specific concepts (e.g., how to view any predicate as a set, and vice versa). An entry on
the <u>Analogies</u> facet is a mapping from a set of concepts to a set of concepts
(e.g., between {bags, bag-union, bag-intersection,...} and {numbers, addition,
minimum,...}; or between {primes, factoring, numbers...} and {simple groups,
factoring into subgroups, groups...}). <u>Intuitions</u> deals with transformations
between a bunch of concepts and one of a few large, standard scenarios (e.g.,
intuit the relation "$\geq$" as playing on a see-saw; intuit a set by drawing a
Venn diagram). Intuitions are characterized by being (i) opaque (AM cannot
introspect on them, delve into their code), (ii) occasionally fallible, (iii)
very quick, and (iv) carefully handcrafted in advance (since AM cannot pick
up new intuitions via metaphors to the real world, as we can).

Since "Sets" is a static concept, it had no Algorithms facet (as did,
e.g., "Set-union"). The Algorithms facet of a concept contains a list of
entries, a list of equivalent algorithms. Each algorithm must have three separate parts:
1. Descriptors: Recursive, Linear, or Iterative? Quick or Slow?
   Opaque or Transparent? Destructive?
2. Relators: Is this just a special case of some other concept's
   algorithm? Which others does this one call on? Is this similar
   to any other algorithms?
3. Program: A small, executable piece of Lisp code. It may be used
   for actually "running" the algorithm; it may also be inspected,
   copied, reasoned about, etc.

There are multiple algorithms because different ones have different properties:
some are very quick in some cases, some are always slow but are very cleanly

written and hence easier to reason about, etc.

Another facet possessed only by active concepts is Domain/range. It is a list of entries, each of the form <D1 D2...Di → R>, which means that the concept takes a list of arguments, the first one being an example of concept D1, the second of D2,..., the last argument being an example of concept Di, and if the algorithm (any entry on the Algorithms facet) is run on this argument list, then the value it returns will be an example of concept R. We may say that the Domain of the concept is the Cartesian product D1xD2x...xDi, and that the Range of the concept is R. For example, the Domain/range of Set-union is <Sets Sets → Sets>; Set-union takes a pair of sets as its argument list, and returns a set as its value.

Once the representation of knowledge is settled, there remains the actual choice of what knowledge to put into the program initially. One hundred elementary concepts were selected, corresponding roughly to what Piaget might have called "prenumerical knowledge". Figure 1 presents a graph of these concepts, showing their interrelationships of Generalization/Specialization and Examples/Isa's. There is much static structural knowledge (sets, truth-values, conjectures...) and much knowledge about simple activities (boolean relations, composition of relations, set operations,...). Notice that there is no notion of proof, of formal reasoning, or of numbers or arithmetic.

2.2 Top-Level Control: The Agenda

AM's basic activity is to find new entries for some facet of some concept. But which particular one should it choose to develop next? Initially, there are over one hundred concepts, each with about twenty blank facets; thus the "space" from which to choose is of size two thousand. As more concepts get defined, this number *increases*. It's worth having AM spend some time deciding which basic task to work on next, for two reasons: most of the tasks will never get explored, and only a few of the tasks will appear (to the human user) rational things to work on at the moment.

Much informal expert knowledge is required to constrain the search, to quickly zero in on one of these few very good tasks to tackle next. This is done in two stages:

1. A list of plausible facet/concept pairs is maintained. No task can get onto this *"agenda"* unless there is some reason why working on that facet of that concept would be worthwhile.
2. Each task on this agenda is assigned a priority rating, based on the number (and strengths) of reasons supporting it. This allows the entire agenda to be kept ordered by plausibility.

The first of these constrainings is much like replacing a *legal* move generator

AUTOMATED THEORY FORMATION IN MATHEMATICS   295

with a *plausible* move generator, in a heuristic search program. The second
kind of constraint is akin to using a heuristic evaluation function to select
the best move from among the good ones. Here is a typical entry on the agenda,
a task:

```
ACTIVITY:        Fill in some entries
FACET:           for the GENERALIZATIONS facet
CONCEPT:         of the PRIMES concept
REASONS:         because
    (1)  There is only 1 known gen'l. of Primes, so far.
    (2)  The worth rating of Primes is now very high.
    (3)  Focus of attention:  AM just worked on Primes.
    (4)  Very few numbers are primes; a slightly more
         plentiful concept may be more interesting.
PRIORITY:        350   [on a scale of 0 - 1000]
```

  The actual top-level control policy is to pluck the top task (highest
priority rating) from the agenda, and then execute it. While a task executes,
some new tasks may be proposed (and merged into the agenda), some new concepts
may get created and (hopefully) some entries for the specified facet of the
specified concept will be found and filled in. Once a task is chosen, the
priority rating of that task now serves a new function: it is taken as an
estimate of how much computational resource to devote to working on this task.
The task above, in the box, might be allotted 35 cpu seconds and 350 list cells,
because its rating was 350. When either resource is exhausted, work on the
task halts. The task is removed from the agenda, and the cycle begins anew
(AM starts working on whichever task is now at the top of the agenda).

2.3 Low-Level Control: The Heuristics

  After a task is selected from the agenda, how is it "executed"? A con-
cise answer would be: AM selects relevant heuristics and executes them; they
satisfy the task via side-effects. This really just splits our original question into two new ones: How are relevant heuristics located? What does it
mean for a heuristic to be executed and to achieve something?

2.3.1 How Relevant Heuristics are Located

  Each heuristic is represented as a condition/action rule. The condition
or left-hand-side of a rule tests to see whether the rule is applicable to the
task on hand. The action or right-hand-side of the rule consists of a list of
actions to perform if the rule is applicable. E.g.,

```
        IF the current task is to check examples of a concept X,
            and (Forsome Y) Y is a generalization of X,
            and  Y  has at least 10 known examples
            and all examples of  Y  are also examples of  X,
        THEN conjecture:  X  is really no more specialized than  Y,
            and add that conjecture as a new entry on the
                Examples facet of the Conjecs concept,
            and add the following task to the agenda:
                "Check examples of  Y"
                for this reason:  Y  may analogously turn out to be
                    equal to one of its supposed generalizations.
```

It is the heuristics' right hand sides which actually accomplish the selected task; that process will be described in the next subsection. The left sides are the relevancy checkers, and will be focussed on now:

Syntactically, the left side must be a predicate, a Lisp function which always returns True or False. It must be a conjunction $P1 \wedge P2 \wedge P3 \ldots$ of smaller predicates P_i, each of which must be quick and must have no side effects. Here are some typical conjucts which might appear inside a left hand side:

```
        Over half of the current task's time allotment is used up;
        There are some known examples of Structures;
        Some known generalization of the current concept (the
            concept mentioned as part of the current task) has
            a completely empty Examples facet;
        A task recently worked on had the form "Fill in facet  F
            of  C", for any  F, where  C  is the current concept;
        The user has used this program at least once before;
```

It turned out that the laxity of constraints on the form of the heuristic rules proved excessive: it made it very difficult for AM to analyze and modify its own heuristics.

From a "pure production system" viewpoint, we have answered the question of locating relevant heuristics. Namely, we evaluate the left sides of all the rules, and see which ones respond "True". But AM contains hundreds of heuristics, and repeatedly evaluating each one's condition would use up tremendous amounts of time. AM is able to quickly select a set of *potentially* relevant rules, rules whose left sides are then evaluated to test for *true* relevance. The secret is that each rule is stored somewhere apropos to its "domain of applicability". The proper place to store the rule is determined by the first conjunct on its left hand side. Consider this heuristic:

> IF the current task is to find examples of activity F,
>   and a fast algorithm for computing F is known,
> THEN one way to get examples of F is to run F on
>   randomly chosen examples of the Domain of F.

The very first conjunct of a rule's left side is always special. It specifies the domain of applicability (potential relevance) of the heuristic, by naming a particular facet of a particular concept to which this rule is relevant (in the above rule, the domain of relevance is therefore the Examples facet of the Activity concept). AM uses such first conjuncts as pre-preconditions: A *potentially* relevant rule can be located by its first conjunct alone. Then, its left hand side is fully evaluated, to indicate whether it's *truly* relevant. Here are a few typical expressions which could be first conjuncts:

> The current task (the one just selected from the agenda)
>   is of the form "Check the Domain/range facet of
>   concept X", where X is some surjective function;
> The current task matches "Fill in boundary examples of
>   X", where X is an operation on pairs of sets;
> The current task is "Fill in examples of Primes";

The key observation is that a heuristic typically applies to *all examples of a particular concept* C. The rule above has C = Activity; it's relevant to each individual activity.

When a task is chosen, it specifies a concept C and a facet F to be worked on. AM then "ripples upward" to gather potentially relevant rules: it looks on facet F of concept C to see if any rules are tacked on there, it looks on facet F of each generalization of C, on each of *their* generalizations, etc. If the current task were "Check the Domain/range of Union-o-Union",[2] then AM would ripple upward from Union-o-Union, along the Generalization facet entries, gathering heuristics as it went. The program would ascertain which concepts claim Union-o-Union as one of their examples. These concepts include Compose-with-self, Compose, Operation, Active, Any-concept, Anything. AM would

----

[2] This operation is the result of composing set-union with itself. It performs $\lambda(x,y,z) x \cup (y \cup z)$.

collect heuristics that tell how to check the Domain/range of any composition, how to deal with Domain/range facets of any concept, etc. Of course, the further out it ripples, the more general (and hence weaker) the heuristics tend to be. Here is one heuristic, tacked onto the Domain/range facet of Operation, which would be garnered if the selected task were "Check Domain/range of Union-o-Union":

> IF the current task is "Check the Domain/range of F"
>   and an entry on that facet has the form <D D...D → R>,
>   and concept R is a generalization of concept D,
> THEN it is worth spending time checking whether or not
>   the range of F might be simply D, instead of R.

Suppose one entry on Union-o-Union's Domain/range facet was "<Nonempty-sets Nonempty-sets Nonempty-sets → Sets>". Then the above heuristic would be truly relevant (all three conjuncts on its left hand side would be satisfied), and it would pose the question: Is the union of three nonempty sets always nonempty? Empirical evidence would eventually confirm this, and the Domain/range facet of Union-o-Union would then contain that fact.

Here is another way to look at the heuristic-gathering process. All the concepts known to AM are arranged in a big hierarchy, via subset-of links (Specializations) and element-of links (Isa). Since each heuristic is associated with one individual concept (its domain of applicability), there is a hierarchy induced upon the set of heuristics. Heritability properties hold: a heuristic tacked onto concept C is applicable to working on all "lower" concepts. This allows us to efficiently analogically access the relevant heuristics simply by chasing upward links in the hierarchy. Note that the task selected from the agenda provides an explicit pointer to the "lowest" -- most specific -- concept; AM ripples upward from it. Thus concepts are gathered in order of increasing generality; hence so are the heuristics.

Below are summarized the three main points that comprise AM's scheme for finding relevant heuristics in a "natural" way and then using them:

1. Each heuristic is tacked onto the most general concept for which it applies: it is given as large a domain of applicability as possible. This will maximize its generality, while leaving its power untouched.
2. When the current task deals with concept C, AM ripples upward from C, tracing along Generalization and Isa links, to quickly find all concepts which claim C as one of their examples. Heuristics attached to all such concepts are potentially relevant.

3. All heuristics are represented as condition/action rules. Once the potentially relevant rules are located (in step 2), AM evaluates each's left hand side, in order of increasing generality. The rippling process automatically gathers the heuristics in this order. Whenever a rule's left side returns True, the rule is known to be truly relevant, and its right side is immediately executed.

2.3.2 What Happens When Heuristics Are Executed

When a rule is recognized as relevant, its right side is executed. How does this accomplish the chosen task?

The right side, by contrast to the left, may take a great deal of time, have many side effects, and the value it returns is always ignored. The right side of a rule is a series of little Lisp functions, each of which is called an *action*. Semantically, each action performs some processing which is appropriate in some way to the kinds of situations in which the rule's left side would have been satisfied (returned True). The only constraint which each action must satisfy is that it have one of the following three kinds of side-effects, and no other kinds:

1. It suggests a new task to add to the agenda.
2. It dictates how some new concept is to be defined.
3. It adds some entry to some facet of some concept.

Bear in mind that the right side of a single rule is a *list* of such actions. Let's now treat these three kinds of actions:

2.3.2.1 Heuristics Suggest New Tasks

The left side of a rule triggers. Scattered among the list of "things to do" on its right side are some suggestions for future tasks. These new tasks are then simply added to the agenda. The suggestion for the task includes enough information about the task to make it easy for AM to assemble its parts, to find reasons for it, to numerically evaluate those reasons, etc. For example, here is a typical rule which proposes a new task. It says to generalize a predicate if it appears to be returning True very rarely:

```
IF  the current task was "Fill in examples of  X",
    and concept  X  is a Predicate,
    and over 100 items are known in the domain of  X,
    and at least 10 cpu secs. have been spent so far,
    and  X  has returned True at least once,
    and  X  returned False over 20 times as often as True,
```

THEN add the following task to the agenda:
"Fill in generalizations of X",
for the following reason:
"X is rarely satisfied; a slightly less restrictive
concept might be much more interesting"
This reason has a rating which is the False/True ratio

Let's see one instance where this rule was used. AM worked on the task "Fill in examples of List-Equality". One heuristic (displayed in Sec. 2.3.1, and again in detain in Sec. 2.3.2.3) said to randomly pick elements from that predicate's domain and simply run the predicate. Thus AM repeatedly plucked random pairs of lists, and tested whether or not they were equal. Needless to say, not a high percentage returned True (in practice, 2 out of 242). This rule's left side was satisfied, and it executed. Its right side caused a new task to be formulated: "Fill in generalizations of List-Equality". The reason was as stated above in the rule, and that reason got a numeric rating of 240/2 = 120. That task was then assigned an overall rating (in this case, just 120) and merged into the agenda. It sandwiched in between a task with a rating of 128 and one with a 104 priority rating. Incidentally, when this task was finally selected, it led to the creation of several interesting concepts, including the predicate which we might call "Same-length".

2.3.2.2 Heuristics Create New Concepts

One of the three kinds of allowable actions on the right side of a heuristic rule is to create a specific new concept. For each such creation, the heuristic must specify how the new concept is to be constructed. The heuristic states the Definition facet entries for the new concept, plus usually a few other facets' contents. After this action terminates, the new concept will "exist". A few of its facets will be filled in, and many others will be blank. Some new tasks may exist on the agenda, tasks which indicate that AM ought to spend some time filling in some of those facets in the near future. Here is a heuristic rule which results in a new concept being created:

IF the current task was "Fill in examples of F"
    and F is an operation, from domain A into range B,
    and more than 100 items are known examples of A,
    and more than 10 range items (examples of B) were
        found by applying F to these domain elements,
    and at least one of these range items 'b' is a distin-
        guished member (especially, an extremum) of B,

THEN for each such 'b' ε B, create the following concept:

> NAME: F-inverse-of-b
> DEFINITION: λ(a)F(a) is a 'b'
> GENERALIZATIONS: A
> WORTH: Average(Worth(A), Worth(B),
>   Worth(b), ‖ Examples(B) ‖)
> INTEREST: Any conjecture involving both
>   this concept and either F or Inverse(F)

and the reason for this creation is: "It's worth
    investigating A's which have unusual F-values"
and add five new tasks to the agenda,
    of the form "Fill in facet x of F-inverse-of-b"
    where x is Conjectures, Generalizations,
      Specializations, Examples, and Isa's;
    for the following reason:
      "This concept was newly synthesized; it is crucial to find where it 'fits in' to the hierarchy"
The reason's rating is just Worth(F-inverse-of-b).

One use of this heuristic was when the current task was "Fill in examples of Divisors-of". The heuristic's left side was satisfied because: Divisors-of is an operation (from Numbers to Sets of numbers), and far more than the required 100 different numbers are known, and more than 10 different sets of factors were located altogether, and some of them were in fact distinguished by being extreme kinds of sets (e.g., singletons, empty sets, doubletons, tripletons,...). After its left side triggered the right side of the rule was executed. Four new concepts were created immediately. Here is one of them:

> NAME: Divisors-of-Inverse-of-Doubleton
> DEFINITION: λ(a) Divisors-of(a) is a Doubleton
> GENERALIZATIONS: Numbers
> WORTH: 100
> INTEREST: Any conjecture involving both
>   this concept and either Divisors-of or Times

This is a concept representing a certain class of numbers, in fact the numbers we call "primes". The heuristic rule is of course applicable to any kind of operation, not just numeric ones. As another instance of its use, consider

what happened when the current task was "fill in examples of Set-intersect".
This rule caused AM to notice that some pairs of sets were mapping over into
the most extreme of all sets: the empty set. The rule then had AM define the
new concept we would call "disjointness": pairs of sets having empty intersection.

There is just a tiny bit of "theory" behind how these concept-creating
rules were designed. A Facet of a new concept is filled in immediately iff
both (i) it's trivial to fill in at creation-time, and (ii) it would be very
difficult to fill in later on. The following facets are typically filled in
right away: Definitions, Algorithms, Domain/range, Worth. Each other facet
is either left unmentioned by the rule, or else is explicitly made the subject
of a new task which gets added to the agenda. For instance, the heuristic rule
above would propose many new tasks at the moment that Primes were created, including "Fill in conjectures about Primes", "Fill in specializations of Primes",
etc.

2.3.2.3 Heuristics Fill in Entries for a Specific Facet

If the task plucked from the agenda were "Fill in examples of Set-union",
it would not be too much to hope for that by the time all the heuristic rules
had finished executing, some examples of that operation would indeed exist on
the Examples facet of the Set-union concept. Let's see how this can happen.

AM starts by rippling upward from Set-union, looking for heuristics
which are relevant to finding examples of Set-union (there are no such rules),
relevant to finding examples of Set-operations, of Operations, of any Activity,
of any Concept, of Anything. Here is one rule applicable to any Activity:

```
IF the current task is to fill in examples of F,
   and  F  is an operation, say with domain  D,
   and there is a fast known algorithm for  F
THEN one way to get examples of  F  is to run  F's
     algorithm on randomly chosen examples of  D.
```

Of course, in the Lisp implementation, this situation-action rule is not coded
quite so neatly. It would be more faithfully translated as follows:

```
IF CURR-TASK matches (FILLIN EXAMPLES F ← anything)
   and  F  isa Activity,
   and the Algorithms facet of  F  is not blank,
THEN carry out the following procedure:
```

1. Find the domain of F, and call it D;
2. Find examples of D, and call them E;
3. Find a fast algorithm to compute F; call it A;
4. Repeatedly:
    4a. Choose any member of E, and call it E1.
    4b. Run A on E1, and call the result X.
    4c. Check whether <E1,X> satisfies the definition of F.
    4d. If so, then add <E1 → X> to the Examples facet of F.
    4e. If not, then add <E1 → X> to the Non-examples facet of F.

Let's see exactly how this rule found examples of Set-union. Step (1) says to locate the domain of Set-union. The facet labelled Domain/range, on the Set-union concept, contains the entry (SET SET → SET), which indicates that the domain is a pair of sets. That is, Set-union is an operation which accepts (as its arguments) two sets.

Since the domain elements are sets, step (2) says to locate examples of sets. The facet labelled Examples, on the Sets concept, points to a list of about 30 different sets. This includes {A}, {A,B,C,D,E}, {}, {A,{{B}}},...

Step (3) involves nothing more than accessing some entry tagged with the descriptor "Quick" on the Algorithms facet of Set-union. One such entry is a recursive Lisp function of two arguments, which halts when the first argument is the empty set, and otherwise pulls an element out of that set, Set-inserts it into the second argument, and then recurs on the new values of the two sets. For convenience, we'll refer to this algorithm as UNION.

We then enter the loop of Step (4). Step (4a) has us choose one pair of our examples of sets, say the first two {Z} and {A,B,C,D,E}. Step (4b) has us run UNION on these two sets. The result is {A,B,C,D,E,Z}. Step (4c) has us grab an entry from the Definitions facet of Set-union, and run it. A typical definition is this formal one:

    (λ(S1 S2 S3)
        (AND
            (For all x in S1, x is in S3)
            (For all x in S2, x is in S3)
            (For all x in S3, x is in S1 or x is in S2))))

It is run on the three arguments S1 = {Z}, S2 = {A,B,C,D,E}, S3 = {A,B,C,D,E,Z}. Since it returns "True", we proceed to Step (4d). The construct <{Z}, {A,B,C,D,E} → {A,B,C,D,E,Z}> is added to the Examples facet of Set-union.

At this stage, control returns to the beginning of the Step (4) loop. A new pair of sets is chosen, and so on. The loop ends when either the time or space allotted to this rule is exhausted. AM would then break away at a "clear" point (just after finishing a cycle of the Step (4) loop) and would move on to a new heuristic rule for filling in examples of Set-union.

3. RESULTS

3.1 Excerpt of the 'AM' Program Running

Repeatedly, the top task is plucked from the agenda, and heuristics are executed in an attempt to satisfy it. AM has a modest facility that prints out a description of these activities as they occur. Here is a tiny excerpt:

** Task: ** Fill in Examples of the concept "Divisors-of".
   3 Reasons:
      (1) No known examples of Divisors-of yet.
      (2) Times (related to Divisors-of) is now v.int.
      (3) Focus of attention: AM just defined Divisors-of.
26 examples found, in 9 secs. e.g., Divisors-of(6) = {1,2,3,6}.

** Task: ** Consider nos. having small sets of Divisors-of.
   2 Reasons:
      (1) Worthwhile to look for extreme cases.
      (2) Focus: AM just worked on Divisors-of.
Filling in examples of numbers with 0 divisors.
   0 examples found, in 4.0 seconds.
   Conjecture: no numbers have precisely 0 divisors.
Filling in examples of numbers with 1 divisors.
   1 examples found, in 4 secs. e.g., Divisors-of(1) = {1}.
   Conjecture: 1 is the only number with exactly 1 divisor.
Filling in examples of numbers with 2 divisors.
   24 examples found, in 4 secs. Divisors-of(13) = {1,13}.
   No obvious conjecture. May merit more study.
   Creating a new concept: "Numbers-with-2-divisors".
Filling in examples of numbers with 3 divisors.
   11 examples found, in 4 secs. Divisors-of(49) = {1,7,49}.
   All nos. with 3 divisors are also Squares. Unexpected!.
   Creating a new concept: "Numbers-with-3-divisors".

** Task: ** Consider square-roots of Nos-with-3-divisors.
   2 Reasons:
      (1) Numbers-with-3-divisors unexpectedly turned
             out to all be Perfect Squares as well.
      (2) Focus: AM just defined Nos-with-3-divisors.
   All square-roots of Numbers-with-3-divisors seem to be
      Numbers-with-2-divisors.
         E.g., Divisors(169) = Divisors(13) = {1,13}.
   Even the converse of this seems empirically to be true.
   The chance of coincidence is below acceptable limits.
   Boosting the Worth rating of both concepts.

** Task: ** Consider the squares of Nos-with-3-divisors.
   3 Reasons:
      (1) Squares of Nos-with-2-divisors were v.int.
      (2) Square-roots of Nos-with-3-divisors were int.
      (3) Focus: AM just worked on Nos-with-3-divisors.

3.2 Overall Performance

Now that we've seen how AM works, and we've been exposed to a bit of "local" results, let's take a moment to discuss the totality of the mathematics which AM carried out. AM began its investigations with scanty knowledge of a hundred elementary concepts of finite set theory (see Fig. 1). Most of the obvious set-theoretic concepts and relationships were quickly found (e.g., de Morgan's laws; singletons), but no sophisticated set theory was ever done (e.g., diagonalization). Rather, AM discovered natural numbers and went off exploring elementary number theory. Arithmetic operations were soon found (as analogs to set-theoretic operations), and AM made rapid progress in divisibility theory. See Fig. 2. Prime pairs, Diophantine equations, the unique factorization of numbers into primes, Goldbach's conjecture -- these were some of the nice discoveries by AM. Many concepts which we know to be crucial were never uncovered, however: remainder, gcd, greater-than, infinity, proof, etc. These "omissions", *could* have been discovered by the existing heuristic rules in AM. The paths which would have resulted in their definition were simply never rated high enough to explore.

All the discoveries mentioned (including those in Fig. 2) were made in a run lasting one cpu hour (Interlisp+100k, Sumex PDP-10KI). Two hundred jobs in toto were selected from the agenda and executed. On the average, a job was granted 30 cpu seconds, but actually used only 18 seconds. For a typical job, about 35 rules were located as potentially relevant, and about a dozen actually fired. AM began with 115 concepts and ended up with three times that many. Of the synthesized concepts, half were technically termed "losers" (both by the author and by AM), and half the remaining ones were only marginal.

Although AM fared well according to several different measures of performance (see Section 3.4), of great significance are its *limitations*. As AM ran longer and longer, the concepts it defined were further and further from the primitives it began with. E.g., "prime-pairs" were defined using "primes" and "addition", the former of which was defined from "divisors-of", which in turn came from "multiplication", which arose from "addition", which was defined as a restriction of "union", which (finally!) was a primitive concept that we had supplied (with heuristics) to AM initially. When AM subsequently needed help with prime pairs, it was forced to rely on rules of thumb supplied originally about *union*ing. Although the heritability property of heuristics did ensure that those rules were still valid, the trouble was that they were too general, too weak to deal effectively with the specialized notions of primes and arithmetic.

For instance, one general rule indicated that $A \cup B$ would be interesting if it possessed properties absent both from A and from B. This translated into the prime-pair case as "*If $p + q = r$, and p,q,r are primes, Then r is interesting if it has properties not possessed by p or by q.*" The search for categories of such interesting primes r was of course barren. It showed a fundamental lack of understanding about numbers, addition, odd/even-ness, and primes.

The key deficiency was the lack of adequate *meta*-rules [Davis 76]: heuristics which reason about heuristics: keep track of their performance, modify them, create new ones, etc.

Aside from the preceding major limitation, most of the other problems pertain to missing knowledge: Many concepts one might consider basic to discovery in math are absent from AM; analogies were under-utilized; physical intuition was hand-crafted only; the interface to the user was far from ideal; etc. A large effort is underway this year at Carnegie-Mellon University, comprised of Greg Harris, Doug Lenat, Elaine Rich, Jim Saxe, and Herbert Simon, to overcome these limitations.

3.3 Experiments with 'AM'

One valuable aspect of AM is that it is amenable to many kinds of experiments. Although AM is too ad hoc for numeric results to have much significance, the qualitative results of such experiments may have some valid implications for math research, for automating math research, and for designing "scientist assistant" programs.

3.3.1 Must the WORTH Numbers be Finely Tuned?

Each of the 115 initial concepts had, supplied by the author, a rating number (0-1000) signifying its overall worth. The worth ratings affect the overall priority values of tasks on the agenda. Just how sensitive is AM's behavior to the initial settings of the Worth numbers?

To test this, a simple experiment was performed. All the concepts' Worth facets were set to 200 initially. By and large, the same discoveries were made as before. But there were now long periods of blind wanderings (especially near the beginning of the run). Once AM hooked into a line of productive developments, it advanced at the old rate. During such chains of discoveries, AM was guided by massive quantities of symbolic reasons for the tasks it chose, not by nuances in numeric ratings. As these spurts of development died out, AM would wander around again until the next one started.

3.3.2 How Finely Tuned is the Agenda?

The top few candidates on the agenda almost always appear to be reasonable things to do at the time. But what if, instead of picking the top-rated task, AM selected one randomly from the top 20 tasks on the agenda? In that case, AM's rate of discovery is slowed only by about a factor of 3. But the apparent "rationality" of the program (as perceived by a human onlooker) disintegrates.

3.3.3 How Valuable is the Presence of Symbolic 'Reasons'?

Only one effect of note was observed: When a task is proposed which already exists on the agenda, then it matters very much whether the task is being suggested for a new reason or not. If the reason is an old, already-known one, then the priority of the task on the agenda shouldn't rise very much. But if it is a brand new reason, then the task's rating should be boosted tremendously. The importance of this effect argues strongly in favor of having *symbolic justification* of the rank of each task in a priority queue, not just "summarizing" each task's set of reasons by a single number.

3.3.4 What if Certain Concepts are Excised?

As expected, eliminating certain concepts did seal off whole sets of discoveries to the system. For example, excising Equality prevented AM from discovering Cardinality. One surprising result was that many common concepts get discovered in several ways. For instance, multiplication arose in no fewer than four separate chains of discoveries.

3.3.5 Can AM Work in the New Domain of Plane Geometry?

One demonstration of AM's generality (e.g., that its "Activity" heuristics really do apply to any activity) would be to choose some new mathematical field, add some concepts from that domain, and then let AM loose to discover new things. Only one experiment of this type was actually carried out on the AM program.

Twenty concepts from elementary plane geometry were defined for AM (including Point, Line, Angle, Triangle, Equality of points/lines/angles/triangles). No new heuristics were added to AM.

AM was able to find examples of all the supplied concepts, and to use the character of such empirical data to determine reasonable directions to proceed in its research. AM derived the concepts of congruence and similarity of triangles, plus many other well-known concepts. An unusual result was the repeated derivation of the concept of "timberline": this is a predicate on two triangles, which is true iff they share a common vertex and angle, and if their opposite sides are parallel. AM also came up with a cute geometric interpretation of Goldbach's conjecture: Any angle (0-180°) can be approximated to within 1° as the sum of two angles each of a prime number of degrees.

3.4 Evaluating the 'AM' Program

We may wish to evaluate AM using various criteria. Some obvious ones, with capsule results, appear below:

1. By AM's ultimate achievements. Besides discovering many well-known useful concepts, AM discovered some which aren't widely known: maximally-divisible numbers, numbers which can be uniquely represented as the sum of two primes, timberline.
2. By the character of the differences between initial and final states. AM moved all the way from finite set theory to divisibility theory, from sets to numbers to interesting kinds of numbers, from skeletal concepts (none of which had any Examples filled in) to completed concepts.
3. By the quality of the route AM took to accomplish this mass of results. Only about half of AM's forays were dead-ends, and most of those looked promising initially.
4. By the character of the human--machine interactions. AM was never pushed far along this dimension.
5. By its informal reasoning abilities. AM was able to quickly "guess" the truth value of conjectures, to estimate the overall worth of each new concept, to zero in on plausible things to do each cycle, and to notice glaring analogies (sometimes).

6. By the results of experiments -- and the fact that experiments could be performed at all on AM.
7. By future implications of this project. Only time will tell whether this kind of work will impact on how mathematics is taught (e.g., explicit teaching of heuristics?), on how empirical research is carried out by scientists, on our understanding of such phenomena as discovery, learning, and creativity, etc.
8. By comparisons to other, similar systems. Some of the techniques AM uses were pioneered earlier: e.g., prototypical models [Gelernter 63], and analogy [Evans 68], [Kling 71]. There have been many attempts to incorporate heuristic knowledge into a theorem prover [Wang 60], [Guard 69], [Bledsoe 71], [Brotz 74], [Boyer & Moore 75]. Most of the *apparent* differences between them and AM vanish upon close examination: The goal-driven control structure of these systems is a compiled form of AM's rudimentary "focus of attention" mechanism. The fact that their overall activity is typically labelled as deductive is a misnomer (since constructing a difficult proof is usually in practice quite *in*ductive). Even the character of the inference processes are analogous: The provers typically contain a couple of binary inference rules, like Modus Ponens, which are relatively risky to apply but can yield big results; AM's few "binary" operators have the same characteristics: Compose, Canonize, Logically-combine (disjoin and conjoin). The *deep* distinctions between AM and the "heuristic theorem provers" are these: the underlying motivations (heuristic modelling *vs.* building tools for problem solving), the richness of the knowledge base (hundreds of heuristics *vs.* only a few), and the amount of emphasis on formal methods.

Theory formation systems in *any* field have been few. Meta-Dendral [Buchanan 75] represents perhaps the best of these. But even this system is given a fixed set of templates for rules which it wishes to find, and a fixed vocabulary of mass spectral concepts to plug into those hypothesis templates; whereas AM selectively enlarges its vocabulary of math concepts. Also, AM must gather its own data, but this is much easier in math than in organic chem.

There has been very little published thought about "discovery" from an algorithmic point of view; even clear thinkers like Polya and Poincare' treat mathematical ability as a sacred, almost mystic quality, tied to the unconscious. The writings of philosophers and psychologists invariably attempt to examine human performance and belief, which are far more manageable than creativity *in vitro*. Amarel [1967] notes it may be possible to learn from "theorem finding" programs how to tackle the general task of automating scientific research. AM has been one of the first attempts to construct such a program.

3.5 Final Conclusions

→ AM is a demonstration that a few hundred general heuristic rules suffice to guide an automated math research as it explores and expands a large but incomplete knowledge base of math concepts. AM demonstrates that some aspects of creative research can be effectively modelled as heuristic search.

→ This work has also introduced a control structure based upon an ordered agenda of small research tasks, each with a list of supporting reasons attached.

→ The main limitation of AM was its inability to synthesize powerful new heuristics for the new concepts it defined.

→ The main successes were the few novel ideas it came up with, the ease with which a new task domain was fed to the system, and -- most importantly -- the overall relational sequences of behavior AM exhibited.

ACKNOWLEDGEMENT

This research was initiated as my Ph.D. thesis at Stanford University, and I wish to deeply thank my advisers and committee members: Bruce Buchanan, Paul Cohen, Edward Feigenbaum, Cordell Green, Donald Knuth, and Allen Newell. In addition, I gladly acknowledge the ideas I have received in discussions with Avra Cohn and with Herbert Simon.

Bibliography

1. Amarel, Saul, *On Representations and Modelling in Problem Solving, and on Future Directions for Intelligent Systems*, RCA Labs Scientific Report No. 2, Princeton, N.J., 1967.

2. Bledsoe, W. W., *Splitting and Reduction Heuristics in Automatic Theorem Proving*, Artificial Intelligence 2, 1971, pp. 55-77.

3. Boyer, Robert S., and J S. Moore, *Proving Theorems about Lisp Functions*, JACM, V. 22, No. 1, January, 1975, pp. 129-144.

4. Brotz, Douglas K., *Embedding Heuristic Problem Solving Methods in a Mechanical Theorem Prover*, Stanford U. Report STAN-CS-74-443, August 1974.

5. Buchanan, Bruce G., *Applications of Artificial Intelligence to Scientific Reasoning*, Second USA-Japan Computer Conference, published by AFIPS and IPSJ, Tokyo, 1975, pp. 189-194.

6. Davis, Randall, *Applications of Meta Level Knowledge to the Construction, Maintenance, and Use of Large Knowledge Bases*, SAIL AIM-271, Artificial Intelligence Laboratory, Stanford University, July, 1976.

7. Evans, Thomas G., *A Program for the Solution of Geometric-Analogy Intelligence Test Questions*, in (Minsky, Marvin, ed.), Semantic Information Processing, The MIT Press, Cambridge, Massachusetts, 1968, pp. 271-353.

8. Feigenbaum, Edward, B. Buchanan, and J. Lederberg, *On Generality and Problem Solving: A Case Study Using The DENDRAL Program*, in (Meltzer and Michie, eds.) Machine Intelligence 6, 1971, pp. 165-190.

9. Gelernter, H., *Realization of a Geometry-Theorem Proving Machine*, in (Feigenbaum and Feldman, eds.) Computers and Thought, McGraw-Hill Book Company, New York, 1963, pp. 134-152.

10. Guard, James R., et al., *Semi-Automated Mathematics*, JACM 16, January, 1969, pp. 49-62.

11. Kling, Robert Elliot, *Reasoning by Analogy with Applications to Heuristic Problem Solving: A Case Study*, Stanford AI Memo AIM-147, August, 1971.

12. Lederberg, Joshua, Review of J. Weizenbaum's Computer Power and Human Reason, W. H. Freeman, S.F., 1976.

13. Lenat, Douglas B., *AM: An Artificial Intelligence Approach to Discovery in Mathematics as Heuristic Search*, SAIL AIM-286, Artificial Intelligence Laboratory, Stanford University, July, 1976.

14. Newell, Allen, and Herbert Simon, Human Problem Solving, Prentice-Hall, Englewood Cliffs, New Jersey, 1972.

15. Nilsson, Nils, Problem Solving Methods in Artificial Intelligence, McGraw-Hill, N.Y., 1971.

16. Shortliffe, E. H., *MYCIN -- A rule-based computer program for advising physicians regarding antimicrobial therapy selection*, Stanford AI Memo 251, October, 1974.

17. Wang, Hao, *Toward Mechanical Mathematics*, IBM Journal of Research and Development, Volume 4, Number 1, January, 1960, pp. 2-22.

18. Winston, Patrick, *Learning Structural Descriptions from Examples*, TR-231, MIT AI Lab, September, 1970.

FIGURE 1: Concepts Initially Given to AM

Below is a graph of the concepts which were present in AM at the beginning of its run. Single lines denote Generalization/Specialization links, and triple lines denote Examples/Isa links.

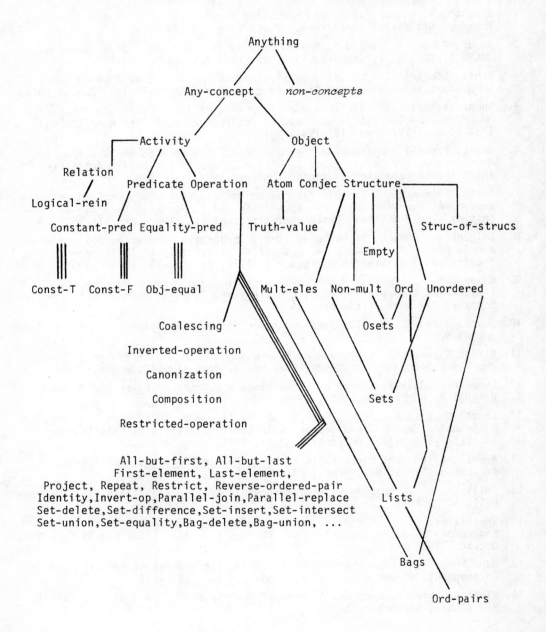

FIGURE 2: Concepts Discovered by AM

The list below is meant to suggest the range of AM's creations; it is far from complete, and many of the omissions were real losers. The concepts are listed in the order in which they were defined. In place of the (usually awkward) name chosen by AM, I have given either the standard math/English name for the concept, or else a short description of what it is.

Sets with less than 2 elements (singletons and empty sets).
Sets with no atomic elements (nests of braces).
Bags containing (>1 copy of) just one kind of element.
Superset(contains).
Doubleton bags and sets.
Set-membership.
Disjoint bags.
Subset.
Disjoint sets.
Same-length.
Same first element.
Count (Length).
Numbers (in unary).
Add.
Minimum.
SUB1 ($\lambda(x)x-1$).
Subtract (except: if x<y, then x-y results in zero.
Less than or equal to.
Times.
Compose F with itself (form F-o-F).
Insert structure S into itself.
Try to delete structure S from itself (a loser).
Double (add 'x' to itself).
Subtract 'x' from itself.
Square ($\lambda(x)$ Times(x,x)).
Coalesced-repeat: ($\lambda(S,F)$ takes true S,
op F, and repeats F(s,t,S) all along S).
Coalesced-join: append together F(s,s), for each sϵS.
Coalesced-replace: replace each element sϵS by F(s,s).
Compose three operations: $\lambda(F,C,H)$ F-o-(C-o-H).
Compose three operations: $\lambda(F,C,H)$ (F-o-C)-o-H.

FIGURE 2: (Continued)

Add^{-1}(x): all ways to repr. x as the sum of nonzero nos.
C-o-H, s.t. H(C(H(x))) is always defined (wherever H is).
Insert-o-Delete; Delete-o-Insert.
Size-o-Add^{-1}.(λ(n) The number of ways to partition n).
Cubing.
Exponentiation.
Halving (in natural numbers only; thus Halving(15)=7).
Even numbers.
Integer square-root.
Perfect squares.
Divisors-of.
Numbers-with-0-divisors; Numbers-with-1-divisor.
Primes (Numbers-with-2-divisors).
Squares of primes (Numbers-with-3-divisors).
Squares of squares of primes.
Square roots of primes (a loser).
Times^{-1}(x): all ways of repr. x as the product of nos.(>1).
All ways of representing x as the product of primes.
All ways of representing x as the sum of primes.
All ways of representing x as the sum of two primes.
Numbers uniquely representable as the sum of two primes.
Products of squares.
Multiplication by 1; by 0; by 2.
Addition of 1; of 0; of 2.
Product of even numbers.
Sum of squares.
Sum of even numbers.
Pairs of squares whose sum is also a square ($x^2+y^2=z^2$).
Prime pairs ({(p,q,r)|p,q,r are primes $\wedge$ p+q=r}).

STUDENT USE OF AN INTERACTIVE THEOREM PROVER

James McDonald and Patrick Suppes[1]

This article focuses on the use of an interactive theorem prover by students in a course on axiomatic set theory. The framework of use is described and then a number of student proofs are presented with comments and explanations. The last part of the article concentrates on feasible ways of improving the theorem prover, especially in its term-finding capability.

1. Introduction

The purpose of this article is to describe student use of an interactive theorem prover that has been developed and used over a number of years in the teaching of logic and axiomatic set theory at Stanford. A more detailed account is to be found in various articles in Suppes (1981a).

Because the uses of the theorem prover are more elaborate in the course in axiomatic set theory we shall restrict ourselves almost entirely to this course, but there is extensive use of the theorem-proving machinery by several hundred students a year in the elementary logic course.

Set-theory curriculum. The curriculum of the course in set theory is classical. It follows closely the content of the second author's textbook (Suppes, 1960). Chapter 1 of the curriculum

---

[1]Institute for Mathematical Studies in the Social Sciences, Stanford University, Stanford, CA 94305.
This research was supported in part by NSF Grant MCS-8011975.

surveys the historical background of Zermelo-Fraenkel set theory and justifies the need for developing the subject axiomatically. Chapter 2 is concerned with general developments, namely, definitions and elementary theorems about fundamental concepts such as those of inclusion, union, and intersection of sets, Cartesian product, power set, and so on. Chapter 3 develops the general theory of relations and functions from a set-theoretical standpoint. Chapter 4 is concerned with equipollence and the concept of finite and infinite sets. Chapter 5 develops the theory of cardinal numbers. Students taking the course at a Pass level are expected to go no farther than Chapter 5. Chapter 6 develops the general theory of ordinal numbers, including the standard representation theorem for well-ordered sets in terms of ordinal numbers. Chapter 7 analyzes the axiom of choice, its equivalents and their consequences.

Student enrollment. To give a sense of the enrollment in the course, which has been taught entirely at computer terminals as a computer-assisted instruction course since 1974, Table 1 shows the

Table 1

Enrollment in CAI Course in Axiomatic Set Theory

Academic year	Quarter			Total
	Autumn	Winter	Spring	
1974-75	8	3	8	19
1975-76	5	4	5	14
1976-77	4	7	5	16
1977-78	4	8	14	26
1978-79	9	6	12	27
1979-80	6	12	11	29
1980-81	7	3	2	12
1981-82	7	6	4	17
1982-83	7	5	9	21
TOTAL	57	54	70	181

data through the academic year 1982-83. The enrollments in the course are not large. It is a specialized course taken by a small number of undergraduates each term but it is a regular staple of the curriculum. The total number of students taking the course each year is larger than it would be if it were offered as a lecture course given once a year. (For quantitative data on this point, see Tables 1 and 2 of Suppes and Sheehan, 1981.) Each student in the course proves between 25 and 50 theorems depending upon the grade he is working for in the course. The main requirement of the course is the proving of theorems using the interactive theorem prover.

Structure of program. An overview of the running EXCHECK program can be seen in the following diagram, which shows its configuration for a typical user. Lesson authors use the entire structure, whereas students use some subset of the graph descending from the course driver. Each labeled object is a separate process running in a TENEX fork on a PDP-10. Together they comprise over two megabytes of code with another two megabytes of data structures.

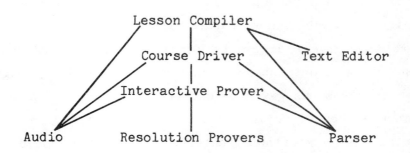

The discussion here is concerned almost entirely with the interactive theorem prover and adjunct routines. Other components parse formulas, speak, present lessons, and perform other pedagogical chores. See McDonald (1981) for a more detailed view of the overall system, and other articles in Suppes (1981a) for extensive discussions of each of the components in EXCHECK.

Structure of proofs. Proofs constructed in the EXCHECK theorem prover are done under the guidance of a command facility that embodies a number of natural proof strategies. These are implemented within two broad sets of commands: those that reduce a pending goal to some set of subgoals which jointly imply it, and those that introduce new lines based upon earlier ones. In both cases, there are facilities for citing prior axioms, definitions, and theorems.

The ability to cite preceding theorems by name is an important feature of EXCHECK. There are more than 600 theorems in the set-theory course, and the students necessarily must use preceding theorems in order to prove the theorems assigned to them. In our judgment, too much work in automatic theorem proving has been concerned with direct proofs from axioms. In any established and developed mathematical subject only a small number of theorems are ever proved directly from the axioms. It is essential that the use of preceding theorems be natural and easy in any theorem-proving environment in which a substantial development of a given mathematical topic is expected.

In section 2 and the appendix we give a detailed presentation of a session in which Cantor's diagonalization theorem is proved. In section 3 we give examples of final reviews for other student proofs and some quantitative data on the variable lengths of proofs for theorems, as given by different students. In section 4 we present the theorem prover as a strategic system, and describe some of the more important features we believe could improve considerably our interactive theorem prover. We concentrate in this discussion on facilities for term finding.

2. Sample Interactive Session

The frames below illustrate the beginning of a rather straightforward session which generates a proof of Cantor's theorem that every set is strictly less pollent than its power set. Frames for the complete session are included as an appendix. The characters in each frame are shown exactly as they would appear to the student, except that student input has been underlined, and the two non-printing characters ESC and Control-\ are shown as $ and !, respectively. Wherever no explicit termination character is shown, the ESC key was pressed after the last underlined character. The notation of an equal sign followed by two colons is the symbol used for equipollence, and correspondingly for the pollent inequalities.

As the proof proceeds, EXCHECK maintains a goal-structured representation of it and makes suggestions based on that structure. Prior to accepting a major command, the program typically offers two suggestions, each selectable by a single keystroke. The suggestion shown in square brackets will perform a complete command with the described effect. The suggestion in parentheses will initiate a command, offering further suggestions as the explication of the command progresses. As shown here, ! selects the major suggestion, while $ selects the minor.

The review mechanism has a number of options. The normal interactive review shown below displays the pending goal and just those lines which are available in the current context. This may include lines inherited from a superior context. Other reviews would show the path of goals down to the currently selected one, or the tree structure of the proof starting at some goal and proceeding down through any specified number of levels.

In the first three frames the student merely accepts the program's suggestion (note that ρA is the power set of A):

Goal G1:   ($\forall A$) A <:: ρA

[Reduce the current goal with a universal reduction]
   (reduce) *!
Doing universal reduction

Goal G2:   A <:: ρA

[Reduce the current goal with an introduction reduction
 using the definition of less power]
(reduce) *!
Doing introduction reduction

Goal G3:   A ≤:: ρA and not ρA ≤:: A

[Reduce the current goal with a conjunction reduction]
(reduce) *!
Doing conjunction reduction

In the fourth frame the student begins to exert control by using the sufficiency condition given in theorem 4.2.1, which reads ($\forall$ A,B)(A ≤:: B iff ($\exists$ F) F: A inj B), rather than expanding the definition of leq power.

Goal G4:   A ≤:: ρA

[Reduce the current goal with an introduction reduction
 using the definition of leq power]
(reduce) *$reduce
Which proof procedure? (introduction) *theorem <Name> *4.2.1

The next goal shown will be labeled G6 since G5 was created at the same time as G4, when the conjunction G3 was reduced. In this frame, the student does a short proof using forward-chaining commands. Since the set theory course uses a sorted language, it is necessary to prove in line 2 that the indicated term is actually a function before it can be existentially generalized in line 3. The command VERIFY invokes a resolution theorem prover used as a black box by the student. TM:1:1 denotes the first term on line 1, and is used to simplify typing. EXCHECK does partial recognition of input strings, so "injection" indicates that the student typed "inj$", the program extended that to "injecti" (which is ambiguous between injection and injective), and the student then typed "o$" to resolve the ambiguity. Remember that $ indicates the ESC key. Typing ? instead of "o$" would have displayed the ambiguity. In the command for line 3, "occurrences (1-1)" is a mildly awkward way of indicating all occurrences, in this case the first through the first.

Goal G6:   ($\exists$ F) F: A inj ρ A

[Use establish to infer the current goal]
  (establish) *theorem <Name> *4.1.10
    ($\forall$ A) {z: ($\exists$ x)(z = <x,{x}> & x ϵ A)}: A inj ρ A

Do you want to specify? (yes) *$yes
Substitute for A? (A) *$A
theorem 4.1.10
        (1)   {z: ($\exists$ x)(z = <x,{x}> & x ϵ A)}: A inj ρ A
[Use establish to infer the current goal from line 1]
  (establish) *1verify   (2) *func(tm:1:1)
Using *definition <Name> *injection
Using *definition <Name> *map
Using *ok

[Use establish to infer the current goal from line 1, line 2]
  (establish) *1eg
Replace term *tm:1:1 Variable *F
F and {z: ($\exists$ x)(z = <x,{x}> & x ϵ A)} aren't in the proper sort relation.
Line justifying a sort for {z: ($\exists$ x)(z = <x,{x}> & x ϵ A)} *2
Occurrences (1-1) *$1-1
1 eg   (3) ($\exists$ F) F: A inj ρ A
Goal 6 fulfilled by line 3.
3 implies using theorem 4.2.1
        (4)   A $\leq$:: ρ A
Goal 4 fulfilled by line 4.

An analysis of the entire session in the appendix shows that students do relatively little typing to create proofs. We wish to emphasize that this was a major design goal of EXCHECK, and in fact, students' initial orientation to the theorem prover emphasizes a minimization of typing. Everyone recognizes the tediousness of technical typing. If the students had to construct in informal mathematical language a complete printed proof it is doubtful very many students would accept this requirement. On the other hand, when the control language can be used in an easy way and most of the typing is done by the program, the students can feel that they are doing that which is intellectually essential and the routine is being left to the program as it should be. This matter of the amount of input may seem like a trivial matter but operationally it is of great importance in order to make the use of the interactive theorem prover attractive to users. The degree of terseness feasible for commands is limited by the set theoretic knowledge of the theorem prover and the quality of its model of the user's intentions.

A point worthy of note is that in the study of structural proofs little if any attention has been given to such interactive proofs. The literature on the structure of proofs is almost entirely devoted to the cleaned-up systematic proofs we put under the label of review in our program. From a psychological standpoint it would in many ways be more fundamental and interesting to examine in a deeper way the structure of the interactive versions of proofs. For example, the final reviews shown here depict perhaps overly detailed proofs, but the simple concluding steps which give this impression are actually generated automatically by the goal machinery for completeness, and in fact appear for the user in their alternate guises of pending or satisfied goals.

We try to encourage students to use the goal machinery, at least for the top few levels of the proof. Besides reflecting good mathematical practice, it makes it easier for EXCHECK to provide the suggestions noted above, lets the reviews be smaller by suppressing irrelevant lines, and helps the teaching assistants understand a proof in progress should the student need help. The last point is important. It is hard for machines or humans to offer assistance when presented with eighty lines representing a mistaken strategy for a proof which the assistant has perhaps never before attempted.

3. Student Proof Data

A small number of proofs will be shown here for illustrative purposes. Extensive data have been kept and analyzed for the entire corpus of student proofs, however, so the comments made here are based upon evidence that goes well beyond these examples.

Of the rules of proof we will mention, the data show that the call on preceding theorems is the most frequently used rule of inference. Closely associated with the use of preceding theorems is the call on preceding definitions. When preceding theorems and definitions are lumped together they are by far the most frequent steps in proofs, as might well be expected.

The second most used class of rules of inference consists of standard elementary logical rules of inference of the kind familiar from natural deduction systems for first-order logic. Examples are the rules for conditional proof, reductio ad absurdum, tautological inference, universal generalization, existential generalization, universal specification, existential specification, forming a conjunction, inferring a member of a conjunct, modus ponendo ponens, etc.

A third class of rules lets the student draw on a resolution theorem prover labeled VERIFY. The student is permitted to call VERIFY, which will run for a fixed number of seconds. If at the end of this short period of time the desired inference is not found, the student must adopt another strategy. It is repeatedly pointed out to the student that when VERIFY does not complete a desired inference it is not necessarily because the inference is not valid. It may be that not enough time was available, but it also may be the case that the inference was not correct. VERIFY is used extensively in the course. The way in which it is used by students is shown below in some sample proofs. What is important is that students do become familiar with it and do make extensive use of it. In order to complete the proofs in set theory in reasonable time it is essential that students not try to make all the inferences proceed by explicit elementary logical steps. Students are continually informed of this fact and become very much aware of it as they undertake the more difficult proofs they are assigned.

Closely connected with VERIFY is the command ESTABLISH, which also calls the resolution theorem prover but does some additional things such as analyzing the formula to be derived by expanding terms and relations using the definitions and theorems available to the student. There are a number of other rules that there is not space to review here but that are available to the student. For example, the command BOOLE calls a Boolean decision procedure.

We now present in final review format several theorems proved by students during the fall of 1982. EXCHECK can present proofs in a number of formats--the final review format is simply a linearized version of the proof with all goal structure suppressed. We avoid printing the goal structure in the final review since the program does a poor job of formatting large and deeply nested structures within 80 columns. However, we could provide both the linear review and a goal structured summary terminating about five levels down from the top. As Lee Blaine (1981) shows, such a review can do a good job of revealing the main outline in a proof.

Again, we emphasize that the proofs we show are exactly those produced by the program. We have not introduced any "manual" modifications of any sort.

The first proof is of the elementary theorem that a set is identical to the empty set if and only if it has no members. The theorems often have names. In this case we call the theorem the empty theorem. Immediately after the theorem is stated or numbered or named, what the student must derive or prove is shown as in the following instance:

Derive:

$(\forall A)((\forall x) \, x \neg \epsilon A \leftrightarrow A = 0)$

```
assume   (1)    (∀ x) x ¬ε A
         (2)    Set(A)
2 eliminate using definition set
         (3)    (∃ x) x ε A or A = 0
1 implies using 3
         (4)    A = 0
1, 4 cp
         (5)    If (∀ x) x ¬ε A then A = 0
assume   (6)    A = 0
theorem 2.3.1
         (7)    x ¬ε 0
6, 7 teq
         (8)    x ¬ε A
8 ug     (9)    (∀ x) x ¬ε A
6, 9 cp
         (10)   If A = 0 then (∀ x) x ¬ε A
5, 10 fc
         (11)   ((∀ x) x ¬ε A → A = 0) and (A = 0 → (∀ x) x ¬ε A)
11 lb    (12)   (∀ x) x ¬ε A iff A = 0
12 ug    (13)   (∀ A)((∀ x) x ¬ε A ↔ A = 0)
```

*** QED ***

We believe that this proof is quite understandable without any explanation. There are of course the usual abbreviations, for example, cp for conditional proof as used to construct line 5. The command teq used to obtain line 8 from lines 6 and 7 permits substitutions on the basis of the rules for tautology and equality. The rule of inference 1b used to obtain line 12 is just the law for the biconditional. Notice finally that line 3 is obtained by eliminating the notation of line 2 by using the definition of a set. The point is we believe that the proof is readable with very little explanation and without any reliance on programming concepts. We emphasize that the students using the interactive theorem prover are not required to have any knowledge whatsoever of programming, and no knowledge of programming of any kind is taught in the course.

The next proof is of theorem 2.10.6 of Chapter 2. This is the standard theorem stating that if A is a subset of B then the union of the sets in A is a subset of the union of the sets in B.

Derive:
$(\forall A,B)(A \subseteq B \rightarrow UA \subseteq UB)$

```
assume     (1)   A ⊆ B
assume     (2)   x ε UA
2 theorem using theorem 2.10.1
           (3)   (∃ C)(x ε C & C ε A)
3 es       (4)   x ε C and C ε A
1 definition using definition subset
           (5)   (∀ x)(x ε A → x ε B)
5 us       (6)   If C ε A then C ε B
4 simp     (7)   C ε A
6, 7 boole
           (8)   C ε B
4, 8 tautology
           (9)   x ε C and C ε B
9 eg       (10)  (∃ C)(x ε C & C ε B)
10 theorem using theorem 2.10.1
           (11)  x ε UB
```

2, 11 cp
      (12)   If $x \in UA$ then $x \in UB$
12 ug (13) $(\forall x)(x \in UA \rightarrow x \in UB)$
13 introduction using definition subset
      (14) $UA \subseteq UB$
1, 14 cp
      (15)   If $A \subseteq B$ then $UA \subseteq UB$
15 ug (16) $(\forall B,A)(A \subseteq B \rightarrow UA \subseteq UB)$

*** QED ***

The rule es used to obtain line 4 is of course existential specification and the rule us used to obtain line 5 is the rule of inference for universal specification. The command simp used to obtain line 7 from line 4 is the rule of simplification for propositional forms. But of course the simplification here is not an equivalence but an implication.

 The next theorem, again a standard elementary theorem (2.12.9 of Chapter 2), is the theorem that the union of the power set of A and the power set of B is a subset of the power set of A union B. Although this proof contains 25 steps, the flow of steps is quite transparent and natural. All that is needed for complete amplification is a listing of some of the prior theorems that are used, but the content of these prior theorems is already pretty obvious from the proof itself.

Derive:
$(\forall A,B)\ \rho A \cup \rho B \subseteq \rho(A \cup B)$

assume (1) $C \in \rho A \cup \rho B$
assume (2) $x \in C$
1 eliminate using theorem union
      (3) $C \in \rho A$ or $C \in \rho B$
theorem 2.12.2
      (4) $C \in \rho A$ iff $C \subseteq A$

theorem 2.12.2
  (5) $C \in \rho B$ iff $C \subseteq B$
3 replace using 4, 5
  (6) $C \subseteq A$ or $C \subseteq B$
assume (7) $C \neg \subseteq A$
7, 6 tautology
  (8) $C \subseteq B$
8 eliminate using definition subset
  (9) $(\forall x)(x \in C \rightarrow x \in B)$
2 implies using 9
  (10) $x \in B$
7, 10 cp
  (11) If $C \neg \subseteq A$ then $x \in B$
assume (12) $C \subseteq A$
12 eliminate using definition subset
  (13) $(\forall x)(x \in C \rightarrow x \in A)$
2 implies using 13
  (14) $x \in A$
12, 14 cp
  (15) If $C \subseteq A$ then $x \in A$
11, 15 tautology
  (16) $x \in A$ or $x \in B$
16 introduction using theorem union
  (17) $x \in A \cup B$
2, 17 cp
  (18) If $x \in C$ then $x \in A \cup B$
18 ug (19) $(\forall x)(x \in C \rightarrow x \in A \cup B)$
19 introduction using definition subset
  (20) $C \subseteq A \cup B$
20 introduction using theorem 2.12.2
  (21) $C \in \rho(A \cup B)$
1, 21 cp
  (22) If $C \in \rho A \cup \rho B$ then $C \in \rho(A \cup B)$
22 ug (23) $(\forall C)[C \in \rho A \cup \rho B \rightarrow C \in \rho(A \cup B)]$

23 introduction using theorem 2.4.7
       (24)   ρA ∪ ρB ⊆ ρ(A ∪ B)
24 ug (25)   (∀ B,A) ρA ∪ ρB ⊆ ρ(A ∪ B)

*** QED ***

   In 1978-79 we examined 9 valid proofs of this theorem. They had a mean length of 12.1, the median was 11, the minimum was 5, and the maximum was 19, which is shorter than the proof given here. The reason is that not all the automatic elementary steps shown above were included in the earlier proof. The closely related theorem that the power set of the intersection of A and B is equal to the intersection of the power set of A and the power set of B had much longer proofs. The mean of 13 proofs was 20.3, the median was 18, the minimum was 5, and the maximum was 50.

   As the final sample theorem from Chapter 2 on general developments we give the proof of Theorem 2.15.3. It is the familiar theorem that uses especially the axiom of regularity to prove that given any two sets it cannot be the case that each is a member of the other. One point of notation in the theorem does need comment. Because of the need for linear notation in the existential specification in line 11, the notation that would ordinarily be subscripted is not, so "x0" is used rather than "x_0". This notation occurs in subsequent lines.

Derive:
(∀ A,B) not (A ε B & B ε A)

assume   (1) A ε B and B ε A
theorem 2.15.2
         (2) If {A,B} ≠ 0
              then (∃ x)[x ε {A,B} & (∀ y)(y ε x → y ¬ε {A,B})]
1 simp   (3) A ε B
1 simp   (4) B ε A
theorem 2.9.1
         (5) A ε {A,B} iff A = A ∨ A = B

theorem 2.9.1
          (6)   B ε {A,B} iff B = A v B = B
5 teq   (7)   A ε {A,B}
6 teq   (8)   B ε {A,B}
7 eg   (9)   (∃ x) x ε {A,B}
9 theorem using theorem 2.3.3
          (10)   {A,B} ≠ 0
10 implies using 2
          (11)   (∃ x)[x ε {A,B} & (∀ y)(y ε x → y ¬ε {A,B})]
11 es   (12)   x0 ε {A,B} and (∀ y)(y ε x0 → y ¬ε {A,B})
12 simp
          (13)   x0 ε {A,B}
12 simp
          (14)   (∀ y)(y ε x0 → y ¬ε {A,B})
13 theorem using theorem 2.9.1
          (15)   x0 = A or x0 = B
assume   (16)   x0 = A
14 replace using 16
          (17)   (∀ y)(y ε A → y ¬ε {A,B})
17 us   (18)   If B ε A then B ¬ε {A,B}
4 implies using 18
          (19)   B ¬ε {A,B}
16, 8, 19 ip
          (20)   x0 ≠ A
15, 20 tautology
          (21)   x0 = B
14 replace using 21
          (22)   (∀ y)(y ε B → y ¬ε {A,B})
22 us   (23)   If A ε B then A ¬ε {A,B}
3 implies using 23
          (24)   A ¬ε {A,B}
24, 7 fc
          (25)   A ¬ε {A,B} and A ε {A,B}

```
1, 25 raa
        (26)  Not (A ε B and B ε A)
26 ug   (27)  (∀ B,A) not (A ε B & B ε A)

*** QED ***
```

The proof of Theorem 2.15.3 depends not directly on the axiom of regularity but on the preceding Theorem 2.15.2 which we have labeled in fact the regularity theorem. It states that if a set A is not empty then there exists a member x of A such that no member of x is also a member of A.

In 1978-79 we examined 25 proofs of this theorem. The mean length was 18.7, the median was 19, the minimum was 9, and the maximum was 26 lines.

To show in a striking way how much the length of proofs can vary we contrast the straightforward mathematically transparent proof given above for Cantor's theorem with a very much shorter second proof. This proof of Cantor's theorem does no simple goal reductions and uses VERIFY much more powerfully. As a result, it consists of only nine lines, instead of 34. This proof was done many years ago in a version of EXCHECK that did not yet support special characters. Thus, "(Ax)" is used in place of "(∀ x)", "pow(A)" for "ρA", "<=" for "≤::", etc.

```
revIEW
A < pow(A)
VERIFY Using:  Th. 4.1.10, Th. 4.2.1, Df. map, Df. injection
        (1) A <= pow(A)
ASSUME (2) pow(A) <= A
2 LET using:  Th. 4.2.1, Df. map, DF. injection
        (3) Inj(f) and dom(f) = pow(A) and rng(f) sub A
LET Using:  Ax. separation Instance:  x not in inv (f)(x) for FM
        (4) (Ax)(x in D <-> x in A & x not in inv (f)(x))
3,4 VERIFY Using:  Th. powerset, Df. subset
        (5) D in dom(f)
```

```
3,5 VERIFY Using:  Df. subset, Th. range, Th. 3.10.9
      (6) f(D) in A
3,5 Th. 3.10.58
      (7) Inv(f)(f(D)) = D
4,6,7,2 CONTRADICTION
      (8) Not pow(A) <= A
1,8 Df. lesspower
      (9) A < pow(A)
```

This proof is harder to follow because its steps are much more compact. On the other hand, it is quite clear what the line of argument is. While this proof is much shorter than the 34 line proof described earlier, many of the lines in the goal structured proof were generated by a single keystroke. In the proof immediately above, the student needed to type seven formulas with a total of 148 characters. In the goal structured proof, the student also needed to enter seven formulas, with a total of 108 characters. We examined 24 student proofs from 1978-79 of Cantor's theorem. The mean number of lines was 29.9, the median was 25, the minimum was 25, and the maximum was 70.

We give just one example of a theorem about cardinal numbers. Theorem 5.4.2 states a standard necessary and sufficient condition for a cardinal number to be a transfinite cardinal, namely, that it be identical to itself plus one. Note that there is a slightly subtle matter involved in the distinction between infinite cardinals and transfinite cardinals. An infinite cardinal is the cardinal of an infinite set. A transfinite cardinal is the cardinal of a Dedekind infinite set. Recall that a Dedekind finite set is a set that is equipollent to a proper subset of itself. In the sense used here, an infinite set is one that is not equipollent to any natural number, although this is not the formal definition which is given independent of the concept of natural number. Standard proofs of the equivalence of Dedekind infinity and ordinary infinity require the use of the axiom of

Interactive Theorem Prover

choice. Thus the difference between infinite and transfinite cardinals is nontrivial.

Derive:
($\forall$ n)(tcard(n) $\leftrightarrow$ n = n+:1:)

assume   (1)   Tcard(n)
1 definition using definition tcardinal
        (2)   ($\exists$ A)(dinfinite(A) & |A| = n)
2 es   (3)   Dinfinite(A) and |A| = n
3 simp   (4)   Dinfinite(A)
3 simp   (5) |A| = n
4 theorem using theorem 4.4.6
        (6)   A =:: A $\cup$ {A}
6 axiom using axiom cardinal
        (7) |A| = |A $\cup$ {A}|
5 theorem using theorem 5.2.38
        (8) |A $\cup$ {A}| = n+:1:
7 replace using 5, 8
        (9)   n = n+:1:
1, 9 cp
        (10)   If tcard(n) then n = n+:1:
assume   (11)   n = n+:1:
theorem 5.2.1
        (12)   ($\exists$ A,B)(A$\cap$B = 0 & |A| = n & |B| = m)
12 es   (13)   C$\cap$B = 0 and |C| = n & |B| = m
13 simp
        (14) |C| = n
14 theorem using theorem 5.2.38
        (15) |C $\cup$ {C}| = n+:1:
15 replace using 11, 14
        (16) |C $\cup$ {C}| = |C|
16 axiom using axiom cardinal
        (17)   C $\cup$ {C} =:: C

17 theorem using theorem 4.1.2
      (18)   C =:: C U {C}
18 theorem using theorem 4.4.6
      (19)   Dinfinite(C)
19, 14 fc
      (20)   Dinfinite(C) and |C| = n
20 eg   (21)   (∃ A)(dinfinite(A) & |A| = n)
21 definition using definition tcardinal
      (22)   Tcard(n)
11, 22 cp
      (23)   If n = n+:1: then tcard(n)
10, 23 fc
      (24)   (tcard(n) → n = n+:1:) and [n = n+:1: → tcard(n)]
24 lb   (25)   Tcard(n) iff n = n+:1:
25 ug   (26)   (∀ n)(tcard(n) ↔ n = n+:1:)

*** QED ***

Note a blemish in the printing of this proof by the program. The "t" for transfinite cardinal is capitalized when it begins a line and otherwise not.

We examined 7 valid standard proofs of this theorem given by students in 1978-79. The mean number of lines was 20.0, the median was 21, the minimum was 13, and the maximum was 24. The very explicit student proof given here is longer than the maximum of this earlier group.

Two theorems that require the axiom of choice and are often assigned to students as the part of their work from Chapter 7 are the law of trichotomy and the Teichmüller-Tukey Lemma that any set of finite character has a maximal element. Examination of 14 valid student proofs of the law of trichotomy yielded the following data: mean number of lines was 24.6, median was 24, minimum was 13, and maximum was 40. Nine proofs of the Teichmüller-Tukey Lemma had a mean length of 41.8, a median of 44, a minimum of 25, and a maximum of 61.

The variability in length of proofs of a given theorem would have been greater if we had examined a larger sample. In our experience the main source of variation is the extent to which students use the powerful and complex commands, especially VERIFY and ESTABLISH, about which we have something more to say in the next section.

4. Strategic Considerations

The proofs reviewed above give a reasonable sense of how the current program behaves, although of course the full range of its capabilities cannot be described so briefly. Students typically spend the first two or three weeks becoming familiar with the machinery while proving some elementary theorems, and may learn more advanced commands or features as the need arises later in the course. We receive occasional complaints about the time needed to learn to use EXCHECK, and thus have striven to keep the use of it as simple as possible.

The suggestions made by EXCHECK can be best understood by viewing them as a collection of strategies. The simplest of these strategies, from both the students' viewpoint and the complexity of our programming effort, is the reduction of goals using sentential operations and the replacement of universally quantified sentences with sentences containing free variables. Among the related synthetic commands, TAUTOLOGY and BOOLE provide useful decision procedures. Almost all proofs use such strategies, especially at the top level, but perhaps none of them can proceed to completion without using other, more complex strategies.

As mentioned earlier, the most useful of these is simply the citation of a previous axiom, definition, or theorem. This can be thought of as goal reduction via standard sufficiency conditions.

In general, there may be alternative sets of sufficiency conditions, in which case the program must somehow decide which reduction to suggest. An analysis from first principles would be prohibitively complex. Instead, EXCHECK makes use of strategies encoded by an expert user of the system, namely Lee Blaine. When proving A = B, for example, the program suggests a reduction to $(\forall x)(x \in A \leftrightarrow x \in B)$. When proving F = G, however, it instead suggests a reduction to dom(F) = dom(G) and $(\forall x) F(x) = G(x)$.

This part of EXCHECK can fairly be described as an expert system, roughly in a category with programs such as MYCIN and PROSPECTOR, even though it differs from them in some major respects, and was not created or envisioned in such terms. Articles on EXCHECK, MYCIN, PROSPECTOR and several related systems can be found in Barr and Feigenbaum (1982). EXCHECK, like other expert systems, emphasizes the need for explicitly encoded expert knowledge, as opposed to a priori principles. Blaine made many local decisions as to what the best rule would be in a given situation, using general notions about the availability of theorems, the power of the existing machinery, and pedagogical clarity. The theorem list went through many revisions in this process, and represents a significant improvement over the original list. We cannot emphasize too strongly that good pedagogical programs will not be able to do everything from first principles. A lot of knowledge and experience must be encoded.

There are limits of course to the guidance that EXCHECK can provide. Sometimes it simply makes an inappropriate suggestion, or fails to notice that a pre-existing theorem would be useful. Poorer students can be waylaid in such situations, but typically students recognize when the program is making poor suggestions and ignore them. We try to emphasize to the students that they should make their own decisions about the best approach, and only then make use of the program's suggestions, assuming they agree. Anecdotal evidence indicates that every term a few students have one or two sessions where they inappropriately pursue the program's mistaken strategy. No students seem to have significant problems in this regard.

On a more fundamental level, there are two major situations in which EXCHECK is almost totally unable to offer guidance, namely proofs by contradiction and proofs of existentially quantified goals.

Proofs by contradiction. Proofs by contradiction are the easier ones for which EXCHECK could be improved to offer guidance. Currently there is no integrated notion of a model built into EXCHECK. A mechanism exists for having the student construct examples or counter-examples for simple assertions, and then carry on a graph-directed dialogue, but this machinery is totally separate from the proof checker. The TEQ command (for Tautology + Equality) proceeds by refining a purported model of the equivalence classes generated by the assertions and the denial of the conclusion. This code can be invoked by the proof checker, but functions as a black box written in a separate language and running in a separate process. Thus it would be hard to integrate.

A nice extension of EXCHECK would have it proceed, when the current goal is to find a contradiction, to internally construct a purported model of the given assertions and then proceed to discover new relations to be added to that model, under the direction of the student. Eventually the purported model would break down, or the student would see how the strategy was failing.

Relations would be added by straightforwardly testing all possible relations among all the terms in the purported model. Such a strategy seems to skirt exponentially expensive behavior, but in practice would probably work reasonable fast, since the number of terms would tend to be rather small, and most relations could be quickly rejected. The student would also be able to curtail extravagant behavior.

Relations would also be added by abstracting over terms which satisfy, or fail to satisfy, relations that cannot be explicitly proved or disproved. For example, if the prover could prove neither $A \subseteq f(A)$ nor $A \neg \subseteq f(A)$, then the terms $\{A: A \subseteq f(A)\}$ and $\{A: A \neg \subseteq f(A)\}$ would be generated. It would also generate terms denoting the union and intersection for each such set, as well as the minimal and maximal elements.

In addition to heuristics for adding such abstracted terms, some obvious heuristics would be needed to avoid generating series of relations unlikely to be useful, e.g., $A \in \rho A$, $\rho A \in \rho \rho A$. A simple bound on the complexity of new constructions would be adequate in this regard, at least for a first running version. Other obvious heuristics could be table driven. A few suggested ones might be:

If $A \subseteq B$, check to see if $A = B$.
If $A \subseteq B \wedge x \in B$, check to see if $x \in A$.
If $x \in A \wedge y \in A$, check to see if $x = y$.
If x and y are not ordinals and are not at adjacent levels in the constructable hierarchy, do not attempt to relate them.

A better set could be empirically determined.

Paper-and-pencil observations indicate that the contradiction-seeking strategy described here is sufficient to let EXCHECK generate a well motivated proof of Cantor's theorem, in which not much more than a dozen terms are actively considered in the purported model.

Term Finding. A strategy for solving existential assertions appears to be a much more formidable undertaking, since there are fewer constraints on terms to be considered. In particular, the terms constructed in such proofs are often rather complex. A good example from set theory is the Schroeder-Bernstein theorem, which can be stated as follows:

Goal: $(\forall A,B)(A \leq:: B \wedge B \leq:: A \rightarrow A =:: B)$

Before proceeding, it should be noted that this theorem precedes the axiom of choice in the curriculum, hence must be completed without it. This precludes the "zig-zag" proof sometimes presented for this theorem, which successively picks an unseen element x and then follows the sequence of applications F(x), G(F(x)), F(G(F(x))), etc.

A few straightforward goal reductions using simple logic and the definitions of less-pollence and equipollence soon lead to the following revised situation:

Goal: (∃ H) H: A bij B

1) (∃ F) F: A inj B
2) (∃ G) G: B inj A

In other words, we have an injection from A into B, an injection from B into A, and the goal is to find a bijection between A and B. Unfortunately, there are no obvious clues as to what that bijection might be.

A traditional pedagogical presentation of this proof might introduce the operator Echo(C) = A ~ G"(B ~ F"C) and then show that A* = {C: C ⊆ Echo(C)} is a fixed point with respect to Echo. From there, it is not too difficult to show that F|A* ∪ inv(G)|(A~A*) is the desired bijection, where F|A* means F restricted to A*, and inv(G) is the inverse of G.

Presumably the student can understand that such a proof works, and may further understand that recognition of a fixed point was a useful trick here. Beyond that, however, it is not clear how well the proof can be internalized as anything beyond an ad hoc solution. The claim we make is that general, intelligible heuristics can be used to find the proof sketched above, in such a way that no major leaps are required beyond those explicitly sanctioned by the heuristics.

The strategy employed will amount to a highly constrained search through the Herbrand Universe, using filters derived from the statement of the problem. A number of powerful heuristics will be used, but all of them are motivated by general considerations. They include ordering principles, fixed point properties, Occam's razor, constructions in the geometric sense, the strengthening of some goals to eliminate other goals, and the generation of simple examples and counterexamples within models such as integer arithmetic. The use of a structured database for concepts will also be important.

Given that the goal and lines 1 and 2 are expressed in fairly condensed notation, the first step is to expand definitions to a more explicit statement of the situation. This is done by consulting a database that specifies the interesting properties of a term. Some of these properties will be sufficiency conditions, while others will express equivalence relations. In essence, the database should capture concepts that are lurking beneath the surface in the traditional representation.

Paper does not provide a good medium for expressing database relations. In the interactive EXCHECK environment, the student would be able to view numerous representations. The following representation, for example, shows the original goal and assumed lines, with the goal explicated as a set of jointly sufficient conditions, and with the assumed lines explicated by a set of some immediate consequences.

```
Goal:   H: A bij B
i.e.:   func(x), dom(x) = A, rng(x) = B, 1-1(x)

   1)   F: A inj B
i.e.:   dom(F) = rng(inv(F)) = a
        rng(F) = dom(inv(F))
        rng(F) ⊆ B

   2)   G: B inj A
i.e.:   dom(G) = rng(inv(G)) = B
        rng(G) = dom(inv(G))
        rng(G) ⊆ A
```

A review of 1-1(x) would show that it is equivalent to

$(\forall y,z)(x(y) = x(z) \rightarrow y = z)$. Other formulas could likewise be reviewed in more detail.

Alternatively, the student might view the relations that hold for bijections:

    x: A bij B   =>   func(x)
                             func(inv(x))
                             dom(x) = rng(inv(x)) = A
                             rng(x) = dom(inv(x)) = B
                             x: A inj B
                             x: A surj B
                             inv(x): B bij A
                             inv(x): B inj A
                             inv(x): B surj A
                             ...

Or the student could view sufficiency conditions for bijections:

    x: A bij B   if   func(x) ^ dom(x) = A ^ rng(x) = B ^ 1-1(x)
                             x: A inj B ^ x: A surj B
                             x: A inj B ^ inv(x): B inj A
                             ...

A general heuristic which can be explained to the student at this point is to decompose the goal into a set of conditions which are jointly sufficient. A sub-heuristic is to not decompose statements of equality or type declarations, since there are often efficient ways to test for such conditions.

EXCHECK now has a better representation of the problem, but there is still no clear way to proceed. The central heuristic to be employed now is that any term used in the proof can probably be expressed as a (possibly complex) operation on terms already present in the expanded statement of the problem. A further assumption is that this complex operation can be expressed in terms of operators ranked a priori for expected relevance. For example, functional application is highly relevant for functions, restriction and images are perhaps less relevant, while composition, union, and intersection are of moderate interest. Cartesian products of functions and powersets of functions are below the threshold for immediate consideration.

The first check is to see if any term already present satisfies the goal. A and B quickly fail since there is no reason to believe they are functions. In fact, the program can note that they are unconstrained in this regard, and potentially could construct counter-examples. Type failure is heuristically considered to be a major defect, so A and B are rated very poorly. F and G are more likely candidates. Still, there is no reason to believe rng(F) = B, nor is there any reason to believe that A = B, which is what dom(G) = A would imply. Inv(G) appears about equally successful as F, and depends on the unattainable condition that dom(inv(G)) = A, i.e. rng(G) = A.

As an adjunct to the following process, rng(F) = B, rng(G) = A, and A = B are noted as unattainable conditions which would trivialize the proof. If a future (and hence more complex) term would succeed subject to any of these conditions, it can be rejected from consideration on grounds similar to Occam's razor—simpler terms have already been shown to be about equally effective.

If terms are even partially successful, they are not discarded, but rather form the basis for the construction of more elaborate terms. Those terms with the mildest or fewest number of defects are employed first in such constuctions.

To pursue the Schroeder-Bernstein proof, since F and inv(G) seem most successful, new terms are generated which include them, according to heuristics for combining functions. In this manner, F|A, F|B, inv(G)|A, inv(G)|B, F"A, F"B, inv(G)"A, inv(G)"B, F@G, G@F, F∪G, F∩G, etc. are generated, where @ is functional composition. All of these have defects, and none seem to depend on conditions any weaker than those that would make F or inv(G) succeed.

All of the above has proceeded rather quickly (i.e., in a few minutes) to a closure on a subset of the available strategies. A review would now look like this:

Goal: H: A bij B
i.e.: func(x), dom(x) = A, rng(x) = B, 1-1(x)

1) F: A inj B
2) G: B inj A

Best candidates, with missing conditions:

F   B ⊆ rng(F)
inv(G)   A ⊆ rng(G)

The student would be able to manipulate the entire set of examined candidates, for example to review those not shown above or to suggest new terms and have them examined. The student would also be able to edit the evidence for a term and to move terms into and out of the set of best candidates.

The next major heuristic involves the introduction of new free terms into the set of terms available for construction. This is a direct generalization of the heuristic in geometric proofs where additional points or lines are constructed to facilitate an otherwise stalled proof. The terms added here, though, are theory independent. All of their properties will be solved for in later stages.

The term constructor now has access to new free terms, which will be represented as x*, x**, etc. As their properties become defined, they may be re-represented as A*, F**, etc. to clarify their type. Thus, F|x*, inv(G)|x*, F"x*, etc. are examined to see how well they succeed. Since such terms are expected to combine in useful ways, they are exempt from the heuristic described above which discards terms requiring conditions as strong as those required by previous, simpler terms.

$F|x^*$ fails in almost the same manner that F failed, but the required conditions for $F|x^*$ to succeed are $A \subseteq x^*$, $B \subseteq F"x^*$, and $B \subseteq rng(F)$. Likewise $inv(G)|x^*$ fails unless $A \subseteq x^*$, $A \subseteq rng(G)$, and $B \subseteq inv(G)"x^*$. FUx^*, $G@x^*$, $F"x^*$, etc. all fail for numerous reasons, primarily failure to be a function. FUH^* is eliminated since for it to be a function $F(x) = H^*(x)$ for all x in A, but then $FUH^* = F$. $Inv(G)UH^*$ is eliminated for a similar reason.

The outcome at this stage is that $F|A^*$, $inv(G)|A^*$, and at most a few other terms appear plausibly useful. A review here would look like:

Goal: H: A bij B
i.e.: func(x), dom(x) = A, rng(x) = B, 1-1(x)

1) F: A inj B
2) G: B inj A

Best candidates, with missing conditions:
F $B \subseteq rng(F)$
inv(G) $A \subseteq rng(G)$
$F|A^*$ $B \subseteq rng(F)$, $A \subseteq A^*$, $B \subseteq F"A^*$
$inv(G)|A^*$ $A \subseteq rng(G)$, $A \subseteq A^*$, $B \subseteq inv(G)"A^*$

The next stage proceeds as before and combines these terms with each other and with terms previously deemed useful. This results in terms such as $(F|A^*) \cup (inv(G)|A^{**})$ and $(F|A^*) \cap (inv(G)|A^{**})$. An analysis of these reveals something important. The sufficiency conditions imply that $(F|A^*) \cup (inv(G)|A^{**})$ will succeed if $A^* \cap A^{**} = 0$, $A^* \cup A^{**} = A$, $F"A^* \cap inv(G)"A^{**} = 0$, and $(F"A^*) \cup (inv(G)"A^{**}) = B$. What is interesting is that none of these conditions are easily refutable.

Given such a promising candidate, the original existential goal can now be reduced by a lemma which claims that

$(F|A^*) \cup (inv(G)|A^{**})$: A bij B if the following subgoals can be satisfied:

F: A inj B
G: B inj A
$A^* \cap A^{**} = 0$
$A^* \cup A^{**} = A$
$(F"A^*) \cap (inv(G)"A^{**}) = 0$
$(F"A^*) \cup (inv(G)"A^{**}) = B$

The new situation can be reviewed as:

Goal: $(\exists C,D)(\ C \cap D = 0 \ \wedge \ C \cup D = A \ \wedge \ F"C \cap inv(G)"D = 0$
$\wedge \ F"C \cup inv(G)"D = B\ )$

1) F: A inj B
2) G: B inj A

Given two joint existentials, a standard reduction would be to try to solve one in terms of the other. This is simply a generalization of solving simultaneous equations. It is closer to being a proof procedure than a heuristic. Solving for D in this manner, the following side proof results:

Goals: $C \cap D = 0$
$A \sim C \subseteq D$
$F"C \cap inv(G)"D = 0$
$B \sim F"C \subseteq inv(G)"D$

Simple boolean operations reveal that the first two subgoals will be satisfied if $D = A \sim C$. A weaker assumption might be possible, but this one eliminates goals, so we heuristically proceed with it to the revised situation:

Goals: $F"C \cap inv(G)"(A \sim C) = 0$
$B \sim F"C \subseteq inv(G)"(A \sim C)$

Since G is an injection, we know from the database that

$(\forall A)(A \subseteq rng(G) \rightarrow G"(inv(G)"A) = A)$ and that $(\forall A,B)(A \subseteq B \rightarrow G"A \subseteq G"B)$. From these we get:

Goals: $G"(F"C) \cap A{\sim}C = 0$
$G"(B \sim F"C) \subseteq A{\sim}C$

and then

Goals: $F"C \cap inv(G)"(A{\sim}C) = 0$
$C \subseteq A \sim G"(B \sim F"C)$

Given these two goals, there is no obvious way to derive one from the other. However, if the second goal is strengthened to be $C = A \sim G"(B \sim F"C)$, then assuming it true, the first can be proved in easy steps:

Goal: $F"C \cap inv(G)"(A \sim (A{\sim}G"(B \sim F"C))) = 0$
Goal: $F"C \cap inv(G)"(G"(B \sim F"C)) = 0$
Goal: $F"C \cap (B \sim F"C) = 0$
Goal: $0 = 0$

Thus it suffices to show that $C = A \sim G"(B \sim F"C)$. Again, a weaker goal might suffice, but this one eliminates other goals, so we proceed with it.

Overall, our proof now reviews as:

Goal: $(\exists C)(C = A \sim G"(B \sim F"C))$

1) F: A inj B
2) G: B inj A

In the current EXCHECK course, this goal is in fact presented as a lemma in the theorem list to facilitate the Schroeder-Bernstein proof.

Viewing it as a separate proof, we can proceed with the same term finding strategy as above, after first reducing the universally quantified goal to one with free variables:

Goal: $C = A \sim G''(B \sim F''C)$
   1) F: A inj B
   2) G: B inj A

We consider A, B, F"C, B ~ F"C, G"(B ~ F"C), and A ~ G"(B ~ F"C). Since we are solving for C, all but A and B are ruled out on the grounds that they are not a closed solution. If C = A, then G"(B ~ F"A) must be the empty set, but a counterexample to this can easily be found which is consistent with the current assumptions. For example, let A and B equal the set of integers, let F(x) = 2x, and let G(x) = x. On the other hand, if C = B, then we can assume A = B and obtain the same unattainable goal that G"(A ~ F"A) = 0.

Thus there are no terms even partially successful for the second stage, so we proceed to consider the following abstraction terms, along with the union and intersection of each such set, and their minimal and maximal elements:

   {C: C = A ~ G"(B ~ F"C)}
   {C: C ⊆ A ~ G"(B ~ F"C)}
   {C: A ~ G"(B ~ F"C) ⊆ C}

Fixed-point considerations make ∪{C: C ⊆ A ~ G"(B ~ F"C)} and ∩{C: A ~ G"(B ~ F"C) ⊆ C} heuristically interesting. Given the first, we generate the following situation:

   Abbreviations: Echo(C) = A ~ G"(B ~ F"C)
                     E = {C: C ⊆ Echo(C)}
   Goal: UE = Echo(UE)
   i.e.:   UE ⊆ Echo(UE)   and   Echo(UE) ⊆ UE

If we assume UE ⊆ Echo(UE), then table-driven heuristics for functions generate the following sequence:

   F" UE ⊆ F"Echo(UE)
   B ~ F"Echo(UE) ⊆ B ~ F"UE
   G"(B ~ F"Echo(UE)) ⊆ G"(B ~ F"UE)
   Echo(UE) ⊆ Echo(Echo(UE))

But this implies $\text{Echo}(\cup E) \in E$, hence $\text{Echo}(\cup E) \subseteq \cup E$, so we can eliminate one sub-goal to obtain the following situation:

Goal: $\cup E \subseteq \text{Echo}(\cup E)$

Alternatively, we could have defined $E = \{C: \text{Echo}(C) \subseteq C\}$ and arrived analogously at this situation:

Goal: $\text{Echo}(\cap E) \subseteq \cap E$

From here on, the proof falls well within the framework of proofs already done in EXCHECK. The goal $\cup E \subseteq \text{Echo}(\cup E)$ can be reduced using properties of subset and family, as follows:

Goal: $D \in \text{Echo}(\cup E)$

assume 1) $D \in \cup E$
Expanding 1 gives us
      2) $(\exists C)(D \in C \wedge C \in E)$
es   3) $D \in C \wedge C \in E$
By the definition of E,
      4) $C \subseteq \text{Echo}(C)$
The definition of $\cup$ implies that
      5) $C \subseteq \cup E$
Hence by a short proof
      6) $\text{Echo}(C) \subseteq \text{Echo}(\cup E)$
But then from 3,4,6
      7) $D \in \text{Echo}(\cup E)$

QED

A similar proof is generated using $\cap\{C: \text{Echo}(C) \subseteq C\}$.

The entire proof thus proceeded through many stages, and employed a number of heuristics, but at no point was an unclearly motivated term produced. The proof is possibly too complex for even such a revised EXCHECK to find it unaided, but a student should have no problem guiding EXCHECK to success. The only danger is a loss of faith that such a long procedure will terminate successfully. Presumably the program would keep hidden most of the failed attempts at various stages, so that students would see roughly the material presented here.

The extension described does not exist in any running form, and would represent a substantial effort—more than a year of solid work. In particular, EXCHECK should be rewritten to convert it completely to LISP, and to enhance its ability to handle such things as restricted quantifiers. That effort alone would be months or perhaps a year of work for any of the original programmers for EXCHECK, and might require several years if done by other people. Thousands of pages of code are involved. If this were done, however, the result would be an elegant pedagogical vehicle for introductory set theory and similar axiomatic courses.

Bibliography

Barr, A., and Feigenbaum, E. (Eds.). *The handbook of artificial intelligence*. Los Altos, Calif.: William Kaufmann, Inc., 1982.

Blaine, L. Programs for structured proofs. In P. Suppes (Ed.), *University level computer-assisted instruction at Stanford: 1968-1980*. Stanford, Calif.: Stanford University, Institute for Mathematical Studies in the Social Sciences, 1981. Pp. 81-120.

McDonald, J. The EXCHECK CAI system. In P. Suppes (Ed.), *University level computer-assisted instruction at Stanford: 1968-1980*. Stanford, Calif.: Stanford University, Institute for Mathematical Studies in the Social Sciences, 1981. Pp. 765-790.

Suppes, P. *Axiomatic set theory*. New York: Van Nostrand, 1960.

Suppes, P. (Ed.). *University-level computer-assisted instruction at Stanford: 1968-1980*. Stanford, Calif.: Stanford University, Institute for Mathematical Studies in the Social Sciences, 1981. (a)

Suppes, P. Future educational uses of interactive theorem proving. In P. Suppes (Ed.), *University-level computer-assisted instruction at Stanford: 1968-1980*. Stanford, Calif.: Stanford University, Institute for Mathematical Studies in the Social Sciences, 1981. Pp. 165-182. (b)

Suppes, P., and Sheehan, J. CAI course in axiomatic set theory. In P. Suppes (Ed.), *University level computer-assisted instruction at Stanford: 1968-1980*. Stanford, Calif.: Stanford University, Institute for Mathematical Studies in the Social Sciences, 1981. Pp. 3-80.

Appendix A
A Complete Session for a Proof of Cantor's Theorem

We present here a complete EXCHECK session used to generate a proof of Cantor's theorem that every set is strictly less pollent than its power set. Double lines are used to indicate points at which the screen is cleared and a new frame started. Each frame is shown exactly as the user saw it, except that $, ! and ↑G here indicate the non-printing characters ESC, Control-\ and Control-G, respectively, and all user input is underlined here. ESC accepts an immediate suggestion, Control-\ accepts a high level suggestion, and Control-G tells EXCHECK to guess something reasonable--usually the immediate goal when a formula is expected. See section 2 of the main text for more explanations.

Interactive Theorem Prover

===
Goal G1:   ($\forall$ A) A <:: ρ A

[Reduce the current goal with a universal reduction]
  (reduce) *!
Doing universal reduction
===
Goal G2:   A <:: ρ A

[Reduce the current goal with an introduction reduction
 using the definition of less power]
  (reduce) *!
Doing introduction reduction
===
Goal G3:   A $\leq$:: ρ A and not ρ A $\leq$:: A

[Reduce the current goal with a conjunction reduction]
  (reduce) *!
Doing conjunction reduction
===
Goal G4:   A $\leq$:: ρ A

[Reduce the current goal with an introduction reduction
 using the definition of leq power]
  (reduce) *$reduce
Which proof procedure? (introduction) *theorem <Name> *4.2.1
===
Goal G6:   ($\exists$ F) F: A inj ρ A

[Use establish to infer the current goal]
  (establish) *theorem <Name> *4.1.10
    ($\forall$ A) {z: ($\exists$ x)(z = <x,{x}> & x $\in$ A)}: A inj ρ A
Do you want to specify? (yes) *$yes
Substitute for A? (A) *$A
theorem 4.1.10
        (1)   {z: ($\exists$ x)(z = <x,{x}> & x $\in$ A)}: A inj ρ A

[Use establish to infer the current goal from line 1]
  (establish) *1verify   (2) *func(tm:1:1)
Using *Definition <Name> *injection
Using *Definition <Name> *map
Using *ok

[Use establish to infer the current goal from line 1, line 2]
  (establish) *1eg
Replace term *tm:1:1 Variable *F
F and {z: (∃ x)(z = <x,{x}> & x ε A)} aren't in the proper sort relation.
Line justifying a sort for {z: (∃ x)(z = <x,{x}> & x ε A)} *2
Occurrences (1-1) *$1-1
1 eg   (3) (∃ F) F: A inj ρA
Goal 6 fulfilled by line 3.
3 implies using theorem 4.2.1
        (4)   A ≤:: ρA
Goal 4 fulfilled by line 4.
==
Goal G5:   Not ρA ≤:: A

Hence   (4)   A ≤:: ρA
[Reduce the current goal with a reductio ad absurdum reduction]
  (reduce) *!
Doing reductio ad absurdum reduction.
==
Goal G7:   Contradiction

Hence   (4)   A ≤:: ρA
assume (5)   ρA ≤:: A
Determine which contradiction to derive and use REVISE to replace the current goal formula with that contradiction.
*5,4theorem <Name> *schroeder-bernstein
5, 4 theorem using theorem schroeder-bernstein
        (6)   ρA =:: A

Determine which contradiction to derive and use REVISE to replace
the current goal formula with that contradiction.
*6definition <Name> *equipollent
6 definition using definition equipollent
        (7) (∃ F) F: ρA bij A
Determine which contradiction to derive and use REVISE to replace
the current goal formula with that contradiction.
*7es

      (∃ F) F: ρA bij A
Ambiguous name for F? (F) *$F
7 es   (8) F: ρA bij A
Determine which contradiction to derive and use REVISE to replace
the current goal formula with that contradiction.
*abbreviation
Finished, Introduce, Eliminate, or Print? (finished) *introduce
What do you want to abbreviate?
    *{x: x in A and x not in inv(F)(x)}
What abbreviation do you want to use? *CS
Finished, Introduce, Eliminate, or Print? (finished) *introduce
What do you want to abbreviate?
    *F(CS)
Is the expression to be abbreviated a Term or a Formula? *term
What abbreviation do you want to use? *CE
Finished, Introduce, Eliminate, or Print? (finished) *$finished
Determine which contradiction to derive and use REVISE to replace
the current goal formula with that contradiction.
*revise
Is the contradiction to be a conjunction, an equivalence, or
an identity? (conjunction) *equivalence
Type the formula that is asserted to be equivalent to its negation
*CE in CS
==

Abbreviations:
CS = {x: x ε A & x ¬ε inv(F)(x)}
CE = F(CS)

Goal G7:   CE ε CS iff CE ¬ε CS

Hence   (4)   A ≤:: ρA
assume   (5)   ρA ≤:: A
By 4   (6)   ρA =:: A
So   (7)   (∃ F) F: ρA bij A
So that (8)   F: ρA bij A
[Reduce the current goal with a biconditional reduction]
 (reduce) *theorem <Name> *2.5.10
Schema:
    x ε {x: x ε A & FM(x)} iff x ε A & FM(x)
Replace for FM *x not in inv(F)(x)
Which variable indicates the parameter places?   *x
    (∀ A,x)[x ε CS ↔ x ε A & x ¬ε inv(F)(x)]
Do you want to specify? (yes) *$
Variables (A,x)   *A
Substitute for A? (A) *$A
theorem 2.5.10
         (9)   (∀ x)[x ε CS ↔ x ε A & x ¬ε inv(F)(x)]
[Reduce the current goal with a biconditional reduction]
 (reduce) *9definition <Name> *subset
         (10)   CS ⊆ A
[Reduce the current goal with a biconditional reduction]
 (reduce) *10theorem <Name> *powerset
         (11)   CS ε ρA
[Reduce the current goal with a biconditional reduction]
 (reduce) *!
Doing biconditional reduction

Abbreviations:
CS = {x: x ε A & x ¬ε inv(F)(x)}
CE = F(CS)

Goal G8:   (CE ε CS → CE ¬ε CS) and (CE ¬ε CS → CE ε CS)

Hence   (4)   A ≤:: ρA
assume   (5)   ρA ≤:: A
By 4   (6)   ρA =:: A
So   (7)   (∃F) F: ρA bij A
So that   (8)   F: ρA bij A
By theorem 2.5.10
          (9)   (∀x)[x ε CS ↔ x ε A & x ¬ε inv(F)(x)]
Also   (10) CS ⊆ A
And   (11) CS ε ρA

[Reduce the current goal with a conjunction reduction]
 (reduce) *!

Doing conjunction reduction
==
Abbreviations:
CS = {x: x ε A & x ¬ε inv(F)(x)}
CE = F(CS)

Goal G9:   If CE ε CS then CE ¬ε CS

Hence   (4)   A ≤:: ρA
assume   (5)   ρA ≤:: A
By 4   (6)   ρA =:: A
So   (7)   (∃F) F: ρA bij A
So that   (8)   F: ρA bij A
By theorem 2.5.10
          (9)   (∀x)[x ε CS ↔ x ε A & x ¬ε inv(F)(x)]
Also   (10) CS ⊆ A
And   (11) CS ε ρA

[Reduce the current goal with a conditional reduction]
  (reduce) *!
Doing conditional reduction
==
Abbreviations:
CS = {x: x ε A & x ¬ε inv(F)(x)}
CE = F(CS)

Goal G11:   CE ¬ε CS

Hence   (4)   A ≤:: ρA
assume   (5)   ρA ≤:: A
By 4   (6)   ρA =:: A
So   (7)   (∃ F) F: ρA bij A
So that (8)   F: ρA bij A
By theorem 2.5.10
         (9)   (∀x)[x ε CS ↔ x ε A & x ¬ε inv(F)(x)]
Also   (10) CS ⊆ A
And   (11) CS ε ρA
assume   (12) CE ε CS
[Reduce the current goal with a reductio ad absurdum reduction]
  (reduce) *12,9implies
12 implies using 9
         (13) CE ε A and CE ¬ε inv(F)(CE)
[Reduce the current goal with a reductio ad absurdum reduction]
  (reduce) *8,11theorem <Name> *3.17.10
         (14) Inv(F)(CE) = CS
[Reduce the current goal with a reductio ad absurdum reduction]
  (reduce) *13,14teq   (15) *↑GCE ¬ε CS
Using *ok
Goal 11 fulfilled by line 15.
11,15cp (16) if CE ε CS then CE ¬ε CS
Goal 9 fulfilled by line 16.
==

Abbreviations:
CS = {x: x ∈ A & x ¬∈ inv(F)(x)}
CE = F(CS)

Goal G10:   if CE ¬∈ CS then CE ∈ CS

Hence   (4) A ≤:: ρA
assume   (5) ρA ≤:: A
By 4   (6) ρA =:: A
So   (7) (∃F) F: ρA bij A
So that (8) F: ρA bij A
By theorem 2.5.10
         (9) (∀x)[x ∈ CS ↔ x ∈ A & x ¬∈ inv(F)(x)]
Also   (10) CS ⊆ A
And   (11) CS ∈ ρA
cp   (16) if CE ∈ CS then CE ¬∈ CS
[Reduce the current goal with a conditional reduction]
 (reduce) *!
Doing conditional reduction
===
Abbreviations:
CS = {x: x ∈ A & x ¬∈ inv(F)(x)}
CE = F(CS)

Goal G12:   if CE ¬∈ CS then CE ∈ CS

Hence   (4) A ≤:: ρA
assume   (5) ρA ≤:: A
By 4   (6) ρA =:: A
So   (7) (∃F) F: ρA bij A
So that (8) F: ρA bij A
By theorem 2.5.10
         (9) (∀x)[x ∈ CS ↔ x ∈ A & x ¬∈ inv(F)(x)]
Also   (10) CS ⊆ A
And   (11) CS ∈ ρA

cp   (16)   if CE ε CS then CE ¬ ε CS
assume   (17)   CE ¬ ε CS
Try to prove the appropriate theorem of the form:
$$\text{for every x, x in } \{x: Fm(x)\} \text{ iff } Fm(x)$$
and then use it to get the goal.
*<u>17,9implies</u>
          (18)   Not [CE ε A and CE ¬ ε inv(F)(CE)]
Try to prove the appropriate theorem of the form:
$$\text{for every x, x in } \{x: Fm(x)\} \text{ iff } Fm(x)$$
and then use it to get the goal.
*<u>8eliminate</u>
8 eliminate using definition bijection
          (19)   Func(F) and inj(F) & dom(F) = ρA & rng(F) = A
Try to prove the appropriate theorem of the form:
$$\text{for every x, x in } \{x: Fm(x)\} \text{ iff } Fm(x)$$
and then use it to get the goal.
*<u>11,19teq</u>   (20)   *<u>CS in dom(F)</u>
Using *ok
Try to prove the appropriate theorem of the form:
$$\text{for every x, x in } \{x: Fm(x)\} \text{ iff } Fm(x)$$
and then use it to get the goal.
*<u>20theorem</u> \<Name\> *<u>3.11.4</u>
          (21)   \<CS,CE\> ε F
Try to prove the appropriate theorem of the form:
$$\text{for every x, x in } \{x: Fm(x)\} \text{ iff } Fm(x)$$
and then use it to get the goal.
*<u>21theorem</u> \<Name\> *<u>range</u>
          (22)   CE ε rng(F)
Try to prove the appropriate theorem of the form:
$$\text{for every x, x in } \{x: Fm(x)\} \text{ iff } Fm(x)$$
and then use it to get the goal.
*<u>22,19teq</u>   (23)   *<u>CE in A</u>
Using *<u>ok</u>

Try to prove the appropriate theorem of the form:

    for every x, x in {x: Fm(x)} iff Fm(x)

and then use it to get the goal.

*23,18tautology (24) *CE in inv(F)(CE)

Try to prove the appropriate theorem of the form:

    for every x, x in {x: Fm(x)} iff Fm(x)

and then use it to get the goal.

*19,22theorem <Name> *3.15.15

19, 22 theorem using theorem 3.15.15

  (25) $(\forall x)(inv(F)(CE) = x \leftrightarrow \langle x,CE\rangle \in F$

Try to prove the appropriate theorem of the form:

    for every x, x in {x: Fm(x)} iff Fm(x)

and then use it to get the goal.

*21,25implies

21 implies using 25

  (26) Inv(F)(CE) = CS

Try to prove the appropriate theorem of the form:

    for every x, x in {x: Fm(x)} iff Fm(x)

and then use it to get the goal.

*24,26teq (27) *↑GCS ∈ CS

Using *ok

Goal 12 fulfilled by line 27.

17, 27 cp

  (28) if CE ¬∈ CS then CE ∈ CS

Goal 10 fulfilled by line 28.

16, 28 fc

  (29) (CE ∈ CS → CE ¬∈ CS) and (CE ¬∈ CS → CE ∈ CS)

Goal 8 fulfilled by line 29.

  (30) CE ∈ CS iff CE ¬∈ CS

Goal 7 fulfilled by line 30.

5, 30 raa

  (31) Not ρA ≤:: A

Goal 5 fulfilled by line 31.

4, 31 fc
>   (32)   A ≤:: ρA and not ρA ≤:: A

Goal 3 fulfilled by line 32.

32 introduction using definition less power
>   (33)   A <:: ρA

Goal 2 fulfilled by line 33.

So that (34)   (∀ A) A <:: ρA

Goal 1 fulfilled by line 34.

==

Abbreviations:

CS = {x: x ε A & x ¬ε inv(F)(x)}

CE = F(CS)

Done G1:   (∀ A) A <:: ρA

Hence   (33)   A <:: ρA

So that (34)   (∀ A) A <:: ρA

*qed ok